Inside the Fisher Body Craftsman's Guild

ALSO BY JOHN L. JACOBUS
AND FROM MCFARLAND

---

*The Fisher Body Craftsman's Guild:
An Illustrated History* (2005)

INSIDE THE FISHER BODY CRAFTSMAN'S GUILD

Contestants Recall the Great General Motors Talent Search

Edited by
John L. Jacobus

McFarland & Company, Inc., Publishers
Jefferson, North Carolina, and London

Library of Congress Cataloguing-in-Publication Data

Inside the Fisher Body Craftsman's Guild : contestants
recall the great General Motors talent search /
edited by John L. Jacobus.
p. cm.
Includes bibliographical references and index.

ISBN 978-0-7864-4861-6
softcover : 50# acid free paper ∞

1. Fisher Body Craftsman's Guild — History — 20th century.
2. General Motors automobiles — Models — Societies, etc. — History —
20th century. 3. General Motors automobliles — Design and construction —
History — 20th century. I. Jacobus, John (John L.)
TL237.I57 2012 629.222 — dc23 2011040086

British Library cataloguing data are available

Front cover: Walter O'Neill at the workbench with his 1957 model entry (*Newark Evening News*); winning models *from left* 1961 Senior Division, Second State Award for New York by Paul Bonfilio (Paul Bonfilio); Samuel T. Kjellman's 1962 winner, Senior Division, Second State Award for New Hampshire (Samuel T. Kjellman); Adrian A. Bruno's 1955 winner, Second National Junior Scholarship (Adrian A. Bruno)

Back cover: Tom Covert's 1962 Second National Senior Scholarship winner (Thomas L. Covert)

Manufactured in the United States of America

McFarland & Company, Inc., Publishers
Box 611, Jefferson, North Carolina 28640
www.mcfarlandpub.com

To the memory of the unsung heroes and heroines of the guild,
namely the moms and dads of the Guildsmen,
and to the memory of my brother and best friend,
Bill Jacobus (1938–2011)

Table of Contents

Between pages 112 and 113 are 16 color plates containing 39 photographs

Acknowledgments

Guildsmen

The following Guildsmen contributed to this book essays about their experiences, images from their scrapbooks, images of their models (either new or from family photo albums) as well as image identification data and captions. Thank you for sharing your personal Guild stories with the editor.

Philip J. Rauth (1947), Mahlon V.F. Freeman (1947), Peter Van Dyke (1948), Roger D. Teter (1949), Gordon D. Williams (1951), William A. Keyser (1949–1952), Charles R. Foreman (1950–1952, 1954), Warren M. Bakken (1949–1953), Joseph R. Rauth (1953), Arnold L. Joslin (1951–1953), Raymond P. Wykes (1951–1953), Wilfred C. Keagy (1949–1953), Adrian A. Bruno (1954, 1955), George E. Anderson (1954–1956), Charles H. Stewart (1951–1955), Herman I. Rauth (1955, 1956), Thomas L. De Fazio (1956, 1958, 1960), Robert Janosko (1955–1958), Gary Grabowski (1957–1963), Roger L. Schneider (1957–1962), Walter F. O'Neill (1956–1959), George G. Herzog (1955–1959, 1963), William D. Scott (1957–1959), Felix Roberto Collazo Velez (1959–1961), Geza A. Loczi (1959–1965), Tony Mazzola and Julie Steinhilber for her cousin Ronald R. Steinhilber (1960–1964, deceased May 2000), Robert W. Lawhn (1961–1964, 1966–1967), Ronald J. Will (1959–1961), Thomas L. Covert (1961, 1962), Paul F. Bonfilio (1958–1962), Samuel T. Kjellman (1961, 1962, 1964), Phil Bonine (1962), Tristram Walker Metcalfe III (1963, 1964), Frederick J. "Bud" Magaldi (1958–1960, 1962, 1965), Jeffrey A. Jones (1962–1965), John "Jack" Given (1963–1968), James Hammang (1964–1968), Tom W. Graboski (1966–1968), Terry P. Graboski (1966–1968), Spencer L. Mackay (1966–1968) and Grant L. Onnie (1967, 1968).

Technical Contributors

The editor wishes to thank the following for their assistance:

My good neighbor Susan Heavey is an online journalist, writer, editor, storyteller, essay and content analyst, and consultant in Silver Spring, Maryland.

George R. Chartier served as technical reviewer of the draft manuscript. A Guild technical expert, collector and archivist and a Guildsman in 1953 and 1954 from Green Bay, Wisconsin, he now lives in Northville, Michigan.

Ronald J. Will served as Guild technical consultant, technical expert, collector, archivist and reunion photographer. A Guildsman in 1959, 1960 and 1961 from Hobart, Indiana, he now lives in Fountain Hills, Arizona. Most of the images were contributed by the Guildsmen-essayists from their personal archives. Many were from Ronald J. Will's Guild photo collection or archives; several came from newspaper or magazine sources; others were courtesy Ronald E. Van Gelderen, Elaine M. Anderson, Virginia Tatseos, and Harry E. Schoepf. Much of the credit for the present day portrait images (Guildsmen with model in-hand), however, belongs to Ron Will (1961 National Senior Scholarship winner), who tirelessly worked with his digital camera at the 2004 and 2008 Guild reunions to document both the models and the model builders. Ron's comprehensive archives of Guild images were very helpful in many ways. Ron was also responsible for the enhanced images that appeared in the Museum of Fine Arts, Boston, catalogue. This work required an auto designer's eye and a graphic artist's touch in order to read the compound surfaces of the models. This was not a problem for Ron, the former manager of product planning and design for Subaru of America. Ron also worked with the author in the review and editing the final manuscript and reviewed the proof copy prior to publication.

We also thank professional photographers Kaile Hart Crowell and Geoff Tesch, who worked with our Guild contributors Gordon D. Williams and Williams A. Keyser, Jr., respectively, to take perfectly framed and balanced, close-up color images of the Guild models as well as portraits of the Guildsmen themselves.

Other Contributors

The bio-sketch spiral notebooks developed by Tom W. Graboski for the 2004 and 2008 reunions, containing names, addresses, phone, fax, cell numbers and brief bio-sketches of Guildsmen attendees, were critical and invaluable to the creation of this manuscript. They were the "Fisher Body Craftsman's Guild, Reunion June 26–27, 2004 Detroit" and "Fisher Body Craftsman's Guild 1930–1968, Reunion-Retrospective-Exhibition, Museum of Fine Arts, Boston August 9–10, 2008" by Tom W. Graboski, Tom Graboski Associates, Inc., Design, Coral Gables, Florida.

Others who contributed included Computer Geeks of Washington, D.C., Meghan Simone, Brett Snyder, Andrew F. Johnson Gallery, Virginia Tatseos, and my friend Jeanie West, who stuck by my side during the arduous manuscript preparation process, who structured my day, read the essays from a layman's point of view, asked good questions and edited the manuscript with common sense and passion.

Introduction

In the 29 essays that make up this book, contestants in General Motors' Fisher Body Craftsman's Guild model car competitions discuss their experiences, trials and tribulations, and techniques as they vied for college scholarships between 1947 and 1968. They vividly remember all facets of competing in this high-profile event during the critical adolescent years of their lives: the knowledge required, the strategic planning they had to do, the technical problems and hurdles they encountered, the price exacted (e.g., dropping out of college for a semester, hundreds of hours of labor committed), the moments of crisis, the heartbreaks (a broken model arriving in Detroit to be judged) and finally, the jubilation of success and being among the best model builders in the country.

Seventy percent of the essayists were national scholarship winners. They came from a variety of socio-economic circumstances with broad geographic representation. They discuss what it was like in junior high school, high school or college to earn a coveted GM college scholarship, and what it took to design and build a top ranked, national award winning "dream car" model for the Fisher Body Craftsman's Guild.

About the Guild

The Fisher Body Craftsman's Guild was an extracurricular industrial arts program in which thousands of boys and young men (15 percent were college age; only males could compete) from across America submitted their 1/12 scale, scratch-built "dream car" models annually to General Motors. A guild for young car designers was the inspiration of Harley J. Earl, design legend of GM. For some contestants in the 1950s and 1960s, their adolescent years were consumed by building five or six consecutive entries for the contest. If they were lucky, ingenious and talented, they successfully became ineligible by age 15, having won a national scholarship award. If they were hard-nosed, determined and fiercely competitive, they dropped out of college for the spring semester of that competition year to make that all-out final push to the June 1 deadline. Many risked failing grades as the Guild deadline, the prom, sports commitments and their final exams all seemed to collide in June.

For GM this was a highly successful public relations gambit and bonanza. It tapped into the love Americans, and particularly teenage boys, had for the automobile. The FBCG was so successful as a talent search and recruiting tool that GM eventually exported the program to its subsidiaries in West Germany (Opel), United Kingdom (Vauxhall Motors) and Australia (GM Holden) before phasing it out in the USA in 1968. This was accomplished under the leadership of Charles M. Jordan and William L. Mitchell, GM design executives, who envisioned an International Guild Competition with a big New York City gala event.[1]

The highly successful West German, co-educational Opel Model Builder's Guild operated from 1965 to 1979. Some girls and young women were top prize winners.

Besides the promise of college scholarships (of which 387 were delivered) and cash awards (of which thousands of dollars' worth were distributed), there was tremendous math, science and engineering value in the FBCG. The Guild would have been at the beginning of that math, science and engineering learning curve for teenagers. Guildsmen had to be able to visualize their design ideas in their minds and express them either with sketches or at a drafting table, then translate their ideas into three dimensions.

Guild Discontinued

In 1967 the word came down from corporate headquarters on West Grand Boulevard to FBCG technical staff that "The General" was no longer going to be in the youth outreach and education business and 1968 would be the last competition season. As the data shown in Table 1 indicate, model building interest among teens had waned. The number of "award-worthy" FBCG model entries had dropped off significantly from 88 percent in 1957–58 to 38 percent by 1968. Only 38 percent of GM's state awards were being claimed near the end of the program as the quality of the models being submitted slipped steadily downward. By 1968 there were so many outside competing social activities, teenagers lacked the time needed for a commitment to high quality design and craftsmanship. The original 1930s era "Party by Fisher" had become a party few wanted to attend by 1968.

At a time when co-education was coming into vogue at many institutions across the country, the Guild was male only. Perhaps GM didn't want to make this transition. Meanwhile, colleges and universities were placing greater emphasis on scholastic achievement as indicated by such measures as higher SAT scores, SAT prep testing, advance placement (AP) courses, high grade point averages, and merit scholarships, as well as school and community service. Essentially, what we had known as high school manual arts (a 1930s and 1940s term) and industrial arts (a 1950s and 1960s term), heirs of the 19th century Arts and Crafts movement, by the 1990s had morphed into the "technical arts"—communications, biological-health and computer sciences.

About This Book

The publication of my previous book about the Guild, and my web site, precipitated an outpouring of cards, letters, essays, life stories, phone calls, memorabilia and more from Guildsmen across the country clearly indicating to me that there was an audience eager for more information about GM's famous Fisher Body Craftsman's Guild. After all, millions of us had been exposed to it as teenagers. Whatever happened to the promise of the Craftsman's Guild, what happened to all those national winners and where could their models be seen? What became of them after the abrupt ending of the program? These were among the questions the first Guild book attempted to answer.

I then began thinking about a sequel. It would be a book of essays written by a cross-section of Guildsmen about their experiences in the contest from 1947 to 1968, featuring a variety of awardees from the highest to the lowest, from diverse geographic locations and various socio-economic levels.

As youths, the guildsmen had been bound together only by a network newsletter and

TABLE 1
FBCG Model Car Competition Awardees as a Percent of Total State Awards Available for Selected Competition Years

Year	*Number of State Awards Available*	*Number of Awardees or Award-worthy Models**	*Awardees as a Percent of Total State Awards*
1949	686	469	68.4
1951	686	491	71.6
1957	784	695	88.6
1958	784	692	88.3
1963	816	583	71.4
1966	816	399	48.9
1967	816	328	40.2
1968	816	308	37.7

*Announcement of state, regional and national winners for the years given, published by the Fisher Body Craftsman's Guild, Fisher Body Division, Detroit and Warren, Michigan. The *Guildsman* newsletter reported that about 2,000 model entries were judged and scored in June and July 1958. The above table shows that only 35 percent of the total models submitted to GM were "award-worthy" and that of this group, they consumed 88.3 of GM's planned Guild State awards.

a mountain of hopes and dreams. Some never met or spoke to another Guildsman due to their geographic dispersion. They participated in the Guild program briefly (a few years at most) from their basement or garage workshops and moved on with their lives. By 1984 the Guild, and all it stood for, had been wiped off the corporate books and the Fisher Body collection of historic Guild models sold at auction, as the Fisher Body Division was organized out of existence. Everything pertaining to the Guild had either been given away, scrapped in a dumpster, or sold, leaving few traces of its existence. All seemed lost until the Internet and its many search engines made it possible for the Guildsmen across the country to reconnect, beginning in the 1990s. The eBay auction site was where they got reacquainted as they grappled with what to do with their memorabilia. Guild coaches and model cars were being sold at the online auction site.

With the help of the Internet and the first Guild book, Charles M. Jordan, John M. Mellberg and GM Design staff volunteers hosted a 2004 Guild reunion at the GM Tech Center in Warren, Michigan, and an exhibition at the Edsel Ford Estate in Grosse Pointe, Michigan. The 2004 reunion included a major public display of more than 60 Guild models. Guildsmen appeared from all over the country with their models and memorabilia, similar to an antique car show. Spurred by the success at the Tech Center, I succeeded in reuniting Guild enthusiasts again at the 2008 reunion in cooperation with the Museum of Fine Arts, Boston. This included a two-day *1950s and 1960s Dream Car Model-rama* exhibition and retrospective. The public made it their business to stop by and talk to the Guildsmen and inspect the 65 models on display at this venue. Guildsmen read about it in the paper and stopped by to reconnect.

Essentially, it was never possible to know the whole Guild story because nothing had ever been published until my 2005 book, *The Fisher Body Craftsman's Guild: An Illustrated History*. Looking back over their lives, many Guildsmen carry a deep curiosity about this part of their past. They are nostalgic for information about that past fueled by fond memories, and this new book will attempt to complete the story for them.

This work is intended as a follow-up and companion to the first. These essays document, at the nitty-gritty level, the model building careers of competitors ages 12 to 20 who participated from 1947 to 1968. They relate how this select group of individuals were able to capitalize on their success at a very early age, many of them becoming leaders in design engineering, automotive design, product design, industrial design and other endeavors.

In order for the reader to thoroughly appreciate the designs of the top models built for the Guild competition, I have tried to include multiple images of the same model. These different views (¾ front view, ¾ rear view, side views, etc.) are necessary to adequately illustrate a three-dimensional object, and these models are no exception.

It has always been an objective of mine not only to capture and identify the model images accurately, but also to capture a portrait image of the builder with model in-hand. The Guild was about people and the models they made, not just the models or *objets d'art*. For this reason the essays include photographs of the Guildsmen themselves. Ron Will has generously provided several photographs of the Guildsmen with their models in-hand, taken at the 2004 and 2008 Guild reunions. In addition some period portrait images from newspapers and magazines are included to show the Guildsmen during their contest days.

Disclosure of Originality

In some of the photographs used in this book, the original competition models may have been recently repaired, refurbished or restored either for one of the reunions (2004 or 2008) or specifically for the book. This repair or restoration sometimes consisted of a new paint job or a new color scheme along with the replacement of parts. Other models shown in the book are replicas or reproductions of the original Guild model. Some Guildsmen painstakingly recreated models that had been lost or destroyed over the years due to theft, misplacement, excessive attic or basement humidity, wear and tear or fire. Most importantly, the essayists try to tell the reader whether they are looking at the original model, a refurbished or restored competition model, or a replica or reproduction of the original model. In all but a few cases, the restoration work was done by the Guildsmen themselves.

Notes

1. *The Fisher Body Craftsman's Guild: An Illustrated History* by John L. Jacobus (McFarland, 2005) was dedicated to Charles M. Jordan (1927–2010), the 1947 First National Senior Scholarship ($4,000) award winner in the Fisher Body Craftsman's Guild model car design contest. He championed the Guild within the corporation until 1968, fought to revive and resuscitate the program after its demise and worked to get a new program started at GMC, including an online, cyber net Guild. After studying automotive engineering and design at MIT, he was hired as a designer at GM Styling and became the public persona of the Guild. He was a friend of the Guild and a friend to all Guildsmen. In 1977 he was named director of design of the entire GM Design staff and eventually became the vice president of GM Design from 1986 to 1992. William L. Mitchell became the vice president of GM Styling from 1958 to 1977 (GM Styling was the forerunner of GM Design).

1

Six Dazzling Model Cars: Family Genes Prevailed[1]

Philip J. Rauth (1947), Joseph R. Rauth (1953) and Herman I. Rauth (1955–1956)

We were "dry dirt farm boys" from York, Nebraska, coming from a family of nine children — seven boys and two girls. The main forces shaping our thinking came from being part of a large rural farm family, church, school and a small town. There was no television or big city life experiences to broaden our thinking. We raised corn, milo, wheat, oats, red clover, hay, a small herd of cattle, a flock of chickens and occasionally sheep or hogs. We made our own cream, cottage cheese, butter, and ice cream. Despite this, or because of it, five out of seven Rauth boys successfully designed and built Guild models.

Our eldest brother, John, entered the Fisher Body Craftsman's Guild model car design contest in 1937. Being the three youngest boys, we hardly knew anything about his model-building exploits, let alone what he was doing in the basement shop. As we grew older, we came to understand that John had won a First State Award for Nebraska at age 16, the first year of the Craftsman's Guild model car competition.[2]

The FBCG model car competition was discontinued during World War II and did not resume until 1946, so our other brothers Paul and Tom missed out. Fortunately, our father liked to listen to classical music, and our interest in the FBCG competition was renewed, once again, after hearing a Guild radio advertisement on General Motors' *Symphony of the Air* program. Our brother Vincent entered the car competition that year and won the 1946 Senior Division, First National Scholarship Award, worth $4,000.[3] The success of our two older brothers, John and Vincent, certainly influenced our desire to enter the competition.

Since we were close in age, we shared duties like milking the cows, separating the milk, and feeding the calves. Building Guild models took all the time that was left after school, studies, sports, and farm chores — only then could we put our full attention into building our models.

Philip J. Rauth, 1947 — Second National Scholarship, Junior Division, $2,000 Award (Ages 13–14)

In the fall of 1946, at age 13, I began the eighth grade and my first model entry for the Craftsman's Guild. I knew that as a junior division participant, it would take more than one year to be successful. I considered building a miniature model Napoleonic Coach, but realized the minute coach details were more than I could contend with. So, I decided on a model car.

At that time, I was limited to building a four-door sedan. I played with various designs of front and rear engine models, doodling on paper in some of my classes at school. At one point my thoughts were, "How can I change and improve on the design of the latest cars that were on the road?" and "After all, there are only so many ways to bend sheet metal." After playing with three or four different designs, I finally decided on a traditional front engine design.

Vincent had used Honduran mahogany the previous year and the hardwood worked well for him. My father took me to the shop of a local cabinetmaker. He gave Mr. Green the dimensions needed for the scale model, and two pieces of mahogany were glued together to provide me with a solid, rectangular block out of which I could carve my model. My father told me from the start that if I was going to start this project, he expected me to stick to it until completion.

Basement Workshop and Tools

Our home workshop consisted of a portion of the basement of our large, two story, wood-framed farm house. This is where all five Rauth boys planned, designed and built their Guild models. Our dad was the one who built the shop's work bench, mounted the vise, and put boards on one wall to hang the tools. The work space took up about half of the house basement area, or about 450 square feet, but this space was shared with a furnace, a coal bin, and a hot water heater. We shared the other half of the basement with the milk separator equipment.

As kids, we Rauth boys thought it was a fairly well supplied shop with hand saws, hand planer, egg beater drill, and a brace and bit. It was a great place to learn the art of cutting by hand. The only power tool I can remember was a small wood lathe, and its homemade attachments — a small "table saw," a jig saw, and grinding stone. This marvelous 4-in-1 set-up was powered by a small electric motor on loan from our mother's washing machine. The electric motor had to be returned every Sunday evening and ready to run for Monday morning's weekly laundry. We had to plan ahead to do all our sawing or wait another week.

I went to work, using mostly hand tools, to shape this solid block of mahogany to match my paper design. Most of the shaping was done with a draw knife, chisels and wood rasps, finished off, of course, with lots of sandpaper. The Honduran mahogany, being an even, fine, tight grained wood, worked easily with the well honed hand tools. After roughing out the basic silhouette, I marked the center line, front to back, and went to work shaping one side. When I finally got that to where I was satisfied, the tedious task of duplicating the design on the other side began. I did not use templates but instead used a compass and calipers to shape the second side.

1947 Model Building Process

I had problems with the model painting process. I began by putting on several coats of shellac to seal the wood and to get a nice smooth, hard base coat. Working on a limited budget, I decided to use the paint leftover from our brother Vincent's project the previous year, the same colors he had used on his prize winning model in 1946. It consisted of two different shades of blue Duco automobile lacquer. I didn't have any spray painting equipment in the shop, but used a hand pumped "bug" sprayer contraption to apply the lacquer. Having not yet studied chemistry, I didn't realize that the lacquer would soften the shellac. As a result, I ended up having to remove both the lacquer and the shellac and begin all over. With the careful application of many coats of lacquer, tediously sanding between coats, and finally hand rubbing with rubbing compound, I finally had a pretty decent two-tone, dark blue paint job.

Philip J. Rauth from York, Nebraska, was the 1947 Second National Junior Scholarship $2,000 award winner at age 14. Carved from Honduran mahogany with solid-top construction, this front engine, 4-door sedan was painted two shades of dark blue Duco lacquer (the same paint pots brother Vincent had used for his 1946 national award winner, except he reversed the upper and lower colors). The model featured plexiglass window inserts painted on the inside with aluminum color paint, brightwork shaped from aluminum and wheels/tires homemade from mahogany. Phil Rauth spent more than 220 hours designing and building this model (courtesy the Rev. Philip J. Rauth).

Then, came the "brightwork," all cut out of a quarter-inch thick sheet of aluminum and shaped with the jig saw, files, sandpaper, and finally, compounded and buffed to a high luster. Bumpers and grille work, along with other trim, were all hand worked in this fashion. The windshield and other windows were fabricated from 1/16 inch Plexiglas. Once having cut and shaped the plastic window inserts to size, they were painted on the inside with silver aluminum paint. I had some difficulties heating and shaping them to fit the curvature of the model and getting them to stay fixed in place. I didn't have the advantage of today's high-tech glues.

I turned the homemade tires from mahogany on the small 4-in-1 wood lathe powered by mom's washing machine motor. The wheels were the easiest part of the whole project. Finally, the model was completed, and it was time to crate and ship it to the Guild judges.

I kept fairly accurate records of the time I spent actually building and finishing the model. According to notes I still have in a scrapbook, I began building my model in the middle of February 1947 and spent more than 220 hours to bring it to completion.

The hours were well spent. I learned I had won State and Regional honors and would travel to Detroit for the annual Guild convention and awards banquet. That was almost frightening for a then 14-year-old farm boy who had never been more than 100 miles from home and had never been on a public bus or train. Fortunately, my brother Vince, who had won the previous year, planned to go with me. It was a wonderful trip to Detroit and I thoroughly enjoyed the experience.

1947 Guild Convention

I did some wonderful things with the other Guildsmen on the Detroit trip. I crossed over into Canada, traveling through the Windsor Tunnel and returning on the Ambassador

Bridge — my first trip to a foreign country. I was treated to a baseball game in Briggs Stadium between the New York Yankees and the Detroit Tigers, receiving two baseballs, each autographed by the players of the respective teams. I still have those two treasures in my possession.

I took a boat ride on Lake St. Claire, had lunch at the swank Orchard Lake Country Club, and visited the Cadillac Assembly Plant in Detroit and the GM Styling Section. I also toured Greenfield Village, played shuffleboard at Harson's Island at The Old Club, and experienced test vehicle demonstrations at the GM Proving Grounds in Milford. All of these things were beyond the wildest dreams of a young farm boy from Nebraska.

Then, finally, I went to the magnificent awards banquet with hundreds of attendees. To be named the winner of the junior division, Second National Scholarship award worth $2,000, was almost unbelievable. Thinking back 63 years, it is still almost unbelievable. I had thought I would be building at least two or three more model entries to become a Regional winner, let alone a scholarship winner. The Guild scholarship provided me with a good start on my college education.

Phil's Conclusions

There were so many challenges for a young teen: coming up with an original design, building to specific scale measurements, taking a blank block of hard wood and forming it into a symmetrical shape, dealing with trial and error, finding a way around difficulties, and painting and polishing the finished project. It took a great deal of commitment and dedication to bring it all to a successful conclusion. The many lessons learned by that experience have served me well throughout my life.

I am grateful to my two older brothers who took the initiative to enter the competition and who set an example for me to follow. I am also proud of my two younger brothers who embarked on that same journey, built models and won awards. I didn't become an architectural engineer, cabinetmaker, craftsman or construction contractor like my brothers. Though I worked several summers as a carpenter in the construction industry while in college and graduate school, I was eventually ordained as a priest. I served as a high school administrator for fourteen years, and as pastor of St. James Catholic Church in Mead, Nebraska, St. Anne's in Campbell, Nebraska, St. Mary's in Nebraska City, Nebraska, and St. Joseph's in Auburn, Nebraska, until retiring in September 2006.

Editor's note: Father Philip J. Rauth still has his 1947 award-winning design, but it's in need of some paint and restoration.

Joseph R. Rauth, 1953 — Third State Nebraska, Senior Division, York, Nebraska (Ages 17–18)

I could tell that mom and dad were very proud of Vincent's and Phil's Guild accomplishments, happy that their boys had some financial help to put themselves through college. Schooling had always been costly with nine children, so our parents encouraged the boys to get involved in model building. Also, since the boys were in the basement building cars, Mom knew they were safe. The Guild competition stirred up a good measure of healthy sibling rivalry and a competitive spirit among the brothers.

Father Philip J. Rauth (age 77) with his 1947 Second National Junior Scholarship award winning model. Rauth served as a high school administrator for 14 years, then as pastor of St. James Catholic Church in Mead, Nebraska, St. Anne's Catholic Church in Campbell, Nebraska, and later, at St. Mary's Catholic Church in Nebraska City, and St. Joseph's in Auburn, Nebraska, until retiring in September 2006 (courtesty the Rev. Philip J. Rauth).

1953 Model Building Process

The winter of 1952 to 1953 was my year to take on the challenge. With Vince and Phil winning national scholarships, Dad gave me all the support he could. He loved that return on effort!

In preparation for the project, I had drafted an intricate set of 1:12 scale plans or drawings showing the top, side, front and rear views of my 1953 design. The 1953 model entry was going to be a 4-door sedan, carved from mahogany with solid-top construction, and I envisioned a light green body and a dark metallic green top, shaped plastic window inserts and polished aluminum trim.

By now, Dad knew to take me to Mr. Green, the cabinetmaker, to get my laminated block of carving wood. He had some heavy mahogany plank, maybe four inches thick, wide enough for a Guild model car body. Mr. Green said, "I can glue it up with only one glue joint." We said, "OK, fine!"

When Dad and I went to pick up the block a few days later, Mr. Green had what we had ordered. But, because the glue in 1952 was not as good as now, and because it was in the middle of a Nebraska winter, Mr. Green set the clamped, mahogany block on top of his shop heater to keep it warm. He forgot about it! Well, the glue joint was fine, but the block was scorched and a little darker brown than Honduran mahogany was supposed to be. Mr. Green asked, "What do you think, Joe, will that cut away and still give you solid wood for the car? I am out of more mahogany." I told him I would give it a try.

A month or so into the building of the model using scale drawings I had drafted, one side was pretty well formed. I started to match the other side and a crack showed up between the headlights and mid-hood on the front end. As I continued to work, the crack grew all

Joseph R. Rauth of York won the 1953 Senior Division Third State Award ($50) for Nebraska ($50) with this design. This solid-top construction, 4-door sedan model was made from an intricate set of drawings or blueprint plans showing side, top, front and rear views. Carved from Honduran mahogany, the model was painted with automotive lacquers (light green body and dark green metallic top), with shaped plastic sheet inserts for window glass and polished aluminum parts to serve as chrome trim (courtesy Joseph R. Rauth).

Joseph R. Rauth, a.k.a. Raymund Rauth, holds his 1953 entry model prior to shipping to the Guild judges in Detroit. He had just finished high school and was 17 years and 11 months old at the time (courtesy Joseph R. Rauth).

the way to the windshield. I went back to Mr. Green and asked, "Now, what do I do?" "Well, I guess we'll have to glue and screw it," he replied. This was done and the crack never opened again, but solving this problem created another menacing situation.

But being young and still learning, I used a "Durham" brand, rock hard, water putty to patch or fill the holes where the screws went in. This turned out to be a major mistake because the "rock hard" putty never expanded or contracted at the same rate as wood. When the putty would shrink, a dimple would show through the high gloss automotive paint, and I would have to re-paint the model. I don't remember how many times I had to repair and repaint that left front fender, but more times than I care to remember.

When the model car came back from the judging in Detroit, the "rock hard" putty holes were ready to pop out again, but I had won the Senior Division, Third State Award for Nebraska ($50).

After high school and the Guild, I became a Benedictine monastery lay brother from 1953 to 1970 and ran a cabinetmaking shop producing church and office furnishings. Regrettably, the shop burned down along with my 1953 Guild model entry in 1970. I have spent most of my adult working life as a cabinetmaker and construction contractor. My hobbies include those of an artist and craftsman chiseling sculptures from limestone and building furniture. I attended the 2004 Guild reunion at the GM Tech Center with my older brothers and Guildsmen John and Vincent.

Joseph R. Rauth (shown at age 75) became a Benedictine lay brother from 1953 to 1970 and ran a cabinetmaking shop producing church and office furniture. He spent most of his life as a cabinet maker and construction contractor, but also enjoys a hobby as an artist, craftsman and stone sculptor (courtesy Joseph R. Rauth).

Joe's Conclusions

Through the years as an artist and craftsman, I have always used wood plugs to fill screw holes. Wood plugs don't dimple or shrink—a lifetime lesson learned! Overall, the Guild contest was a great education. It involved conceptualization, scale drawings and scale work, execution, building both sides the same—or as close as young hands could manage—as well as painting, polishing, packaging, shipping, and waiting. All lessons learned and used for life.

Herman I. Rauth, 1955 First State Award and 1956 Styling Scholarship

Entering the Fisher Body Design Contest had become a rite of passage and a matter of family pride, following in the footsteps of my four older brothers. However, I took a low-key approach, as the Guild was something I wanted to do.

Mom and Dad were interested, supportive and sympathetic toward the Rauth boys' ventures, but were busy providing the family's "essentials." I was pretty much left on my own to find solutions to model building problems. Dad would check on my model's progress periodically as he serviced the coal stoker and lifted out the clinkers from the furnace combustion chamber each day during the heating season.

I cobbled together a homemade air compressor using an old refrigerator piston compressor, an electric motor, and a small tank. The purchase of a gauge, air regulator, hose and fittings completed the homespun machine. Although the pounds per square inch of pressure were minimal, it provided enough air to put on a coat of paint on my models before running out of pressure. Quite an improvement over the hand-operated "bug" sprayers my brothers had used.

1955 Model Entry (Ages 16–17)

My 1955 entry was a four door sedan of modest design. It did not have enough of the future in it to warrant recognition above a Senior Division, First State Award for Nebraska of $150. I was very proud and happy to have won the state award, but I was also aware of the letdown feeling that accompanied it. The lesson I learned was that after putting in what I knew were hundreds of hours of labor to complete and finish a competitive Guild model, the hours invested were not enough to get to the top of the competition. I would have to commit more time to the Guild if I wanted to win a college scholarship.

The 1955 model did serve as a great learning experience for planning and building the 1956 entry. The confidence gained through the first model building experience was a benefit of immeasurable quantity and whetted my interest in automotive design and my awareness of design in general.

1956 Model Entry (Ages 17–18)

A sporty red and white convertible seemed the right goal for the next try. The thought that kept running through my mind at this point was, "You haven't heard the last of this model builder quite yet." I began to plot my course of action for the 1956 model entry. I was going into my senior year in high school and getting ready for college. I knew this would be my last opportunity to commit the amount of time that a competitive Guild model required.

By the time I entered the competition the local cabinetmaker, Mr. Green, who had previously supplied the laminated mahogany to my brothers, had retired and closed his shop. My use of poplar for the 1956 model was more because of the convenience and availability than by choice. It was planed and two pieces were laminated together and proved to be quite stable for carving. I found that poplar worked well, but perhaps not quite as well as the mahogany that I'd used for my 1955 model

The design and construction problems of modeling a convertible certainly raised the cost of doing business. Although excited about taking a sporty direction, at times, I questioned making that choice — building a sedan model seemed far easier. Now having to contend with interior space requirements added a larger investment of time and labor. Forming the plastic windshield and trim also took a lot of trial and error before it reached an acceptable level of appearance.

Sometime prior to our Guild years, the Rauth home workshop acquired a Dremel power tool with multiple attachments. I was able to use the very small circular saw blade to put a groove or channel in some aluminum wire for the windshield trim. Many attempts were discarded before I had enough good quality linear inches to wrap fully around the windshield.

All the chrome-like aluminum parts were cut, filed, shaped and fitted, then sanded and buffed to a bright flawless luster. I remember visiting the local flooring and carpeting company to obtain discarded pieces of scrap aluminum which were later selectively reshaped to make bumper and grille parts, light bezels and air vents.

Some of the other metal parts were made from found items. Tubing and plate stock were used to make gauges, taillights and headlights. A pin with a very small glass head was used for the gear shift lever and the pointed end of a pin was used for the dial indicator needle on the speedometer.

The hubcaps were made on a very small metal lathe by first turning and forming the hub profile in wood and then fastening a thin oversized sheet of aluminum over the wood. A shaped hardwood stick was then pressed against the spinning aluminum plate to press and stretch the metal to conform to the wood profiles. Lastly, a shaped fin-like metal cutting bit was pressed into the hub, with the machine at rest, making a series of fin or spoke-like indentations. This was achieved by using the indexed stop pin and cranking the head and bit into the hub over and over again until completing 360 degrees. Many attempts were discarded before I got four hubs that looked alike. Lastly, the hubs were polished on the lathe before being carefully cut off the wooden mandrel to fit the inside diameter of the recess molded in the official Guild rubber wheels.

The double striped whitewalls were done by chucking up the official Guild wheels and painting the stripes as they were slowly rotated. It took more than a few clean-offs to get all four tires to look alike.

Finally, the rubbed and waxed Bolero red and Dover white lacquered model was adorned with the quasi-chrome attachments, wheels and hand-spun hubs as well as other open cockpit interior details. A sturdy wooden shipping box was assembled to protect the spit-polished entry. After shipping by rail from the Burlington Rail Road Depot in York, Nebraska, it seemed that I had nothing to do but wait.

I found out via telegram I had won a Senior Division, First State Award for Nebraska ($150). Later, I was awarded one of the 10 National Styling Scholarships worth $1,000. That news was thrilling, as it meant that I had also won a trip to the Guild Convention. My brother Phil was my escort on the railroad trip to Detroit.

1956 Guild Convention

The new experiences that I was exposed to throughout the Guild Convention were eye-openers: a first class buffet luncheon at the Bloomfield Hill Country Club, the parade of Guildsman down Main Street in Grand Rapids, a furniture factory tour, a visit of the GM design facilities for an evening chat with professional car stylists, the high speed test track at the Milford Proving grounds, and sightseeing at the Henry Ford Museum and Greenfield Village. All were great opportunities.

I cruised on Lake St. Clair and swam along its shores at the famous Metropolitan Beach. While on the cruise, the captain opened the access doors to allow Guildsmen to peer into the engine compartment to see two humongous diesel engines that powered the craft.

With a degree in education from Kearney State College, I enjoyed a 35 year career as an educator in the Nebraska schools, but also enjoyed a part-time venture as an artist in ceramic pottery doing a few street shows and exhibiting in a few local galleries. My wife Jenene and I live near Council Bluffs, Iowa.

Herman's Conclusions

The lessons learned from the Guild competition were numerous and challenging. Model building taught me astute observation skills such as accuracy in measurements, the precise fitting of parts and pieces, extreme care and craftsmanship in the implementation of a model design and a great amount of patience and persistence. The process of reaching

Herman I. Rauth (age 72), with his model that won a 1956 National Styling Scholarship. He also won a 1955 Senior Division First State Award for Nebraska ($150). With a degree from Kearney State College, Herman was an educator in the Nebraska school system for 35 years and enjoyed a hobby as an artist making show quality stoneware pottery (courtesy Herman I. Rauth).

the end goal brought me challenges sometimes requiring research and using unfamiliar materials or techniques. It called for reaching out to others as well being open to their ideas, suggestions and expertise, working with available materials, and then, returning to one's own corner of the world to try them out. Then if something didn't work the first time, I quickly learned to either repeat or delete the process and move in a different direction until my results met my own quality requirements. All of these things were essential to reach acceptable quality in one's model.

The lessons learned from the Guild competition were very useful to me throughout my adult life. I am thankful to my father and mother for their encouragement and support and also to my older Guild brothers for passing on their Guild enthusiasm, knowledge and support to me. I am grateful to the people who had a wealth of foresight in setting up the competition as well as for the support GM provided in implementing the Guild over its long life.

Notes

1. Five Rauth brothers (pronounced Roth) constructed six Guilds models for competitions in 1937, 1946, 1947, 1953, 1955 and 1956.

2. John F. Rauth (deceased), the eldest brother, was the 1937 first State Award winner for Nebraska ($150) at age 16 in the first FBCG model car competition. Carved from balsa wood, the deep, dark blue four-door sedan model was an aerodynamic, fastback design featuring a cab-forward passenger compartment and rear power plant. Polished tin was used to indicate each window area. John Rauth was in Army Air Corps Flight School when Pearl Harbor was attacked. He went on to become a fighter pilot with 52 missions out of Africa. Shot down over the Bay of Sicily, he was taken prisoner a short while before Italy was liberated by Allied forces. He flew with the Missouri Air Guard throughout the Vietnam conflict. John F. Rauth attended the 2004 Guild Reunion at the GM Tech Center with his model and his brothers Vincent and Joe. He and his family operated a high-end home remodeling business that focused on kitchens, bathrooms, and room additions.

3. Vincent J. Rauth (deceased) was winner of the 1946 Senior Division National Scholarship Award worth $4,000 in the FBCG model car competition at age 19. His 1946 model car was carved from Honduran mahogany and featured an aerodynamic, cab-forward design with a rear engine. Cooling scoops, or air intakes, were located in the front of the rear quarter panels. He used two dark shades of blue "Duco" lacquer to paint the model with a light blue lower body and a dark blue upper or roof. Vince Rauth attended Notre Dame University, the same university his father attended, and studied architectural engineering, and worked for Minoru Yamasaki and Associates from 1963 to 1979. One of their projects was the World Trade Center in New York City. Vincent studied art and design at Cranbrook in Birmingham, Michigan, and attended the 2004 Guild Reunion at the GM Tech Center with his brothers, John F. Rauth and Joseph R. Rauth.

2

A Tale of Two Brothers

Roger D. Teter (1949)

In the spring of 1949, I was 14 years old and interested only in shooting my BB gun, climbing trees or going fishing in the small creek behind our barn. The last thing I considered doing was building a model car and entering it in the Fisher Body Craftsman's Guild contest. So how did a boy who was not particularly interested in cars win the 1949 Third National Junior Scholarship and a $2,000 award in the only Fisher Body model car contest he ever entered? It all began with my brother.

During the Second World War, my older brother, Keith, had carved many airplane models from balsa wood. After the war, he'd started to build flying model airplanes and enter them in meets held near our home in Elkins, West Virginia. Early on, Keith saw that he needed someone to run after the airplane models, retrieving them when they had completed their flight. Perhaps he also just wanted company as he built models on the enclosed back porch of our small farmhouse. Whatever the reason, Keith badgered me to help him. He insisted that I was wasting my time playing, fishing or climbing trees and that I should build and fly model airplanes with him.

My first attempts at model-building, however, were less than stellar. The wings drooped, and the paper sagged. Keith was a patient teacher, though, and soon had me completing airplane models that actually flew. Of course, all stages of my model had to earn his approval. If anything was not up to Keith's standards, I had to start over. As Keith learned new modeling techniques and better ways of doing things, he made sure that I learned them as well.

As members of the local model airplane club, Keith competed with the older group and won many prizes while I competed against the younger, less-skilled model builders. It was fortunate that those prizes included model kits and supplies because money for hobbies was scarce. With Keith's instruction and skills leading the way, the two of us became proficient model builders and competitors. In every model airplane meet that we entered, we won in at least one event and usually in many different events. Together we pooled our winnings and continued to build better and less expensive models.

Then, in 1948 Keith carved a model car from laminated blocks of balsa wood and entered the Fisher Body contest, winning the Senior Division First State Award for West Virginia. His rear engine design model was painted light blue and featured a long extended rear end. The "trunk" compartment sat forward of the engine and was accessed through doors on either side of the car with the passenger compartment moved forward.

Afterward, he insisted that I enter the contest with him the following year. Being three and a half years younger than Keith, I would be eligible to enter the contest's less-competitive junior division. So it was at Keith's insistence that, in early 1949, we branched into building model cars. Several local children had contracted polio at that time, leading health officials to advise families to keep their children at home and avoid large groups. Since we were not allowed to travel to model airplane meets, my brother and I focused our attention and energy on building model cars.

Although Keith's 1948 entry had been carved from balsa wood, he seemed to think that it would be easier for me to cast a model out of plaster than to carve it from wood. He decided that we should develop our models in soft clay, make a mold from the clay, and then cast the car from plaster of Paris. This process helped us ensure that one side of the model would be a mirror image of the other. Casting a plaster model, however, turned out to be more difficult than Keith had originally thought. In later years, he reverted to carving his models from wood.

We labored on our cars throughout the spring and summer of 1949. As we approached each new question of what to do or how to make something, Keith would experiment and decide how we would proceed. After he had developed a technique and made his parts, he would then teach me how to proceed. As a result, all my efforts could be focused on simply building the model.

My 1949 model car design had wraparound bumpers and a grille with vertical bars in the center. I struggled to develop a good headlight design and finally made them rectangular. The rear window of my model resembled the back window of the 1950 Pontiac Catalina. That year, Keith designed a fastback 4-door sedan.

When we were satisfied with the design of our soft clay models, we covered them with a thick liquid latex mixture, which hardened into a flexible cover. We then made an outer plaster form to hold the flexible latex covers in place after the clay models were removed. At that point we were ready to cast our models in plaster. We built up the thickness of the plaster to about one-and-a half inches inside the latex mold and then added a fabric reinforcement to strengthen the final product. After it was removed from the latex mold, we soaked each model in a solution designed to strengthen and harden the final product. Then, we set the plaster models aside for many days to slowly dry and harden. Throughout the entire process we were very fearful of breaking our models.

During the curing process, we worked on other parts. Keith had located a low-speed drill that he strapped tightly to our workbench. We then used the drill as a mini lathe to make the wheels. After a couple of tries, Keith perfected the process and taught me how make wheels for my model. We used polished wire for the window trim on our models. We soldered small diameter wires to the larger molding wire, and then, ran the small wires through holes in the plaster with adhesive for firm attachment points. Although we searched for other materials to use, we eventually made the bumpers and grille parts from wood painted silver. Developing the parts for the chrome bumpers and grille proved difficult. Keith explored using different materials, but those, too, were made from silver-painted wood since we were running out of time.

When the plaster models were dry, we sanded, primed, painted and polished them. Because our father was a General Motors car dealer, Keith and I had access to and experience in using tools and facilities that were very helpful in painting and finishing our models. We also had access to paint and other supplies at my father's automobile paint shop; we mixed the paint from various leftover cans until we had a color that suited us. I painted my 1949 stylish 4-door sedan blue, and Keith painted his a light green.

Above: Roger D. Teter (age 14) of Elkins, West Virginia, with his model car that won the 1949 Third National Junior Scholarship ($2,000) award. This photograph was taken a few minutes before brothers Roger and Keith packed their model cars for shipment to enter the Fisher Body design contest in Detroit. Their dad was a Pontiac and GMC dealer in Elkins. ***Left***: Father (with tie) Lloyd S. Teter, Sr., with his son Homer Keith Teter, age 17, who won the 1949 Senior Division First State Award for West Virginia ($150). Constructed from plaster of Paris, this model had hand-carved wooden wheels/tires and polished wire window trim secured in place by multiple soldered pins. It was painted with light green automotive lacquer with hand-carved wooden bumpers and grille painted silver (courtesy Roger D. Teter).

Eventually, we completed our cars and shipped them off before the scheduled deadline. A short time later we received telegrams: Keith had again won the Senior Division First State Award for West Virginia. I had won the Junior Division Regional Award as well as the First State Award for West Virginia and a trip to Detroit to attend the awards ceremony that August.

I took the train to Detroit by myself, where I was fortunate enough to win the $2,000 scholarship in the Junior Division. We referred to it as a two-year scholarship, for it actually paid for two full years at West Virginia University. When I was at the 1949 awards dinner in Detroit, an elderly gentleman (whom Keith later identified from a photograph as Charles F. Kettering, GM's director of research and development) took me aside and commented on how well my car's two sides matched. He said that the modeling techniques I'd used had been one of the reasons that my car had won the scholarship. He had gone on to point out, though, that my model's front windshield posts were too small and would not adequately support the roof of an actual car — a defect, he added, that had taken only a few points off

Sixty years later, Roger D. Teter poses with his model car that won the 1949 Third National Junior Scholarship award. The $2,000 prize financed two full years for Roger at West Virginia University, where he earned both undergraduate and graduate degrees in engineering. There followed a successful, lifelong career at Lockheed Missiles and Space in California (courtesy Roger D. Teter).

my model's overall score. I had no idea, at the time, that I was talking to the famous inventor of the automobile's first electrical self-starter ignition system, first installed in a Cadillac in about 1912.

Later, my model car was put on display by the Guild. During that period, the fragile plaster model was damaged with deep scratches along each side and large chips gouged out of the plaster. Fisher Body informed me of this fact and tried their best to repair the model.

That was my last Guild competition since contestants were no longer eligible to compete after winning a scholarship award. The following two years, although away from home, Keith continued to design, build and enter the Fisher Body competition. In addition to his 1949 award, he won the Senior Division First State Award for West Virginia in 1950, and in 1951 he won both the Regional Senior Division and the First State Awards. He carved these models from balsa wood and moldings, with bumpers and grille parts cast from metal.

In 1953, Keith entered an art contest sponsored by *Mechanics Illustrated* magazine and won a scholarship to the Art Center in Los Angeles. After graduating from there, he went to work for Ford Motor Company in Detroit as an automobile designer — the artistic skills he had honed as a boy coming to full fruition. Later, Keith returned to the Art Center and became the chairman of the Industrial Design Department. Meanwhile, his passion for automobiles had spilled over into restoration. He painstakingly restored, by hand, many Mercedes, BMWs, Fiats and MGs from the 1930s and early 1940s. He also drove those vehicles in his daily commute from his home to the Art Center, which by that time, had moved to Pasadena. The September 1973 issue of *Car and Driver* magazine included an article featuring Keith and two of his restored BMWs.

As for me, winning the Fisher Body scholarship had assured that I would go to college. I attended West Virginia University and graduated with a bachelor's degree in mechanical engineering in 1956 and a master's degree in engineering mechanics in 1962. I headed west to California and began working for Lockheed Missiles and Space Division in the newly formed aerospace industry. As a structural engineer I worked on the design and analysis of the first U.S. reconnaissance satellite, called *Discoverer*. My entire career since has been at Lockheed working on various aerospace, missile and re-entry programs.

Recently, I read a copy of Mr. Kettering's August 1949 Craftsman's Guild award banquet address on the meaning of opportunity. He began by saying, "I think you boys have been very lucky in being born at a time when you can't fall down out there on the street without falling upon a couple of opportunities." In addition to my being born at such a time, I was also fortunate to have had a very talented older brother who took the time and had the patience to teach me many skills that have served me well throughout my life. Keith died much too soon in 1994 at age 62 without my ever thanking him for his guidance and encouragement.

3

An Artist Emerges from a Hotbed of Design

Gordon D. Williams (1951)

Growing up in Highland Park in the northeastern corner of Los Angeles, California, I had no idea it was a hotbed of automotive design. It was a place where consecutive state, regional and national award winners from the Fisher Body Craftsman Guild hailed, but I had no clue about this rich local history and environment for designing and building model cars. For most of my childhood, I showed only the usual kid interest in cars and model building, not the passion that led multiple men from the area to build award-winning model cars. Charles Pelly, who became my buddy in sixth grade (and would eventually become another Highland Park Guild winner and a lifelong friend), was the only person I knew with any interest in car design. Then it literally fell into my lap.

It wasn't until I fell sick in the winter of 1949 (age 13) that my sister's boyfriend, Bob Cadaret, thought I would be interested to see his FBCG winning models and his state and regional prizes. He set them on my bed, and my circuit breaker switch was thrown to "ON." My interest level was instantly catapulted into the design and model building world. Everything was different from that moment on: I simply had to design and build a Fisher Body competition model car. Bob, who was also from Highland Park, was delighted and encouraged me to read the *Guild News* and learn everything I could about starting a model entry. He had already begun working on his next model for the 1950 Guild competition and introduced me to Ron Hill, a neighbor who was also building a car that year. They both prevailed to win First Grand Nationals that summer, Bob in the senior division and Ron in the junior division.

After all that summer's excitement, I began work on my 1951 model entry — a 2-door sedan made of balsa wood and with a metallic green body and white window areas. I ordered a beautiful block of balsa wood with its distinct wonderful smell and began trying out design profiles on the scaled template drawings provided by the Guild. The design process went on for quite a while until I was satisfied. Bob guided me, and I would ask his advice along the way. But I never allowed him or anyone else to work on the actual model with the exception of the initial band sawing for the original side-view profile.

I was familiar with most hand tools, as my dad was a "fix-it man" and I had taken the usual junior high school shop classes. Luckily, my parents allowed me to work in the garage and stay up late. My friends couldn't understand why I wouldn't hang with them more often, but I was locked into this project and wouldn't allow myself too many breaks. I

Two views of the Guild model entry made by Gordon D. Williams (age 14) of Highland Park, California, which won the 1951 First National Junior Scholarship ($4,000) award (courtesy Kaile Hart Photography).

learned how to honor the center line and how to make templates and use files, sanding blocks, chisels and X-Acto knives to shape the contours. There were so many obstacles and dilemmas. Inaccuracies always had to be compensated for and made right — and there were many. "Duratite" surfacing putty was my friend, and sometimes it got so thick that it would crack with the application of lacquer primer. Side-to-side symmetry was very challenging, as was chiseling and sanding balsa wood end grain.

Hugh D. von Delden (left) and Gordon D. Williams (right) of Highland Park, California, won the 1951 Fourth Senior and First Junior National Scholarships worth $1,000 and $4,000, respectively. Other 1951 Highland Park winners were Stanley Mott, Third State Senior, and William Ritter, honorable mention. Two other Guildsmen from Highland Park won national scholarships in 1950, Robert A. Cadaret and Ronald C. Hill, who won First Senior and Junior National Scholarships, respectively, worth $4,000 each. Two pairs of national scholarship winners from the same community, back-to-back (1950 and 1951), was unprecedented in Guild history. Stanley Mott, Hugh von Delden and William Ritter also won state awards in 1950 (courtesy Mr. Pluria Marshall, owner/publisher, and Terence Lau, controller, *Los Angeles Wave*, Los Angles).

All the chrome trim, bumpers and the grille were cut, bent, filed, sanded and polished from brass prior to chrome-plating. Tires were turned on a wood lathe from maple. My dad had an old Shopsmith multiple tool that I used as much as possible in the building process. Lacquer spray painting was a real experience with many primer coats, a lot of very accurate sanding and finally the Carlotta Brown undercoat required for the 1951 Olds metallic Palm Green finish coat. The best part: rubbing it out with Dupont #7 rubbing compound, then Vista Simoniz wax. Nothing compared with the joy of the final polishing and fitting in the completed chrome parts, wheels, windshield wipers and door handles.

When all was done, delicate, thin chrome-plated trim was inlaid around each designated window area and the windshield perimeter. The headlamps were recessed into the front fenders without chrome trim. There was a simple front chromed bumper and grille assembly, and the rest of the model lacked a lot of metallic brightwork. The front and rear wheels were covered with sheet metal down below the centerlines of each axle consistent with the overall body design lines, which were rounded and aerodynamic. It was a static model, but it looked like it was in motion.

Building the shipping box was another hurdle, one I let my dad help with as we were running out of time. We needed photos of the finished design, and my dad's public relations photographers at the Los Angeles Department of Water and Power took some beautiful 4-inch square transparencies. Then we raced to the post office just beating the deadline. I was just 14 years old, but I was confident that I had built a winner.

While my model project was going on another older Highland Park modeler, Hugh D. von Delden, was working on his own. We were in some contact with each other to see what was going on. We soon found out I had won the 1951 Junior Division First State Award for California and the Regional Award, while Hugh had won the Senior Division First State Award for California and Regional Award. He went on to win the 1951 Fourth National Senior Scholarship (worth $1,000). We were written up together in the local newspapers, and it was quite a blast in the summer of 1951 as we traveled by train together to Chicago and Detroit for the Guild's national convention.

On the train trip to Chicago, I had waited in line forever to get seated in the dining car for dinner. I was so young, inexperienced and starving. It took forever for my order to be taken, then it was another long wait for my dinner to be served. I thought I had ordered a hot shrimp dinner, but it turned out to be an awful cold salad with little tiny shrimp in a dressing (Shrimp Louie). Things like that are very disappointing to a kid. Still, once I arrived, the convention experience was a really an exciting adventure with some apprehension thrown in, as I really didn't know anyone well enough to have a real buddy in the midst of this grand adventure. There were factory tours, wonderful speeches, good food, outing events with swimming, boating, running, prizes for everything — and, oh yes, I fell in love with our nurse.

Of course, I was nervous, apprehensive and walking on air the evening of the banquet where the national awards were to be announced, where my dreams were to be realized winning First National Junior Scholarship award (worth $4,000). More importantly, the convention's inspirational speeches by business and educational leaders and their advice to young men helped form the ethic and attitudes that I have incorporated and practiced throughout my life.

Due to parental planning my trip home was diverted to Yellowstone in order to join them and my sister on a camping trip. Under other circumstances this might have been ok, but this time I really wanted to be home celebrating with my friends. My parents prevailed. Certainly, I was a very lucky kid to have won first State, Regional and First National Junior Scholarship awards on my first try. It had a huge impact on me and was a wonderful platform to begin high school and plan for college. I followed in the footsteps of Bob Cadaret, and after graduating from high school I applied to the Art Center School for their four-year course in transportation and industrial design — another huge continuation of the Fisher Body Craftsman's Guild process.

One thing that helped propel me into the winners' circle were my peers (Bob Cadaret, Ron Hill, Hugh von Delden and Chuck Pelly) who were Guild participants and winners.[1] Exposure to their ideas and seeing their models made a big difference for me. I also discovered skills I never knew I had — a good design eye and craftsmanship abilities. Now I am glad that I didn't have a clue as to the huge field of talented Guild contestants that I was up against.

Gordon D. Williams, industrial designer and sculptor in Altadena, California, shown with his 1951 award winning model. The rear of the model is facing the camera. While growing up in Highland Park, he knew other highly successful Guild model builders like Robert A. Cadaret, Ronald C. Hill (neighbor) and Charles W. Pelly (a 6th grade buddy and lifelong friend) (courtesy Kaile Hart Photography).

At the time of my national award, I had just graduated from junior high and was not yet in high school. The Guild model builders who were already in Franklin High School when they built their models had an extra boost, as the woodshop teacher, Hugh Barrett, encouraged the guys to use woodshop class to build their Guild entries. Highland Park treated the Guild model builders with some delight. Several years later, my friend Charles Pelly, also of Highland Park, went on to win the junior division, First State award for California, the Regional award and 2nd National Junior Scholarship award in 1954.

After graduation from the Art Center School, I headed for Detroit and then Sweden before returning to California and forming an industrial design office. Later, at a time of economic distress, I responded to a wonderful opportunity in interior design and eventually opened a commercial interior design office in San Francisco called Gordon Williams Design. This was a 35-year experience with office and education projects locally, and banking and publishing projects in Juneau, Alaska, and Athens, Georgia. Upon returning to Los Angeles In 1995, I worked as a consultant to industrial design firms as a designer and model builder, also known as a "prototype fabricator." I streamlined my home shop for model building prototype work as well as abstract sculpture, which I have been doing all along but now enjoy even more.

I have come full circle now since my Guild days. The next chapter in my life began in March 2010 by our relocating from the Los Angeles area to Albuquerque, New Mexico, for the dramatic beauty and so my wife Binah and I could better pursue our artistic goals, she in fiber arts and mixed media painting and I in abstract sculpture. We live in the scenic Sandia Heights area, 6000 feet elevation, with views of the Sandia Mountains and of Albuquerque. We completed a joint art exhibit, in September 2010, as part of the Sandia Heights Art Tour. My work is mostly smooth flowing abstract forms from wood, but also in stone, metal and resin. We are very happy here. Amazingly, my first art project still perseveres, 60 years since completion of my 1951 FBCG model. It looks very good despite some cracks and slight color and surface deterioration and is currently on display in our glass book case.

Notes

1. As noted above, Gordon Williams was the 1951 Junior Division First National Scholarship Award winner and Hugh D. von Delden was the 1951 Senior Division Fourth National Scholarship Award winner. Both young men were from the town of Highland Park, Los Angeles, California. Now retired from a career as an aerospace engineer, von Delden resides in the town of Sierra Madre, California. Von Delden and Williams recently reconnected for the first time since their Guild days just at the moment Williams was relocating in New Mexico. Hugh D. von Delden worked as an aerospace engineer at Cal Tech's Jet Propulsion Laboratory on the world's first satellite (*Explorer I*) launched in 1958, the first interplanetary space craft to Venus launched in 1962, as well as other firsts, such as the *Surveyor* moon lander and the Mars *Viking* land rover.

There were other Guild scholarship winners from Highland Park, California: Robert A. Cadaret (1950 First National Senior Scholarship winner), Ronald C. Hill (1950 First National Junior Scholarship winner) and Charles W. Pelly (1954 Second National Junior Scholarship winner). The success of these five men, all living in the Highland Park community, was unequalled in the history of the famed FBCG auto design contest. Cadaret and Hill went on to become famous automobile designers at General Motors Styling and beyond. Pelly became a successful independent industrial and automotive designer and founder of Designworks USA, now BMW Designworks USA, and is the principal of The Design Academy, an industrial design consultancy. Cadaret, Hill and Pelly were featured in John L. Jacobus' earlier book, *The Fisher Body Craftsman's Guild: An Illustrated History.*

4

An Artisan's Ancestry: From Cars to Craft

William A. Keyser, Jr. (1949–1952)

My family is full of craftsmen going way back — machinists, toolmakers, plumbers, iron mongers, electricians and blacksmiths. My father, as a young boy, worked in his father's plumbing business but later took up carpentry, cabinet making and furniture making. He was self-employed with a shop adjacent to our home. My mother was an excellent seamstress and tailor, making most of her and her children's clothes. But none of my relatives had any training in design or the visual arts. That was where I would make my mark, and it all began with the Fisher Body Craftsman's Guild.

At an early age I became interested in working on my own projects. My parents gave me a bench and a small area behind the family car in our garage to use as my own workshop. Christmas and birthday gifts usually consisted of tools: a toolbox with basic hand tools, a small disc sander that my Dad had made, a jig saw. Combined with access to the hand and power tools in my father's shop, I thought I could do most anything.

I tried model airplane building and then model railroading as hobbies but lost interest. The two industrial arts activities I enjoyed the most as a youth were the All-American Soap Box Derby and the Fisher Body Craftsman's Guild competition — two scholarship programs sponsored by General Motors. The former consumed my time from 1947 through 1951, as I designed and built a brand new racer each year. Almost simultaneously, I was intently engaged in the FBCG model car building from 1949 to 1952. For the Derby I enjoyed local success in the Pittsburgh area, but for the Guild I was a national scholarship winner. Both of these extracurricular programs had a profound impact on my life and led me to a successful career in design and the visual arts, but I give most of the credit to the FBCG.

Sometime in 1948, as a 12-year-old seventh grader, I saw a magazine advertisement in *Mechanics Illustrated* for the Fisher Body Craftsman Guild competition. I was immediately excited by the thought of designing and building a model car from scratch. Already committed to participating in the Soap Box Derby and working part-time with my father as well as having a grass-cutting job, I knew that in order to succeed in both the Derby and the Guild, I would have to really apply myself, plan carefully and use all my spare time.

I began my entry for the 1949 FBCG competition — a red 4-door sedan — by collecting photos of existing cars. Then, I tried to think of different ways of treating the grille, bumpers, and body sections, exploring alternatives through small, crude sketches. As suggested by the information I received from the Guild, I began a clay model but quickly became very impatient. It was new for me, and the malleability of the material was difficult to control. And it seemed like a waste of time to labor over making both sides of the clay model identical. As a result I didn't make a complete clay model before moving on.

The body was laminated from several blocks of pine, some added on when I suddenly ran out of material due to poor planning. The grille and window trim were made from 1/16 inch thick aluminum sheet, cut, filed and sanded to shape. Bumpers were re-worked sections of aluminum wallboard moldings. Headlights were round steel furniture glides mounted within steel bezels from political campaign buttons. Door and trunk handles were indicated by the round heads of brass escutcheon pins. Wheels were lathe-turned with tires painted black, hubcaps painted silver. All the metal trim was glued onto the bare wood body and masked off prior to painting. The car was then painted red with brushed-on enamel: white for the windows, red for the body. The paint seeped under the masking tape in spots and needed to be scraped off the aluminum with a knife, which left the trim uneven and lack-luster.

In retrospect, I now realize the design lacked a coherent concept. It was a collection of disparate automobile details, largely because I did a slipshod job on the clay model and did not integrate the diverse details into a cohesive design. Some transitions between body parts did not flow together smoothly and many curves were not graceful. The workmanship on the model was also lacking, with the shaping and sanding of both the metal and wood parts troublesome. So I was pleasantly surprised on August 1, 1949, when I received a Western Union telegram informing me I had won Second State Award in Pennsylvania. I was elated because it represented success in competing with other boys from across the entire state, whereas my Soap Box Derby competitors were only from the Pittsburgh area.

My 1950 model car, a green 4-door sedan, was only a slight improvement over the previous year. The design and clay modeling phases, as well as the construction methods, were very similar. The craftsmanship improved somewhat, but the necessary precision still

This two door sedan designed and built by William A. Keyser, Jr., was carved from wood, painted with black enamel and featured carved aluminum bumpers, sculpted Lucite headlamp lenses, die-stamped aluminum hubcaps, and a brushed, multi-coat enamel finish. The final color coats were meticulously hand rubbed with oil and rottenstone. The design captured the 1951 Junior Division First State ($150) and Regional (Pennsylvania and Maryland) Awards that included a trip to Detroit for the Guild Convention and Awards Banquet (courtesy Geoff Tesch Photography).

fell short. It was made from carved wood with aluminum trim, and brushed enamel paint, and featured a ventilation air scoop in the roof. When the telegram arrived on August 1, 1950, I was disappointed to learn that I had only taken third place in the state. At this point my mother consoled me and later wrote in a letter to GM: "Personally I think [that was] the best thing that happened because it taught Bill that we must take these disappointments in life without becoming discouraged and [it] gave him an incentive to try harder in the next year's competition." Fortunately, it was a huge consolation when my Soap Box Derby racer was awarded a trophy for the Best Designed Racer in Pittsburgh that summer.

I responded to my mother's inspirational advice the next year in the 1951 FBCG competition with a 2-door sedan made of carved wood and painted with black enamel. The design featured a front and rear-end that were much more three dimensional than the previous cars and better integrated into the overall design. The headlights, instead of being simple rounds, had lenses three dimensionally carved from blocks of clear Lucite. Bumpers were sculpted from thicker aluminum and became more dimensional. The hubcaps were die-stamped out of aluminum. After several coats of enamel paint, the model surfaces were hand rubbed using oil and rottenstone. This created a satin smooth finish. While the entire design flowed well, there were still some flat, uninteresting areas. The workmanship was better than in previous years but still not perfect. Brushing on the enamel finish continued to give me problems, but it was the best car I had made to date.

William A. Keyser, Jr., a 15 year old sophomore at North Catholic High School in Pittsburgh, holds his 1951 Junior Division First State and Regional (Pennsylvania and Maryland) Award winning model following his return from the Fisher Body Craftsman Guild Convention in Detroit. This photograph, along with a story of Keyser's quest, appeared in a fall 1951 issue of the school newspaper, *The Trojan News* (courtesy Frank M. Orga, president, North Catholic High School).

I was overjoyed when I learned my entry won First State and Regional Awards for the Pennsylvania and Maryland area, including a trip to Detroit for the national competition. It was my first train ride and my first time out of Pennsylvania. The trip was awesome! We toured the GM Styling Section, Chevrolet Forging Plant, Pontiac Stamping Plant, and other automobile assembly lines. While these tours were very interesting, the most important part of the trip was seeing the other models, especially the National Scholarship winners. It was amazing how beautifully designed and how wonderfully crafted they were. The paint jobs — almost all sprayed and rubbed lacquer—were impeccable, and the metal trim really sparkled! The secret seemed to be using chrome-plated brass. I knew immediately that I had no chance of being a national winner that year and that I would have to raise the bar considerably on both my designing and my craftsmanship if I were to advance to such a high level.

When Guild registration opened for the 1952 competition, I immediately began designing my next car, a 2-door sedan carved from poplar. I spent hours sketching designs of overall body shapes and of grilles, headlights and bumpers. Next, I made a rough, preliminary clay half-model, to quickly see how my ideas worked. A second complete clay model was sculpted, this time to exact scale for both sides of the car, followed by full-size elevation and plan view drawings to finalize the design. This design phase took at least five or six months. But now I had a very precise, clear and detailed picture of how the car would look, and I knew exactly how I would construct it.

I was determined to eliminate the masking of trim work prior to painting and the tedious scraping of paint seepage afterward. The design of my car featured a horizontal rub molding that ran the length of the car and integrated at either end into the front and rear bumpers. This trim divided the lower body into two horizontal sections. Likewise, a thin molding under the windows separated the cab and roof from the lower body. This design suggested a layered assembly scheme. The horizontal rub moldings and the window trim could be sandwiched between the separate wooden body and roof segments. The entire assembly could then be screwed together from underneath, after the wood parts were spray painted.

I started actual construction by making the bumpers from 1-inch by 1-inch by 6-¼ inch brass bar stock, hack-sawed, filed and finally sanded and polished by hand. The grille components, head and taillight receptacles and the license plate frame were sweat soldered to their respective bumpers. These were tapped and screwed to two thinner brass bars, the edges of which would be visible as the horizontal rub moldings. Hubcaps were made in two pieces from sheet brass, cut to shape, contoured with turned hard maple stamping dies and

Three-quarter front view of the 1952 Second National Junior Scholarship ($3,000) award winning model made by William A. Keyser, Jr., at age 15 (courtesy Geoff Tesch Photography).

William A. Keyser, Jr., at age 73 in 2009, holding his model car which won the 1952 Junior Division First State Award for Pennsylvania, Regional Award (Pennsylvania and Maryland), and 1952 Second National Junior Scholarship Award, a $3,000 college scholarship. Keyser attributes the FBCG with developing his lifelong passion for designing and making beautiful objects in his career as a craftsman, artist and teacher (courtesy Geoff Tesch Photography).

then mounted on turned maple wheels. Two side air vents and the door handles were formed from brass stock and eventually recessed into the wood body sections. All the brass trim parts were then chrome-plated and polished by a commercial metal plating company.

The wooden body parts were made of poplar, shaped and fitted around the chrome bumpers, rub molding and window trim. The separate components made shaping quite simple. The pieces were then carefully sanded, primed, sprayed with 15 coats of black automotive lacquer and rubbed out with rubbing compound. The windows were made from 1/16 inch thick opaque white Plexiglas, heated in my mother's kitchen oven and draped over a wooden mold that had been covered with thin stainless steel sheeting. The plastic windows and handmade wheels/tires were added after the painting. I estimated about three hundred hours were spent constructing the model, not counting the design phase.

My parents were always very supportive of all my youthful activities. My father worked very hard making a living at his own woodworking business, and while he furnished me with all the materials, tools and advice I needed, he was frequently not available when there was a crisis. It was my mother who calmed me when things went wrong and who picked me up when I was emotionally down. And so at the televised 1952 National FBCG Convention, when I was asked to comment after being announced as the Second National Junior Scholarship winner, all I could think to say in my excitement was "Hi Mom!" It was a special moment for me and a special tribute to her as well.

Believing I wanted to be an engineer, I earned a bachelor of science degree in mechanical engineering from Carnegie Institute of Technology in 1958. But longing for the passion I had experienced designing and building for the Fisher Body competitions, I then pursued a master of fine arts degree in furniture design at the School for American Craftsman at the Rochester Institute of Technology (RIT). Subsequently I taught furniture design at RIT for 35 years while operating my own studio designing and building custom furniture for the corporate, religious and residential market. My work has been exhibited nationally and internationally, has won many awards and is in the permanent collections of numerous museums, corporations and private collectors. I received a National Endowment for the Arts Grant (1975), was named a Distinguished Alumnus of Rochester Institute of Technology (1987), was elected to the College of Fellows of the American Crafts Council (1997), was cited for a Merit Award from Carnegie Mellon University (1998) and received the Award of Distinction from the Furniture Society (2003). When I stopped teaching in 1997, I returned as a student and earned another master's degree, this time in painting and sculpture. Today, I continue doing furniture, painting and sculpture.

I believe I have been true to my heritage, continuing the lineage of craftsmanship inherited from my ancestors and passing it on to the next generation. While earning his living as an engineer, my own son's passion is working with his hands. He has built much of his own furniture as well as furniture for clients on commission. He is actively preparing to build a woodworking studio and a timber frame house, harvesting the logs, milling the timbers, and doing other work himself. From my teaching at RIT, I can count hundreds of former students who are following their dreams and earning their livings creating furniture. I get immense satisfaction watching the propagation of my craftsman's heritage.

I attribute my successes to what I first learned in the Fisher Body Craftsman's Guild competitions. Choosing from among many diverse interests as a youth taught me to prioritize my time and energy. The necessity to plan was first experienced when faced with the numerous and complex steps involved in creating a finished model, and I certainly learned patience while executing that plan. The attention to detail required precision work at every turn. Rebounding from mistakes and working at a job for months on end as well as striving year after year to do a better job taught persistence. But perhaps the greatest thing I derived from the Guild was finding the passion for my work and my life that has been with me since I first competed with that red sedan.

5

An Engine, a Dungeon and a Lot of Lacquer

Warren M. Bakken (1949–1953)

When I was eleven I found a drawing in an old book of Henry Ford's first gas powered internal combustion engine. It looked like an interesting project and I needed something a little different from the model airplanes and cars that I had been building since the third grade. My little scratch-built model so impressed my sixth-grade teacher that she insisted that I bring the engine to class and give a little talk. After much coaxing, she even convinced me to donate the model to the school, so she could use it as a history teaching aid. Building it, and subsequent Fisher Body models, though, wasn't easy. My family home in Grand Forks, North Dakota, in 1948 was so small that my brother Wally, two years my senior, and I shared a bedroom with just enough space for a desk and a ¾ sized bed. My oldest brother Wendell had already left home to serve in World War II and then work at Boeing Aircraft Company. My father was in the painting business and hand-painted both signs and automobiles.

I could have worked in the garage, but it was too cold in the winter, and too hot (and buzzing with mosquitoes) in the summer. The workshop I later shared with Wally was under a trap door in the middle of the kitchen floor, down some almost vertical steps, in a dugout cellar. I spent hundreds of hours each year working in that windowless, dead air space. I couldn't work on laundry days though, because my mom pulled out the wringer washing machine and two rinse tubs, taking up the whole kitchen floor.

The local Chevrolet dealer promoted the first Fisher Body Craftsman's Guild design contest I entered in 1949. They handed out the instructions and held a local competition before shipping the model cars to Detroit for the "official" judging. The rules stated that you had to be 12 on September 1, 1948, to be eligible for the Junior Division (age 12 to 15). I was only 11, but I entered anyway. I carved my first Guild model entry, a two-door sedan, from laminated balsa wood. The balsa wood bumpers were brushed with aluminum paint, and, because I didn't own a drill or a lathe, the wheels were hand-carved from balsa wood. I used a hacksaw blade to score the tire treads. The only material I had for trim on my 1949 model car entry was a rolled up sheet of very thin polished aluminum, about the thickness of heavy paper, courtesy of Wendell and the Boeing Company. There was very little of it, and I treated the stuff like platinum. I used it to make the grille and the chrome-looking strip down the center of the hood as well as window area inserts. My dad, who was a sign painter, showed me how to brush on the lacquer primer and sand it down with progressively smoother sandpaper until the final coat was sanded with 600 grit. With the model all primed and sanded, I let dad take over and he brushed on the flat green blackboard paint. When that was dry, a nice thick coat of marine spar varnish was applied. It looked stunning!

My best source of materials was my brother-in-law, Jerry, who worked in a dental lab. He supplied me with buffing wheels and abrasive compounds, and later on in my Guild career, with some half-round, stainless steel dental bridge wire that was just perfect for window trim. Jerry tried to drum up interest in using dental wax, instead of sculptor's clay, to work out design ideas. He was also my best friend and was always willing to help in any of my modeling endeavors.

The car salesmen and the local Chevrolet dealer, who were judging our 1949 models, awarded me first prize and $10 for beating four other contestants in town, mostly because I had the best paint job. In addition, I was officially notified by the Guild that I had placed third in the Junior Division in North Dakota and won $50. If I hadn't won all that cash, I might have given it up as a lost cause. But the reward jump-started my enthusiasm, and I was ready to try again.

I entered the Guild competition again in 1950 as a legitimate 12 year old. This time I focused on a Sherwood Green two-door sedan design also with hand-carved balsa wood wheels. I employed my dwindling supply of polished thin sheet aluminum for window trim, which was carefully cut with an X-Acto knife and glued in place using Duco Cement™, the only glue I could find that would adhere to aluminum. I made window area inserts from 1/64 inch thick celluloid, after painting them gray on the inside so that the outside surfaces had the highly-polished look of glass.

My dad did a lot of gold and silver leaf appliqué on window glass and this could have a very smooth and lustrous finish. He showed me how to apply silver leaf to the bumper surfaces of my 1950 model. It looked pretty good for a while, but after a short time the silver leaf tarnished and looked just like old neglected silverware. All the lustrous finish had to do, though, was last through the Guild judging process. It did, and I won the Junior Division First State Award for North Dakota and $150.

My brother Wally became interested in the contest once he saw I was winning prize money and having a good time doing it. We would both work on our models together when we could and often shared the expense of tools and materials. When the weather was nice in the spring he would take his model outside to work in the sun since it was kind of crowded in the "dungeon" cellar workshop.

About this time I saw an article in *Model Airplane News* on how to chrome-plate model airplane pistons. It sounded like the magic bullet I needed to win a scholarship. The article told what chemicals to buy, how to set up the plating tank and so forth. This looked like the perfect solution to my chrome trim problem. All that was needed was a car battery, a window shade roller spring for a rheostat, a teaspoon of sulfuric acid, some chromium trioxide crystals and two lead plates. Just heat it to "coffee cup" temperature, the article said, and you are in business. I tried and tried but couldn't get it to plate or "throw," as it was called. The only thing that seemed to help was to add a cup of sugar to the tank. This started a vigorous rolling boil and I quickly rushed the whole experiment outside because I thought it was going to blow up. After it had cooled down a bit, I tried it again and amazingly, it had plated a gray finish that could be polished to a high luster, but only on very small parts.

For my third try at the Guild competition, I spent less time on the actual building and more time designing. And since my chrome plating wasn't working, I decided to build a

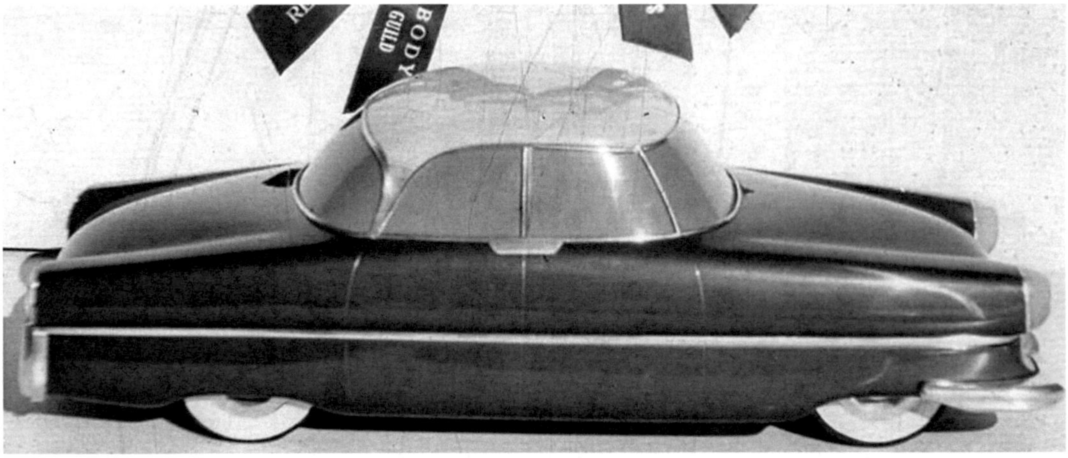

Side view of the four-door sedan made by Warren M. Bakken (age 14) of Grand Forks, North Dakota, for the 1952 Guild competition. This was his fourth entry in the Guild model car competition with a dark metallic Cypress Green lacquer body and light Sportsman's Green lacquer top and light blue indicated window areas. The construction featured a carved balsa wood body, resin cast headlights using the "Lost Wax" process, sculpted aluminum bumpers and polished aluminum hubcaps. The model won the Junior Division First State Award for North Dakota ($150), Regional Award and a trip to the Guild convention in Detroit (courtesy Warren M. Bakken).

bright red lacquered two-door sedan, with integral bumpers which were also painted red so I wouldn't have to deal with the chrome problem. I used aluminum paint on the window areas and I cut elliptical-shaped headlights out of my ever-dwindling supply of aluminum sheet. This was scribed with rectangular cross-hatching and glued on for headlamps. The wheels were again carved from balsa and hubcaps were painted aluminum. I didn't bother with chrome around the windows. This model won a respectable Junior Division First State Award for North Dakota and $150.

On my fourth attempt in 1952, I was still working with balsa wood and made a 4-door sedan, spray painted with dark metallic Cypress Green for the body and light Sportsman Green for the top, with windows indicated with a light blue. I used only as many coats of lacquer as were necessary to sand smooth with 600-grit sandpaper, and just thick enough to rub and polish with fine rubbing compound. I used very fine powdered calcium carbonate (chalk) called whiting for the final polishing before waxing the finish. My painting skills had improved. The big bulbous headlamps were shaped initially in wax, and then cast from acrylic resin. The bumpers were carved from a solid block of aluminum. I was still using my highly prized, but even more limited, supply of polished sheet aluminum — this time for hubcaps. The half-round stainless steel wire around the windshield and the polished aluminum bumpers impressed the judges enough to give me Junior Division First State and Regional Award with a trip to the national awards banquet in Detroit.

When I left home to attend the National Convention, I didn't have far to go to catch the train since our house was just a block from the station. Like other regional winners, I had to change trains in Chicago. Since I was one of the first to arrive early that morning, I got to see a bit of the Windy City — even a big league ball game at Wrigley Field. A couple of Guild representatives met the trains as they arrived. They could've passed for FBI agents in their tailored suits as they stood on the platform with photos in hand to identify each lost-looking kid. Our train for Detroit didn't leave until late in the evening. The GM "babysitters" left us in the Conrad Hilton while they made the short trip to the train station to meet new arrivals. About fifteen or sixteen of us were sitting around the hotel lobby watching a little round 6-inch television, the first one I had ever seen. Then, sometime in the evening, Stan Mott bounded in.

I was an impressionable 15-year-old who had never been away from home, and I was just in awe of this guy from California. He was 19, dressed in low-cut worn blue jeans, a white T-shirt, and the latest California fashion — a ducktail haircut. He walked into the room with this old beat up leather suitcase and set it down near the wall, looked at everyone, picked it up again, turned it around and said, "I was showing you the good side!" After he arrived we could no longer be wallflowers.

At 10 P.M. four taxicabs dropped our lot off at the train station. We were soon in our bunks in a sleeper car for the overnight ride to Detroit, but Mr. California would have none of that. He said, "Who needs to sleep anyway? Let's play some blackjack ... for money!" I wasn't new to card playing, but I was to blackjack. My new friends were all senior division winners, as the sensible kids my age had gone to sleep. We played in the middle of the aisle, using a suitcase as a table until well past midnight, the first of many late nights for a regular group of us.

Once we arrived, I soon realized how much going to Detroit gave modelers an edge that could be the difference between winning a scholarship or not. For example, during one of the side trips at the convention, we stopped for about an hour at the General Motors Building in Detroit where all the Guild model car entries were on display. We just swarmed over the place trying to pick up ideas for next year. I noticed many of the model cars had their tires filed flat on the bottom to look just like a real tire. But it was never mentioned in any of the *Guild News* publications. Such detail made a lot of difference because anything that helped make a model look longer and sleeker without falling outside the Guild design rules offered an aesthetic plus. It solved a lot of design problems for me on my next entry when I could drop the wheels down and lower the fender and roof lines for a more streamlined visual effect.

My last year of eligibility in the Junior Division was 1953. From my trip to Detroit last year, I knew how perfect the cars had to be to win in the Senior Division. I really didn't want to keep building cars year after year; I just wanted to win a scholarship. Besides, I would have been competing with my brother since we would both be seniors. So I went all out to try and win a scholarship in 1953.

The year started out good. My dad came home with an old electric motor one day, and I made it into a lathe by fastening a two inch diameter metal faceplate to the end of the motor shaft with a setscrew. I could now fasten little flat pieces of brass to it with plaster and turn out taillights and headlamp rims, using a sharpened file resting on a piece of wood as a cutting tool.

To make the solid aluminum wheels/tires, a more substantial holding fixture was required. I bolted a round piece of wood about three inches in diameter and an inch thick to the same faceplate. A shallow circular recess was cut into the wood that was half as deep as the tire width, and just under the two and seven sixteenths inch diameter of the aluminum bar stock. The one inch thick disks of aluminum were press-fit into the fixture, and the outside half of the tire sidewall, hubcaps, and tread were shaped and grooved. I burnished the final shine on the hubcaps during the spinning operation with balsa wood impregnated with rubbing compound. The piece was then turned around and the inside of the wheels and the other half of the tire treads were cut. I only needed to make three wheels because

Warren M. Bakken (age 15) in the early stages of designing his 1953 model entry in clay. His mother allowed him to do this work in the bedroom, as it was both noise-free and dustless. Once the model was cast in plaster, however, work had to done outside or down in the cellar or "the dungeon," as brother Wallace called it (courtesy Warren M. Bakken).

the rear ones were cut in half. To paint the whitewall tires, I used a drafting compass with an India ink attachment to draw the perfect line of separation from the white which was painted on the wheel first. Then all I had to do was brush the flat black up to that line. The lucky people that entered after 1953 had their wheels supplied by GM.

The *Guild News*, which we looked forward to receiving every few months, showed how easy it was to make a clay model and cast it in plaster. Both Wally and I wanted to try our luck building the models out of plaster, instead of wood, but the clay was very expensive and not something we could readily buy at the store. After experimenting with some of Jerry's dental wax, we broke down and sent away for some good quality modeling clay. I think having the clay model to work with was the step that put me over the top — it made it so easy to try different looks and designs. I had mine sitting on a board so it looked a lot like a derelict car body sitting out in an empty field. I think it would have been better to have the wheels in place on the clay model to get the right proportions for the wheel openings.

My clay model was cast in plaster in April. It took weeks for it to dry out. I was afraid to put it in the oven for fear it would crack. This two-door sedan entry was painted with lacquer colors — namely Colonial Blue for the body, Honey Cream for the top, and black to indicate the windows. The window trim was the same stainless steel alloy dental wire that both Wally and I used the previous year that was so easy to form and dazzled after being polished. My exhausted stash of prized paper thin, polished aluminum came to the rescue covering the butt-joints of the dental wire trim used on both the windshield and the

Close-up view of the sculpted brass front bumper made by Warren M. Bakken for his 1953 plaster cast Guild model entry shown during one of many "cut and fit" assemblies. This was shaped from a solid brass bar (2 inches by 2 inches by 7 inches) using a hack saw, various hand metal files and a vise. Brass extensions were soldered on as the bumper wrapped around the sides toward the front wheel wells. Cardboard templates were used to achieve symmetry and the bumper was permanently attached to the plaster cast body with a one-quarter inch diameter threaded rod soldered to the back of the bumper assembly. The brass bumper and other brass parts for the model were sent to Minneapolis for chrome-plating for a cost of $12.50. Note the brass headlight rim (courtesy Warren M. Bakken).

rear window. About 60 small brads had to be soldered to the back of these alloy trim pieces to act as fasteners, and it amazed me that soft solder would bond so well. The headlamps were the viewing lenses from a box camera, and behind each lens was an aluminum reflector with a scribed rectangular grid. I was still masking and spraying the final coats of lacquer just three days before the June 30 deadline. The finished model weighed 15 pounds complete with the solid turned aluminum wheels/tires, chrome-plated brass bumpers and trim.

I had visions of another trip to Detroit. Everything was going along just fine until the unthinkable happened. As my dad and I were placing the model into the shipping box, the metal case holding his reading glasses slipped out of his breast pocket, a sharp corner landing on top of the left front fender, knocking off a chip as big as a pencil eraser. After nine months and more than 700 hours of work, I had serious doubts about my chances of winning. My dad was devastated. We didn't have time for any repairs and tinkering with it would only make it look like bad workmanship. So we decided to ship it that way and hope the judges would think it happened in transit or was damaged when the model was unpacked.

Now I wasn't so sure I'd be going to the Guild convention. How many points were the judges going to deduct for the chip? What was my competition doing this year? Their learning curve was going up too. I was so relieved to get a telegram in July stating that I had won the Regional Award again. My brother Wally also won the Senior Regional Award, so we traveled together to Detroit. We took the train to Chicago and spent the day walking around the loop area sightseeing and the evening waiting for the other winners to arrive. Stan Mott showed up again, having won the senior division in California in his last year of eligibility. Some of other winners who arrived included Tom McDonnell from Stockton, California, who became the Third National Junior Scholarship winner, and Robert Leger

Two views of the two-door plaster model made by Warren M. Bakken (age 15) of Grand Forks, North Dakota which won the 1953 First National Junior Scholarship ($4,000) award (courtesy Warren M. Bakken).

from Sulphur Springs, Louisiana, who became the Fourth National Junior Scholarship winner. During our stay in Detroit, regional winners had GM employees with former Guild connections as chaperones. Our group leaders included Gordon Drummond, a 1932 Grand National Winner in the Guild's former Napoleonic Coach Competition, and Ted Mandel, a top national model car builder in 1937.

At the 1953 Guild convention, my group leader kept pressuring me to give one of the famous luncheon speeches, but I just couldn't. Usually GM executives and their wives were in attendance, as well as prominent college presidents and other dignitaries, and my palms got sweaty just thinking about it. He said they would help me write the speech, but I flatly refused. Stan, on the other hand, was self-confident, making speeches and successfully promoting himself to GM officials as someone they should hire. A Guild photographer accompanied us on our sightseeing trips, and Stan was always aware of that. Our souvenir photo album shows Stan Mott in just about every picture. He was eventually hired by GM Styling and moved on to become a cartoonist for a popular automotive magazine.

Despite my aversion to the limelight, I was most relieved at the banquet to hear my name called as the First National Junior Scholarship ($4,000) award winner. Anything less would have haunted my dad forever.

My brother Wally (Wallace E. Bakken) had entered models when he was fourteen and fifteen, in the Junior Division, and won two second place State awards. The next year he moved up to the Senior Division and built four more models and won three trips to Detroit, the first one with me in 1953. Wally's 1954 design which won Senior Division, First State and Regional Award, was cast in plaster from a clay model pattern. It was a two-door hardtop with a metallic maroon lacquer paint job, black lacquer for windows and stainless steel trim. He had an elaborate chrome-plated brass grille and bumper combination, and chrome-plated brass hubcaps. All this was done with the most rudimentary metal working tools—hacksaw, Swiss needle files, polishing wheels, and the most used tool of all: a hand vise to hold the tiny parts.

Wally's 1955 model was also cast in plaster from a clay model. This was a beautiful 2-door hardtop with a metallic blue body, white top, and black windows. The concave sculptured doors were a rather new innovation at the time and showed a lot of originality. Most of the trim pieces were chrome-plated brass, except for the window frames which were made from half round dental wire. Other than the bumpers, which were attached last, all of the trim, except for the hood ornament, was secured during the primer paint application stage. The door handles were integral with upper decorative side chrome trim pieces. This made the fine sanding between paint coats extremely difficult, as he had to avoid scratching the chrome surfaces. All these parts had to be masked and re-masked before the final coats of color paint were applied.

His reward for all this hard work was a First State Award, and a Fourth National Senior Scholarship of $1,000. He and John E. Holding from Medford, Massachusetts, may have been the only contestants without a parent working for GM to actually tie for an award, and both received scholarships. During the 1955 awards banquet broadcast on national radio, Charles F. Kettering, director of research and development for General Motors, introduced the winners. When Wally's name came up, he said: "Mr. Bakken here is somewhat of a philosopher. When I asked him how many hours he had put into building his model car, he replied, 'It's not how many hours you put in, it is what you put into the hours.'"

Wally was always pressed for time when it came to the Guild deadlines because once school was out, he spent his summers working long hours on our uncle's farm 90 miles away. He went on to earn a master's degree in geology at the University of North Dakota and became a consultant in oil exploration, specializing in the Williston Basin Area of western North Dakota. He passed away in 1994 in Denver, Colorado, at age 60.

Many scholarship winners were asked if they wanted to sell their models to GM and a number of them did. Since I decided to keep my model, they asked for paint colors, trim details, and other various construction techniques so they could make a duplicate. I'm glad I didn't sell my model, as other Guild members later told me they regretted their decision. GM used the purchased models in tours across the country to promote the Guild. I know they made a duplicate of my 1953 model because when it was returned to me the blue body color had been repainted and the chip in the front fender was almost invisible. There was also evidence that the wheels had been removed and replaced.

A recent photograph of Warren M. Bakken and his 1953 national award winning Fisher Body model. The blue blazer he is wearing is similar to the ones given all the regional winners in Detroit in 1953. The Fisher Body Craftsman's Guild patch, with diamond-shaped logo, had been preserved and was worn to the 2008 Guild reunion at the Museum of Fine Arts, Boston, held August 9–10, 2008 (courtesy Warren M. Bakken).

I received a bachelor's degree in mechanical engineering in 1959 from the University of North Dakota and went to work for Lockheed in Burbank, California. I later moved to Plattsburg, New York, to work for General Dynamics on the construction of underground missile sites and then worked for Lockheed Missiles and Space Division in Sunnyvale, California. In 1970, I concluded my engineering career at Boeing Commercial Aircraft in Seattle as a model designer at the wind tunnel.

From my student days of building balsa engines, planes and cars to a profession of designing scale models for the wind tunnel, my love of modeling has remained strong. My little museum of past modeling projects resides in the basement along with my 1953 Fisher Body model. I think the condition of my model today isn't too bad considering it is more than 50 years old. There are now plenty of other little chips and dings to keep company with that first one from my Dad's eyeglass case.

6
From Chicken Coop to Cars

Wilfred C. Keagy (1949–1953)

When I was almost seven years old, my parents thought I needed a workshop. They tore down the old chicken coop, and my uncle used the same materials to build me a small workshop area. I was always out there in my spare time making some gadget or another while growing up the only child of Clark and Ethel Keagy in a small Appalachian dairy farming town named Lafayetteville, Pennsylvania, about 75 miles east of Pittsburgh.

Sometime during the eighth grade, I found a newspaper ad about the Fisher Body Craftsman's Guild automotive model contest. I said to myself, I'd better find out what this is all about, as it sounded interesting. I had always enjoyed building model airplanes, especially during the war years of the early 1940s. So, I mailed in the Guild coupon and before long received their membership package including a Guild pin, membership card and their famous "how-to" booklet, *Designing and Building a Model Automobile.*

At age 15, my first Guild passenger sedan model entry in the 1949 Junior Division competition was made from one piece of plaster. I poured it using a flexible mold made from room temperature vulcanizing rubber taken from a clay model that I had sculpted. I made the wheels the same way. The body was painted blue and the windows were indicated using black paint. Thin pieces of aluminum were glued on the paint to simulate body trim or chrome. I won a Junior Division Honorable Mention award for Pennsylvania, an X-Acto brand tool set and a certificate.

A year later, my second passenger sedan model in the 1950 Senior Division competition did not turn out as well. About this time, two-part liquid plastics became available and a catalyst or chemical hardener was required. I thought I could make a model using this plastic material the same way I poured the plaster. I sculpted the clay model, made the rubber mold and poured in the liquid plastic mixed with the hardener. When I removed the rubber mold, the surface finish of what would be the body of the model was not shiny and smooth. I mixed up another batch of liquid plastic and painted it on to mask these imperfections for a shiny, finished surface worth painting. This didn't do much good and things just got worse. The front and rear bumpers were cast out of Babbitt metal using plaster molds and painted silver to simulate chrome. The body was painted dark red and the windows were indicated with white paint. Time was running out, but I shipped it any way, this time receiving a senior division Certificate of Design and Craftsmanship and a Guild T-shirt.

I continued to experiment with liquid plastic. Soon, I discovered I could cast black-colored plastic in a soup can by adding black dye. I put these round black plastic discs in

Wilfred C. Keagy (age 17) from Lafayetteville, Pennsylvania built this model for the Guild competition and won the 1951 Senior Division Third State Award for PA ($50). This 2-door passenger car model was designed in clay and used to make a mold or mask from room temperature vulcanizing (RTV) rubber. The resulting plaster of Paris model was painted medium green with a yellow top and medium gray indicated window areas. A liquid plastic metal called "Chrome-a-Lac" was painted on carved wood bumpers to simulate bumper chrome (courtesy Wilfred C. Keagy).

a wood lathe and could turn fairly good-looking scale rubber tires from them. This is how I made the wheels for the rest of my models.

In 1951, I returned to the technique used for my first Guild entry — sculpted 1:12 scale clay model, rubber mold and plaster-of-Paris as prescribed by the Guild — to make my third passenger sedan model entry as a 17-year-old in the Senior Division. The body was painted a medium green with a yellow top and medium-gray windows. For the simulated chrome bumpers, I used a paint-on plastic metal called "Chrome-a-Lac" advertised by the Guild and sold by a Fisher Body vendor.[1] This model won a Senior Division Third State Award for Pennsylvania ($50) and a bronze Craftsman's Guild signet ring.

That year I graduated from high school and began working full-time in a Ford body shop. I wanted to learn how to spray paint automobile bodies, and while the job didn't give me enough time to build a model car, I did learn how to paint. With money from my next job as a part-time milkman in a nearby town, I was able to buy a few tools and build a table to mount them. I had one motor and moved it from tool to tool: 8-inch table saw, grinder, wood lathe, jig saw and small metal lathe.

With no idea what I wanted to do in life, I thought the electronics my uncle dabbled in might be interesting. A magazine advertisement for a home study course on television and radio repair caught my eye, and my Dad promised to pay for it if I completed the program. Soon I was doing my radio and TV repair lessons, but I still hoped I could make another model good enough to get me to the Guild convention. So in 1952, I entered my fourth model.

This passenger sedan for the Senior Division was made of plaster and my description of its construction problems and solutions was published in the *Guild News*.[2] The general idea of the design for my 1952 Guild model came from the picture of a car of the future I saw in a magazine. I changed the parts of the car I didn't like until they suited me. I started transferring these design ideas into clay.

While pouring the plaster for the 1952 model I designed in a structure to soundly and reliably attach the wheels. As I topped off the liquid plaster in the rubber mold, I embedded a varnished wooden armature for a solid, wheel/tire mounting surface. The wood block was barbed with screws and nails to hold it in place. After the plaster model had dried, about a month, I scraped and sanded the body surfaces of the model so they were smooth. This made up for the imperfect clay model. Using the template method, I matched the two sides of the plaster model car for symmetry.

My scraping and sanding had changed the original design to some extent so I had to build up the roof about ⅛ inch with plastic wood. The plaster model was smoothed with No. 320 sandpaper, and 3 coats of sealer were applied, then 15 coats of "wet sanding" undercoat or lacquer primer were applied.

Then the trouble began. The primer started to rise away from various spots on the surface of the model creating bubbles. I naively tried to apply masking tape to the primed surfaces to prevent the bubbles from rising to the surface. I cut through the primer, and using a hypodermic needle, I ran primer thinner under these raised surface blemishes. This softened the primer and seemed to solve the problem. After sanding this smooth, I applied six coats of color lacquer and sanded in between coats with No. 500 Garnet paper. Masking the model, I sprayed two tones of blue body color, then gray to indicate the window areas. But, afterward, I did not like this color combination, and some of those earlier blemishes re-appeared as the primer began to sink into the small holes of the plaster. Then one evening a friend suggested I paint it red and yellow, which suited me fine. So the model car was again sanded and primed numerous times and given a two-tone, red and yellow paint job picked from an automotive paint book. The window areas were again indicated with light gray paint.

The wheels and headlight rims were turned out of plastic. The wheels were painted black then masked to add the whitewalls popular in the 1950s. The wheels were mounted by counter-sinking screws into the flat part of the wheel and into that wooden armature locked in the rock-hard plaster-of-Paris. The headlight lenses (clear plastic) and taillight lenses (red tooth brush handle plastic) were carved, shaped and polished. The carved walnut bumpers were coated with Chrome-a-Lac plastic metal, and match-stick dowels were used to fasten these parts to the plaster model. The window frames were cut and filed out of aluminum at least two times before a perfect set was completed. Model airplane cement didn't seem to work for attaching the aluminum trim. However, pressing the metallic trim parts into hard lacquer softened with lacquer thinner seemed to hold things tight like an adhesive. The metallic window trim was the most difficult job of all.

However, when I went to polish the final paint job, the red and yellow paint job was not thick enough, and I rubbed through in several places. Not having enough time to repaint it, I had to ship it. Despite the imperfections in the paint job, I won the Senior Division First State Award and Regional award as well as that trip to the Guild convention I'd been hoping for. At the convention, I got a good look at the top winning models and saw how they were made. When I returned home, I decided to make one more model before I got too old for the competition. I wanted to win another trip to the Guild convention and win a national scholarship in the worst way.

My fifth model in 1953 was made out of wood. One day on my milk run I stopped at a lumber yard, bought a poplar board and cut it into length-wise pieces that fit the side view of my tracing paper design. These were glued together to make a block of wood to carve. I used the prescribed Guild method of templates to get both sides of the model car the same shape and used chrome-plated brass for the brightwork and trim. This was not easy since I had to order the brass material from a town 100 miles away. The thin pieces of brass came in sheets. This was cut in to narrow strips using a fine toothed metal-cutting blade on my table saw. The brass parts were filed and shaped to size and sent out of town to be chrome-plated.

I didn't have any fancy spray paint equipment, so I rounded up several spare car tires, took them to a service station to fill them with air, and using a small jar-type sprayer device, painted my model. It took half-dozen trips to the service station to do the job. I wouldn't tell the service station manager why I needed to borrow all the air.

Earlier that year, on Easter Sunday 1953, I got married to my sweetheart, Mildred, and she worked while I designed and built my Guild model. We bought a 22-foot house trailer and pulled it onto my parents' property and set up housekeeping. I quit my milk route and worked on the model car full-time.

The 1953 entry was painted bright red and had clear acrylic plastic inserts for windows that were painted on the inside a light gray color. The model was finished a few days before the deadline and taken to a railway freight depot about 25 miles away. Before I knew it, the Guild notified me I had won the Senior Division First State award for Pennsylvania and the Regional Award and was eligible for another trip to the Guild convention. Fisher Body made first class travel arrangements for me to travel from Altoona to Detroit on a Pullman "sleeper" car. When I arrived, I took a cab to the Book-Cadillac Hotel, found the right floor and discovered I was the last one to arrive.

Those trips to Detroit as a Guild Regional winner took me the farthest away from home I had ever been. I met one other boy at the convention who was married. The Craftsman's Guild employed two buses — one for the Junior Division (ages 12–15) and one for the Senior Division (ages 16–19) — and General Motors stylists served as chaperons to give us guided tours. They had a roll call after each time we stopped. Once they found me on the wrong bus, as I had met a Junior Division winner from my same Region 5 (Pennsylvania and Maryland), and we had struck up quite a conversation. In Pontiac, Michigan, they gave us a parade through town in the latest convertible GM cars, tops down, and the Guildsmen were sitting high up on the rear seat backs at the tonneau level. At the convention, I received a navy blue sports jacket, a light-blue beret or tam-o-shanter ("tam"), a gold signet ring, a FBCG Key Award and tie clip with the FBCG diamond emblem. A few years later, in 1958, I received a Fisher Body silver anniversary plate to commemorate 50 years for the division.

The 1953 Guild Convention was televised nationwide that year and my wife and parents watched the awards program. My fellow Guildsmen and I were ushered into the banquet room in no particular order and took our models to the dinner tables. Nothing was said in advance about who won until Dr. George J. Fisher, honorary Guild president, made the presentation to each of the eight boys.[3] One of the Guild officials had said, "There will be some surprised boys tonight," and there were — me, when they called my name as winner of the Third National Senior Scholarship award, worth $2,000. I really got what I had set out for, and I was 19 years old.

As a national winner in the Guild competition, my high school got a large trophy to remember me. It was made of blocks of maple and walnut wood, a brass name plate, and a gold-colored ball representing the world supported by gold-colored eagle wings. This arrived by train damaged with a few loose pieces, but the school managed to put everything

Wilfred C. Keagy (age 74) from La Verne, California, holding his 1953 Third National Senior Scholarship ($2,000) award winner circa July 2007. Will, who came from a small dairy farming town near New Enterprise, Pennsylvania, worked in a Ford body shop and as milkman and studied electronics. Using his Guild scholarship, he became an electronics technician, moved to Pasadena, California, and worked in the design of electronic test equipment, mainframe computer manufacturing and aircraft controls (courtesy Wilfred C. Keagy).

back together. My Guild legacy was complete and I would be remembered down through the ages.

Many years later, long after my high school had been torn down, I met one of my old high school teachers, Mr. Ross Miller, at the post office in New Enterprise, Pennsylvania. He said he had taken home and saved something very special just for me. It turned out to be the engraved pieces from the front of that 1953 trophy. Not only was I forgotten, but someone had stolen the maple and walnut for a woodworking project or their fireplace. In 2007 I decided to make a new trophy with the preserved antique brass engravings. I had a small picture of a 1953 national award trophy from an old Guild newsletter that I had blown up to make a full-size blueprint.[4] I hand-crafted a new trophy pretty much from scratch using some walnut and maple woods and a can of gold paint. The only replacement part I couldn't find was the smooth ball on top.

While living in our 22 foot trailer in a Chicago trailer park, Mildred worked and I attended DeVry Technical Institute to study electronics using my new FBCG university trust fund. After two years of study, I graduated as an electronics technician. During this time, 1955 and 1956, I returned to the Guild awards banquet in Detroit to relive my glory days as a national winner. A year later, I got a card about a group of Guild models touring the country that would soon be in Southern California. Mildred and I had moved to Pasadena in September 1956; we were living in a trailer park, and I was looking for work. I finally located the traveling Guild model exhibition at a department store in the San Fernando Valley. Fisher Body had made a copy of my 1953 winning model since I wouldn't sell out. When they had asked me to sell my model to GM, I had declined, as I had too much time invested in the model — roughly something like 700 hours. The Fisher Body professional model-makers had duplicated my model very accurately except for the Plexiglas windows with the light gray paint. They literally took an impression of the model using RTV rubber. From what I could see, there were duplicate copies of other Guild models also on display that day.

After I got home from Detroit, I got a postcard about a company in New York that had recorded the NBC television and radio interviews of the 8 national scholarship winners by GM's director of research and development, Charles "Boss Ket" Kettering, at the August 18, 1953, Guild awards ceremony. I bought a copy and years later made cassette copies off that old warped, scratchy 78 rpm record. I sent a copy to fellow Guildsman Warren M. Bakken, the 1953 First National Junior Scholarship ($4,000) award winner, who lived permanently in Seattle. Warren was glad to have it and even made a transcription of it.

Eventually, I went to work for the Pasadena-based Consolidated Electrodynamics, a manufacturer of electronic test equipment. Mildred and I stayed in the area and prospered working at Burroughs (now Unisys) at their mainframe computer manufacturing plant and later with McDonnell Douglas Aircraft controls. We invested substantially in apartment buildings as our retirement plan to hedge against lay-offs. We cleared the land and built our own custom home, brick-by-brick, in La Verne, California. I even included a nice workshop for myself. Looking back, I had come a long way from that customized chicken coop I had in Lafayetteville. Mildred and I have been married 57 years; we have two sons, Alan and Brian, and are blessed with eight grandchildren.

Notes

1. *Guild News*, Vol. 5, No. 3, March 1950.

2. "1952 Regional Winner Gives Work Method (Magazine Photo Prompts Design; Primer Brings Major Trouble)," by Wilfred C. Keagy of New Enterprise, PA, *Guild News*, Vol. 8, No. 3, February 1953.

3. Dr. George J. Fisher was the national Boy Scout commissioner and honorary Guild president, *Guild News*, Vol. 1, No. 1, March 1946.

4. *Guildsman*, Vol. 1, No. 4, 1954.

7
The Son Who Built the Model Car

Arnold L. Joslin (1951–1953)

It is a moment that sticks in my memory. I was sitting in the library at Murphey Elementary School in Raleigh, North Carolina, and reading the latest issue of *Boys' Life* magazine when an advertisement caught my attention. The Fisher Body Craftsman's Guild was seeking boys to design and build a model automobile to enter in their competition. Prizes would be awarded, including university scholarships. I wrote down the address and eagerly sent away for the information packet. When it arrived, however, I realized there was no need to rush: I was only ten years old and would have to wait until I was twelve to be eligible.

The General Motors–sponsored contest was designed to appeal to boys just like me—one who loved to build models, had some drawing talent, and liked cars. My older brother Ivan and I knew every different automobile make and could distinguish one from the other. One of our favorite games to play while on the road was to see who could first correctly identify an oncoming car, which we usually could do from a great distance. We also, along with our dad, eagerly awaited autumn visits to the car dealerships in Raleigh to see their newest models. This was a huge event all through the 1950s. The day for the introduction of next year's model would be announced weeks in advance, and great pains were taken to shield the new cars from view beforehand. The dealership's windows would be covered as they set up their floor display ahead of the huge crowds that would later show up to view Detroit's latest offerings.

Ivan and I also built and raced Soap Box Derby cars in the summer of 1948, and the derby could have possibly sidetracked me from the Guild competition except for the fact that Raleigh discontinued the event for a number of years afterward. Thus, I was eager to embark on a car model building project when I blew out my twelve candles and was finally eligible to enter the Fisher Body Craftsman's Guild.

First Try: 1950–1951 Competition (Age 12–13)

As advised by the Guild instruction booklet, I worked out my design in clay. It was a simple, basic 2 door sedan shape with little originality. The rear fender shape was possibly influenced by Buick, the side trim by Oldsmobile, and the rear end resembled a Packard. The headlight rims were typewriter key tops from the shop at Burroughs Adding Machine Company, where my father worked as a repairman. All the chrome plated parts were simulated with silver paint, and the window and side trim was simply painted on.

My model was made from plaster, and the casting process was perhaps the most challenging yet interesting part of the building experience. A five-part mold was constructed over the clay version, using clay "dams" to separate each part. After the mold was removed from the clay, it was reassembled, cleaned, and coated with Vaseline. Plaster was then poured into the mold to create the model car, and after it hardened, the mold was carefully removed. The model was cleaned up, sanded, and then brush-painted a dark green with black windows and silver bumpers, grille, and trim.

The wheels were formed from plaster as well, in two halves that were then glued together. The outside wheel profile, including the tire and hubcap, was cut and filed from a thin piece of sheet metal, then rotated around a nail in a flat board while plaster hardened underneath. The inside profile had the tire and a simulated brake drum.

I finished the car a few weeks before the deadline and began constructing a shipping box. My dad reviewed the rules and discovered the model was too long by about a half-inch. Afraid it would be rejected without consideration, I sanded a quarter-inch off the front and rear — ending up with flat ends. I painted the repaired areas with little blending into the original bumpers and grille, so it was obvious. My sturdy wood shipping box had a cloth-covered foam rubber pad on the bottom to absorb shock and vibration. We carried it to the Railway Express office at the train station and off it went to Detroit. In August, a telegram came stating I had won the Junior Division Third State Award for North Carolina and that a check for $50 was to come. The announcement also appeared in Raleigh's newspapers following a press release from GM.

Opening New Worlds: 1951–1952 Competition (Age 13–14)

By the time I got the results from my 1950–51 efforts, I had already begun work on my design for next year's model. Looking back, that entry was perhaps my most advanced one in terms of originality. To be honest, it was not altogether my originality, as it was based on a rendering of a "car of the future" I saw in a magazine. With that concept drawing as a guide, I worked out my 2-door sedan design in clay. One feature was ahead of its time: the twin beam headlamps. I envisioned them with two bulbs side by side, one for low-beam and the other for high-beam, with the pair enclosed in a single housing behind an oblong lens. Dual headlamps appeared on cars later in the 1950s but were not incorporated behind a single lens for roughly 30 more years.

For the next several months, I repeated the process of making a five-piece plaster mold from the clay model and then casting a plaster version. The car was brush-painted a 2-tone blue with white windows, with silver paint again for the chrome. Flat toothpicks were cut and glued to form hood and fender ornaments, door handles, and trunk decoration.

I completed my second model shortly after my 14th birthday in January, months before the deadline. I think I was beginning to realize that plaster was a limitation in building a truly competitive car, based on what I was reading in *The Guildsman* magazine, GM's publication for Guild members. Its pictures of winning models looked far superior to what I had done. I didn't have much higher hopes for my second entry than I had had for the first, despite its more advanced design.

So immediately after finishing it — and months before the June 30 deadline — I started my model for the next year in clay and transferred it to white pine. The futuristic design was selected out of the many that I had collected or drawn by then. I band-sawed eight pieces of ¾ inch thick white pine to the profile shape of the car and glued them together. By summer 1952, I had carved, sanded and otherwise shaped the woodblock with basic contours. In the meantime, I had shipped my second model to Detroit. In July, a Western Union telegram from the Guild congratulated me for winning two Junior Division Awards, namely, First State North Carolina Award and the Regional Award (comprised of competitors from North Carolina, South Carolina and Georgia). Local newspapers carried the story once more, this time sending a photographer to take my picture in my basement work area wearing my FBCG T-shirt and holding my 2-door 1951 Third State Award winning model in my lap. The caption under the photo read, "Raleigh's Model Car Expert." The State Award meant a $150 cash prize, but the Regional Award offered something much more exciting: an all-expenses-paid trip to Detroit for the annual Guild convention, where the national awards would be announced. Travel to Detroit was by train, and soon my tickets arrived for the big adventure.

Arnold L. Joslin (age 14) of Raleigh, North Carolina, holds his 1951 entry which won the 1951 Junior Division Third State Award for North Carolina ($50). In addition to the cash award, he received a Guild T-shirt and a bronze Guild ring. This photograph of Arnold in his basement workshop was published by the *Raleigh News and Observer* on August 6, 1952, announcing his 1952 Junior Division First State Award ($150) and Regional Award (North Carolina, South Carolina, Georgia) with an all-expenses paid trip to the Guild convention in Detroit. The photograph caption read, "Raleigh's model car expert." The 1951 dark green two-door sedan shown was constructed (similarly to his 1952 model, which was still in Detroit for the 1952 Convention) using a plaster casting from a clay model, chrome parts simulated with silver paint, window and side trim painted-on, and plaster wheels/tires. As the deadline neared, portions of the front and rear of his 1951 model had to be sanded off for it to comply with overall length requirements (courtesy Ms. Peggy Neal, photograph researcher, *The News and Observer*).

For a shy, 14-year-old kid traveling alone, this was to be an exciting week of new experiences. My family saw me off from the Seaboard Rail Lines station in Raleigh around mid-morning. I had first-class accommodations, with lunch served in the exquisite setting of the dining car with crisp white table linens and uniformed waiters. The train moved through Ashland, Virginia, where my grandparents were waiting to wave at me as I chugged past. Then it was on to Washington, D.C., and a 3-hour layover before boarding another train for the overnight trip to Detroit.

My accommodations for the overnight leg of my trip were on a Pullman sleeper car, which was again a new and exciting experience. After another fine meal in the dining car, the porter came by to convert my private space into a bedroom, lowering a hinged mattress platform from the wall above the window. Drawing back the curtains from the narrow window by the bed, I watched the lights of the countryside pass by until I drifted off to sleep. The train pulled into the Detroit station the next morning, where a Guild representative escorted me to the Sheraton Cadillac Hotel in downtown Detroit.

Upon seeing all the other model cars as each boy picked his up before entering the awards banquet, I suddenly realized that my car came nowhere near matching up to the quality of most of them. In fact, when I compared my model to others around me, I felt a sense of embarrassment for even being there and realized that I was fortunate to have competed in a state and region where there was not a lot of competition—and in the Junior Division (ages 12–15) at that. I had played by the rules, but surely there were many contestants in other states who received honorable mention awards whose models received higher scores than mine. I also immediately knew that the wood model I had begun at home was not going to cut it for the following year's competition, with its painted-on, simulated chrome and brushed-on paint job.

Being out of the running for a scholarship had its advantage, however, as I was able to relax and enjoy the dinner without one bit of nervousness. The music from the band was especially enjoyable, played in the big-band jazz style made famous by such greats as Les Brown and Tommy Dorsey. As we marched into the room carrying our models, I remember the sweet saxophone sounds as the band played the University of Michigan fight song at a slow, walking tempo. Of all the other awe-inspiring and mind-boggling experiences that week, perhaps the most impressive to me was the buffet spread at the Bloomfield Hills Country Club. To this boy from a modest family, the sheer extravagance and quantity of food before me was something I had never experienced and could hardly comprehend.

Success: 1952–1953 Competition (Age 14–15)

Upon arriving home, I set aside the wood model I had begun and started over. In addition to refining my design in the clay model, I began to search out sources of aluminum from which to fabricate metal parts. Some of my prize money went toward the purchase of a small air compressor and paint spray gun from the local Western Auto store.

The body style of this third Guild entry was a 2-door sedan made out of clear white pine. I made four lengthwise templates from the clay model and traced the outlines on the wood, cut the profiles with a band saw, and glued the eight pieces together for a basic shape. I carved my model using the template and station line method; templates were made from the clay model and cutting done with a Dremel rotary tool. After cutting the station lines to shape, I cut away the wood remaining between the station lines with X-Acto carving tools and shaped the body with sandpaper. The grooves to receive the trim pieces were carved in as the trim pieces were constructed.

The front bumper and grille combination was the most complex piece of metal work that had to be designed, sculpted, shaped and finished. I began with a solid block of soft aluminum about 6 and a half inches long and 2 inches square. I made top view and front view templates from the clay model and traced the outline onto the metal block. I cut along these outlines with a metal-cutting band saw to achieve the rough shape. After many, many hours of filing, drilling, more filing, block-sanding and buffing, the bumper/grille was done and fitted to the front of the model.

Little details mattered. The headlight rims were typewriter key top rings, fitted with lenses turned from a flat piece of plastic glued to a grinding wheel from the rotary tool set. The side trim was made out of a bar of aluminum ⅛-inch thick by ¼-inch wide set in grooves cut into the sides of the model. Door handles were made in three parts: the handle itself, a pushbutton, and thin aluminum trim around the handle filed to shape. Front and rear emblems contained a small glass jewel taken from a dime store ring, and the license plate was a miniature cardboard license replica made for the American Veterans Association. I painted over the "North Carolina 53" with black lacquer, cut the edges off to make the plate smaller, and then coated it with clear lacquer. The taillights were cut and filed from ¼ inch thick aluminum and set atop the rear fenders to resemble tail fins. The taillamp lenses were turned from red and clear plastic using the Dremel rotary tool as a lathe and files and sandpaper to shape them.

Even though the model was made of wood, I still made the wheels from plaster turned with templates by hand as the plaster set up. The outside template was formed to make an inset to receive the hubcap, while the inside one simulated a brake drum. When it came time to cut the forms used to turn the wheels from plaster, I got the measurements for the tires and the inside brake drums by measuring the tires and wheels on our family's 1941 Chevrolet. When the car was finished, the tires appeared skinny compared to the balloon tires that had begun appearing on new cars in the early 1950s. I remember my parents feeling bad about not being able to afford a newer automobile when they discovered that I had taken the measurements from our family car. (In the long run, however, it worked to my advantage. The model continues to look more contemporary and less dated than it would have with the then-popular balloon version.) The hubcaps were turned from ¼ inch aluminum on a metal lathe and included red-painted wheel rims. The finished wheels were cut roughly in half and attached with wood screws to the bottom of the car.

Next, I sprayed four coats of Testor's brand sanding sealer on the pine body, sanding between each coat and rubbing out the last. This was followed by six coats of dark grey primer-surfacer, wet-sanded between each coat. At this point the scored lines for the door, hood, and trunk openings were laid out in pencil and cut with a file that had been ground down to a sharp blade. Then nine coats of black lacquer were sprayed on the body, sanded between coats and rubbed down about every three coats. After masking, fifteen coats of cream lacquer were sprayed on the roof of the model, sanded and rubbed out the same way. The windows were painted to simulate tinted glass. First, they were sprayed with seven coats of white lacquer, then with a clear lacquer-based metallic blue-green coating. This was rubbed out and protected with five more coats of clear lacquer.

In my final month, I worked about 13 hours a day and shipped the finished model off with just one day to spare before the June 30, 1953, deadline. This time, when I again won First Place State Award and Regional Award in my division, I felt I would have a good chance at one of the national awards. The 1953 convention was in many ways a repeat of the 1952 trip, with the overnight trip to Detroit and the layover in Washington, with the added apprehension of what might occur at the awards banquet. Indeed, I had such a nervous stomach that my food did not digest properly, and later that night, alone in my hotel room, I had to get out of bed and throw up.

Until then, though, it was sheer excitement. The band again played the beautiful march music as the contestants entered the banquet room to prolonged applause. During the meal, the boys on each side of me thought surely that my car would be one of the scholarship winners. As for other details of that night, if it were not for a photograph of Dr. George J.

Rear view of the Guild model made by Arnold L. Joslin (age 15) of Raleigh, NC. It won the 1953 Junior Division First State Award for NC ($150), Regional Award (North Carolina, South Carolina, Georgia) and Second National Junior Scholarship ($3,000) award. After his 1952 regional trip, Arnold upgraded his shop to include a Dremel rotary cutting tool and compressor with spray paint equipment. This 2-door sedan with black lacquer body and cream lacquer roof was designed using clay for a reference, but the model was carved from clear white pine. Grooves were cut to accept trim pieces and the bumpers were sculpted from aluminum. The wheels were made from plaster and had aluminum wheel covers. A large Guild trophy was awarded, in Arnold's honor, to Needham Broughton High School in Raleigh during his Sophomore year (courtesy Arnold L. Joslin).

Fisher, honorary Guild president, handing me the certificate, I would have no specific recollection of the events. It was one of those times when the mind races so fast that reality seems unreal. I do remember the emcee beginning his announcement with, "The next award goes to a Southern boy from Raleigh, North Carolina." He asked me some question based on what I had written in a questionnaire concerning my future ambitions, what an "engineering designer" was. I couldn't really come up with an answer, but he covered for me and moved on. I was 15 years old and had just won the 1953 second National Junior Scholarship award worth $3,000.

While Guild judges used a points system, their results were subject to approval by an honorary board, primarily GM executives. Later, one judge told me my model had actually scored a higher point total than the first place winner, but that at least one honorary judge hated my downward sloping front-end design; he stated "it looks like the front end of a 1928 Franklin" and refused to have it win the top award. Thus Warren Bakken's model, with a lower score, was put ahead of mine.

The balance of the convention was lots of fun, and I was accorded special privileges as a scholarship winner with a gold satin lapel badge attached to my navy blue Guild blazer. The ticker-tape parade through downtown Flint was especially memorable, although the event was obviously contrived by GM. I felt very special perched on the back of the light blue 1954 Buick Skylark convertible, waving to young ladies who were throwing confetti from second and third story windows as we passed by. Another memorable moment was at a breakfast when the band began to play "Carolina in the Morning." It didn't sink in at first, until the counselor at our table said, "They're playing that for you." I stood up and waved a "thank-you" to the band as everyone applauded.

An experimental 1:12 scale model made to 1954 Guild sports car specifications by Arnold L. Joslin after returning from the 1953 Convention. The 2-door car had a red, white and metallic green color scheme, black indicated window areas, and official Guild rubber wheels/tires available for the first time that year. After hearing the judge's comments at the convention about his 1953 entry, this model was made "just for fun," to incorporate design ideas that would have increased the score of his Second National Junior Scholarship award winning model. The 1954 model featured a simpler oval grille-bumper design (which would have been easier to make), elongated, forward slanted front fenders, wraparound windshield, and fender line raised to the beltline (allowing for a more dramatic color separation). National scholarship winners were no longer eligible for the Guild model car competition, but were free to experiment with new design ideas (courtesy Arnold L. Joslin).

Life went back to normal after the convention, but several "Guild moments" followed. There was the telephone call from Fisher Body asking me if I would sell them my model for the General Motors "Parade of Progress," a traveling exhibition about GM technology which moved from city to city and included a display of recent Guild winners. Then, there was the trophy presentation in my honor to Needham Broughton High School during my sophomore year. The senior class president called me onstage during an assembly, and an official from the Guild brought out a huge trophy and made a short speech about my accomplishment with my dad and mom there to watch.

One souvenir I brought home from the convention was a set of the new molded rubber wheels the Guild furnished the contestants starting in 1954. The Guild also revised the rules to include sports cars and convertibles in addition to 2-door and 4-door full-sized sedans. Already aware of several design upgrades that I wished I would have included on my scholarship winning model — changes I thought would have earned me first place and another $1000 — I cut down and remade the wood model that I had put aside after the 1952 convention. I made it into a sports car and used the new rubber wheels. Of course, that model was never entered into competition but provided many enjoyable hours of model making to me, even after I was ineligible to compete.

Arnold L. Joslin (circa 2009) standing with a display case containing his 1953 national award winning model. He won the Second National Junior Scholarship in 1953 at age 15 with this beautiful two-door sedan. He began his Guild career at a fairly high level of accomplishment, building Regional Award level entries in 1952 and 1953. Encouraged by his Guild experience, in 1959 and 1960 he worked as a styling intern at GM Styling in Warren, Michigan, and later became a product design engineer at Hewlett-Packard in Colorado (courtesy Arnold L. Joslin).

Epilogue

Winning the Guild scholarship had many far-reaching consequences in my life. First and foremost, it paid my tuition through seven years of college, supplemented by my summer jobs and by living at home for five of those years. But it also allowed my parents to use the money they had set aside for my college expenses to make a down payment on a house, the first and only home they ever owned. The notoriety gained by this experience was also the primary factor in our family being named Raleigh's "Family of the Year" in 1956. It became a family joke that people in the offices where my Dad went to fix machines would ask him, "What is your son doing now?" He would counter with, "Which one? I have three sons," and they would invariably ask about "the one who built the model car." The most humorous answer he got was, "The one who built the model airplane and went to Chicago!"

Finally, being a Guild alumnus opened the door for me to successfully apply for GM's styling program for design students during the summers of 1959 and 1960. This in turn played a huge part in forming my career choices in the field of industrial design and has enriched my entire life experience. I have spent most of my adult life in Loveland, Colorado, as a mechanical engineer and industrial designer working on the design of Hewlett-Packard products.

8

If at First You Don't Succeed — Try, Try Again!

Raymond P. Wykes (1951–1953)

My interest in designing and constructing models for competition began with my friends and peers in Boston and the Junior Aviation League (JAL), a model airplane club sponsored by the Jordan Marsh department store and the *Boston Herald Examiner* newspaper. While my interest in airplane design and aeronautics carried me through a successful aerospace career, it was the Fisher Body Craftsman's Guild that helped shape the values and virtues needed to get there.

After reading a Guild advertisement in a hobby magazine, my twin brother, Donald, designed and built a model for the 1950 competition. He won the Senior Division Third State Award for Massachusetts ($50). Encouraged by Donald's success and intrigued by a potential replacement for the JAL Club, which had disbanded, I filled out the forms required to compete in the 1951 FBCG competition.

First Try 1951 (Age 18)

My understanding of automobiles and what was involved in designing them was almost nonexistent. My initial challenge was to familiarize myself with new automobile sales brochures. I used a large pad of tracing paper to outline them from the photos. That helped me become familiar with all the visual exterior components and design-lines of current makes and models. Next, I revised the sketches by over-laying my own ideas on another separate sheet of tracing paper. The evolution of this initial model design attempt was very crude as design lines were traced and re-traced. I met all the requirements to submit a reasonable model of my own design in the 1951 FBCG competition.

For that model car, I designed a two-door sedan with a solid, dark green body with white indicated window areas, using solid top construction. It was carved from laminated white pine. The dark green color was a Ford production paint from a local dealer, and the white-indicated window areas were painted with Testor's model airplane dope.

No clay model was used, and all the surface development and required dimensions were laid out using a ruler and "Rock of the Eye," a method to judge accuracy by eye. The whole procedure was very unsophisticated. The only carving and power tools I had at the time were purchased to construct my model airplanes. They included a hand saw, wood chisel, wood rasp, Dremel tool and X-Acto kit.

I lacked the resources to purchase other essential tools to help build the 1951 model, such as a wood lathe and a band saw to shape the basic wood block. Consequently, the four wheels were laid out on a ¾-inch thick white pine board with a compass and carved entirely with hand tools. The basic laminated block used for carving the body consisted of white pine planks that my father had stored away for safekeeping. These were glued and C-clamped together. The very elemental straight, flat bumpers and other metal parts were made from a 1/16-inch thick sheet of annealed aluminum. The window frames were made from malleable lead solder strips with a 1/16-inch diameter that were glued into grooves outlining the window areas.

I used my mother's vacuum cleaner, along with a hobby shop paint spray attachment, to apply the wood sealer, undercoat, and final five color coats of automotive paint. I was given a stern warning by mother that "no spray painting was to occur anywhere in the house," under any circumstances. Instead, I had to employ a paint spray booth outdoors made from a large corrugated box to prevent the contamination of freshly painted surfaces from dust and debris. The areas of which I was most proud, and considered to have above average craftsmanship, were the wood surface sealing, undercoating, and application of the five color coats of paint.

I used slivers of a clear plastic cylindrical rod to make the headlight lenses and slices of red plastic cylindrical rod to make the taillight lenses. The circular headlight and taillight housings were made from 1/16 inch thick chrome-plated costume jewelry mountings that were purchased from a local department store. The chrome-plated hubcaps were found in a local hardware store and looked a lot like the 1950s Volkswagen-style hubcaps. The whitewalls on the hand-carved white pine tires were masked and spray painted with white Testor's model airplane dope. My model car entry was crated and shipped before the June 30 deadline. I believe that my above average paint job must have impressed the Guild judges, as my 1951 entry won a Senior Division First State Award for Massachusetts of $150 as well as a Regional Senior Award for Region 2, which encompassed Massachusetts and Rhode Island. Regrettably, my twin, Donald, did not enter again after his first successful try.

Second Try, 1952 (Age 19)

Before constructing my "second try" for the 1952 FBCG Model Car Competition, I gave an honest and very critical assessment of my 1951 model to help in planning the next design. At the 1951 Guild convention I had viewed hundreds of models on display, and learned that the competition was extraordinary in terms of quality and perfection. There were also large and obvious differentials in terms of talent, design and craftsmanship among contestants. These facts had to be addressed for my 1952 model in order to advance beyond first state and regional award levels.

A neighbor offered me the use of both his wood and metal lathes as well as his band saw. A shop in Medford Square that stocked tempered bar and sheet aluminum was located. The annealed aluminum I had previously been using did not "buff-up" to a brilliant chrome-like finish. Now, the major challenge was to find the time to complete a more competitive model before the inevitable June 30, 1952, deadline. That was going to be extremely difficult since I was starting my freshman year at University of Massachusetts as an engineering student. All of the construction had to be done at home during major holidays and between semester breaks.

In 1952, I designed a two-door sedan, featuring solid-top construction with laminated white pine. I chose a maroon body and a cream top color scheme from Ford production colors, with white model airplane dope to indicate the windshield and window areas. I personally selected the white pine at the lumber yard to make sure it was free of imperfections. A clay master model was used to work out the design details before transferring critical cross-sectional body information to the carved white pine. "Plastic Wood"™ was occasionally used to refine some of the design features.

Rectangular headlight lenses made from a solid bar of clear plastic were fitted into tempered aluminum housings, and then blended into the front fenders. Side trim, window trim, and headlight and taillight housings were made from 1⁄16 inch thick tempered aluminum. The front and rear bumpers were carved and sculpted from tempered aluminum bar stock. I turned the white pine wheels on my neighbors' wood lathe and designed them so they could be "press fit" onto the aluminum hubs I had machined on the metal lathe. White Testor's model airplane dope was used to spray paint the whitewall tires.

By the time the college year ended, all of the planning and hard work focused on the construction of a more competitive model appeared to have paid off. The car had been built pretty much in the confines of my bedroom "hobby area" at 42 Butler Street, Medford, Massachusetts. My social life and normal activities had been curtailed and sacrificed in order to complete the model before the deadline. It was an "eating-sleeping-working-on-the-model" existence.

I was elated to receive a formal letter from FBCG Secretary J. P. Wines stating that my 1952 model had been judged the Senior Division First State Award winner for Massachusetts ($150) and Regional Award winner. Again, I was given an all-expenses paid trip to Detroit for the 1952 Guild convention.

The trip to Detroit that year did not result in my goal of winning a national scholarship, but I was informed that the model was "in the running." This fact was very encouraging. Hearing from the national winners about all the sacrifices they had made impressed me — it was an incentive to try harder. Some of them had designed, built and entered as many as six unique Guild models over as many years. All of this knowledge increased my determination to "go for broke" in 1953.

Both my twin brother, Donald, and I applied for and transferred to Bradley University after our brother, Robert, highly recommended that school to continue our college educations. He was attending Bradley at the time and told us it was highly rated for its engineering department. This unanticipated move had a major impact on giving me that "go for broke" effort I required to win.

Third Try, 1953 (Age 20)

My 1953 model would require a quantum leap in creativity and a higher level of attention to craftsmanship in order to beat the competition Therefore, it was mandatory that I devote a maximum effort to the design and construction process, way beyond what I had given in 1951 and 1952. This would be my last attempt to win a national scholarship, after which I would be ineligible for the contest due to age — there would be no other chance for a scholarship.

Attending Bradley University as a sophomore in the engineering school was another challenge for budgeting design and construction time. I needed enough time to keep my grade point average within the "I'm serious about obtaining my degree in engineering" range, but I still wanted to compete for a Guild scholarship. Despite the fact that the model design and construction had to be a second priority, I still worked on it while at the university.

Immediately after arriving home from the 1952 Guild convention, I initiated preliminary efforts into the planning and purchasing of the required materials that would have a major impact on the basic body design and construction for my next model. The convention ended August 18, 1952, and that left me until the beginning of my sophomore classes in September to complete my preparation.

The preserved 1952 clay model and its accompanying Guild score sheet were used to study styling improvements or plan changes for the 1953 version. My luggage for the trip to Bradley University in Peoria, Illinois, contained that clay model, a laminated white pine body block to carve, and my inventory of woodworking tools so that I could construct the model in my spare time. This last model entry was to be a two-door sedan, with a Cadillac black body and Ford canary yellow top color scheme, white indicated window areas, and solid-top construction carved from white pine.

1953 Model Built at Bradley University

Based on observations of the national award winning models at the 1952 Guild Convention, I decided against using polished aluminum for the trim and bumpers. Instead, I upgraded to chrome-plated brass — it was the obvious choice to compete at the national level. This choice was confirmed by my impression of Bill Morgan's 1952 First National Senior Scholarship award-winning model. His craftsmanship, pleasing design lines, and choice of materials were extraordinary. I took copious notes and photos of his model to record what it would take to ascend to a higher and more competitive level. Bill was very generous with his time, and tried to answer all of my seemingly unending questions.

As the carving of the white pine progressed through the surface and design-line stages, at least two cans of Plastic Wood were used to revise any non-pleasing surface development areas called "dips and valleys." It also helped blend surfaces into the required dimensions, for example, around bumpers and headlight housings.

Design and Construction Details

Serendipity played a major role in helping me make the serious decisions during the 1952 model construction process, and it was at work again in 1953. Besides offering an excellent four-year engineering program, Bradley University had a well-staffed and highly rated Horology Department that was involved in the design and manufacture of timepieces and intricate jewelry. I obtained permission to use these excellent facilities and spent several evenings on the design and layout of a very sophisticated and detailed front grille. My main objective was to integrate the grille design into the front-end body surface development.This "above average" craftsmanship would surely impress the Guild judges.

Many hours were spent between classes in the horology machine shop cutting out brass parts and silver-soldering them together as an assembly for the grille and the bumpers. The silver-solder required a high temperature to melt and liquefy it. The chrome-plate did not dull at the silver-soldered joints, but remained bright and shiny. The grille, for example, had to be assembled and silver-soldered in a jig/fixture. A vivid and painful memory remains of my lack of respect for the liquefaction temperatures. I absent mindedly forgot to put on protective gloves and paid for it with severe burns to the palm of my left hand. Fortunately, I am right-handed.

Another example of serendipity was locating a model train hobby shop nearby that was advertised in the Peoria phone book. While perusing the miniature railroad track section of the shop, the idea of fashioning brass HO gauge rails into chrome-plated window frames came to mind. The brass model railroad track material turned out to be ideal for filing and bending into shapes that I used for the chrome-plated window framing of the final model.

Some of the highlights and "how-to" details for my 1953 Guild entry were:

Body — Using any number of drawings and sketches, the styling and design of the scale model automobile were worked out in an integrated fashion using modeling clay. The primary design idea was integrating the front and rear bumpers into the body. Templates made from the finished clay model were used to carve the laminated white pine body so that longitudinal symmetry was maintained.

Front bumper and grille — The front bumper was shaped from two pieces of tempered brass bar stock, and then disassembled for smoothing, finishing and polishing. At another stage, it was re-assembled using silver-solder. It was then chrome-plated. The mating joint surfaces were milled so they fit perfectly flush. In the wide-oval cavity of the front bumper, a grille was placed consisting of 22 half-elliptical pieces of brass of varying major and minor axes to create a fluted bar. This piece was also silver-soldered together and chrome-plated. Jigs and fixtures were required for the assembly of these intricate parts. Holes that were earlier drilled in the flat, brass bar stock made housings for the signal lights, which were turned from clear rod stock, and fog lights, which were turned from a yellow plastic screw-driver handle. Bolts hidden behind these lights helped attach the bumper and grille to the wooden body. The brass rear bumper, similarly integrated into the design of the body, was carved, shaped and filed, then smoothed to a satin finish with emery paper, polished and chrome-plated.

Front and rear wheels — The tires were turned on a wood lathe from soft basswood, painted black and masked for the whitewall paint. The hubs or wheels, with integral hubcaps, were turned on a metal lathe and mounted on separate front and rear aluminum support brackets. They were then aligned with the fender wells and the correct fender-well off set distances. The wooden, simulated wheels/tires were press fitted onto the hubs and splinter-like shims were used to lock them in place.

Headlights — The convex outer lens surface was carved and shaped from a clear plastic rod, and straight lines were scored on the reverse-side for the photometric lens pattern. The polished lenses were inserted in turned brass headlight bezels which had been chrome-plated. Small sparkling jewels from an old broach my mother had became the light bulbs and were visible to the Guild judges.

Taillights—Two chromed-plated, half-elliptical housings assembled at 120 degrees were reinforced with a soft solder gusset. The red lenses were carved from translucent Lucite plastic. The red lenses were shaped with a Dremel power-tool, smoothed and polished to a flawless finish.

Side window trim — As opposed to a painted lacquer window surface, semi-opaque plastic sheet was painted white on the inside surface, and inserted into the window cavity. This created a highly polished, durable, glass-like exterior window surface. Then, chrome-plated, annealed, brass model railroad track was used around each window opening as brightwork or trim. The trim was cradled in specially carved grooves.

Front to rear, side chrome strip — Brass, miniature railroad track was carved, shaped, and tinned, and pins were soldered perpendicular every two inches for mounting. This was chrome-plated. The pins held the chrome-plated trim to the wooden body, and the piece was cradled in a specially carved groove.

Lower rear-quarter panel, chrome strip — This was carved and shaped from brass bar stock with perpendicular attachment pins, smoothed, polished and chrome-plated.

Antenna — Telescoping hypodermic needles with the end blunted.

Windshield wipers — Rectangular slivers of white pine were dipped in black automotive lacquer to create the rubber blades, 1/32-inch diameter piano wire was used for the wiper arms, and sections of hypodermic needles (a short tube section) were used for the hinge or attachment points to the windshield cowl. The windshield wipers could swivel back and forth.

Raymond P. Wykes kneels on the sidewalk in front of his boarding house at Bradley University with his unfinished Guild entry about seven weeks before the June 30, 1953, deadline. The model is shown with several coats of primer (note the matte finish) and most of the chrome-plated parts such as front bumper, grille and rear bumper have been temporarily attached. All the chrome parts were removed before all the various primer or color coats were applied. At the time this photograph was taken (May 9, 1953), Ray had just completed his sophomore year at Bradley University School of Engineering in Peoria, Illinois (courtesy Raymond P. Wykes).

Academic Success and Guild Scholarship in Sight

I was still "hanging in there" in terms of a respectable grade point average, and the 1953 model construction was still on schedule. The wood surfaces were sealed, primed and aged for the final color coats of paint. All the brass parts were chrome-plated and then fitted "one last time" into their grooves, and adjusted for attachment for the final assembly process.

My twin brother, Donald, and I were offered a ride back to Massachusetts by a fellow student who was also going home for the summer. The new model was packed into the 1952 model shipping crate for the return trip home along with all of my construction tools.

The time between arriving home and the June 30, 1953, FBCG deadline was spent on a concentrated effort. I applied the final color coats, attached the chrome-plated brightwork, and mounted the wheels/tires and hubs. A production Cadillac black color was used to finish the lower body and a Ford production color, Canary Yellow, was used to finish the top, creating a two-tone paint job. A white lacquer was used to indicate the windshield and rear window areas. I completed the final attachment of all the metal and plastic detailed parts: headlights and tail lights, telescoping antenna, windshield wipers, winged hood ornament, and door handles. The basswood tires — painted lamp black with wide whitewalls — were press fitted onto the aluminum hubs.

It came down to staying up the entire evening of June 29, 1953, to finish the model by 7:20 o'clock in the morning and then pack it for shipping. The model and shipping crate had to be delivered to the Medford Square Post Office at 8 o'clock. The timing was so close that the glue attaching some of the detail parts was still drying as the model was lowered into the shipping crate.

As the summer proceeded, I anxiously awaited word from Guild Headquarters in Detroit. A formal letter finally arrived from Mr. W. S. McLean, the Guild secretary, stating that I had won the Senior Division First State Award for Massachusetts of $150 and the Regional Award. An invitation was enclosed for the Guild Convention and Awards Banquet in Detroit on August 18, 1953.

1953 Guild Convention and Awards Banquet

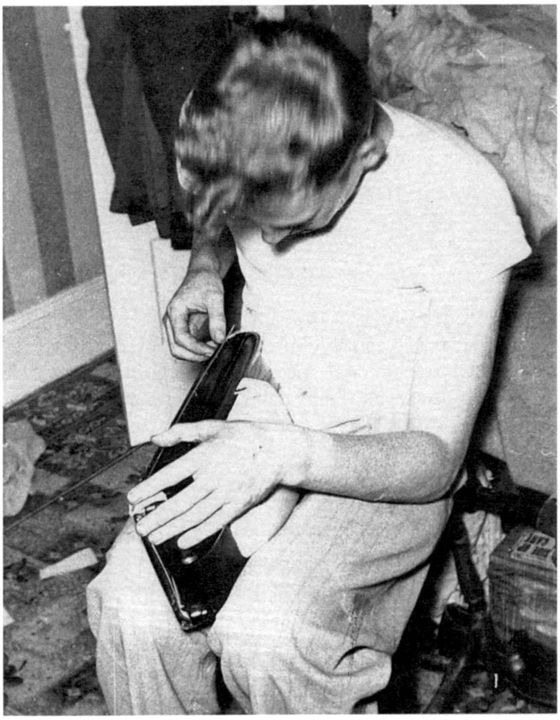

Raymond P. Wykes works on his model at home in Medford, Massachusetts, the night before the shipping deadline. He worked on the model all night and finally completed it by 7:20 the next morning, leaving just enough time to get to the Railway Express Station by 8:00. The glue attaching some of the parts was still drying when he lowered the model into its shipping crate (courtesy Raymond P. Wykes).

My 1953 model entry generated a lot of interest and a cheerleading among family members, relatives, neighbors and friends in my hometown of Medford, Massachusetts. My mother, the head cheerleader, kept them all informed about the planned national broadcast radio and television programs to announce the national scholarship winners on August 28, 1953, at 6:30 in the evening.

All of mom's friends' attention was focused on the radio program and the preliminary speeches by GM and Guild executives. In order to build anticipation, the names of the national scholarship winners were withheld until the conclusion of the broadcast.

All my "try and try again efforts" finally paid off by winning the highest Guild National Award. I took the 1953 First National Senior Scholarship — a $4,000 university trust fund! I was 20 years old and still had college bills to pay at Bradley University. The estimated time invested in the 1953 model's construction was approximately 750 hours, excluding the clay model surface development effort. Transferring the clay master model design to white pine included a lot of sweat, but no tears.

Front elevated view of Raymond P. Wykes' refurbished 1953 First National Senior Scholarship ($4,000) award winning model photographed circa September 1991. The original 1953 national award winning model was heavily damaged in a 6.7 scale earthquake in the Northridge area of California in October 1989. The model fell 5 feet to the floor along with his large flying model airplane competition trophies. The contents of Mr. Wykes' apartment were nearly destroyed. In a labor of love, he spent from December 1990 to September 1991 restoring and reconstructing the car. He repaired and revised some of the chrome and detailed parts and repainted it a two-toned blue similar to his 1968 Mercury Cougar. The original model was painted a Cadillac black with a Ford canary yellow top. The reconditioned model is 95 percent the same as the original (courtesy Raymond P. Wykes).

Mother sent me a telegram while I was still attending the Guild convention in Detroit. Dated August 28, 1953, it read: "Congratulations, son! You finally succeeded. You surely deserve everything because you worked very hard to get it. I am very proud of you. Everyone here is proud and excited for you."

Rebound from Detroit

Preparations for returning to Bradley University were eased with the knowledge that the scholarship would relieve me and my parents from paying the remainder of my college expenses. The scholarship was well depleted when my goal of graduating from Bradley with a bachelor's of science degree in mechanical engineering was achieved.

I received a letter from GM and the Guild asking if I would sell my original, award-winning 1953 model. This presented a very thought-provoking dilemma that had to be approached with diplomacy. My letter of response stated that "no amount of money" could convince me to give up my model. It held too much sentimental value, and I wanted to keep it intact as a physical reminder of my experiences while competing in the FBCG competitions. Looking back today, I don't regret it. This model represented a major experience of my youth and had a huge impact in terms of continued education and professional development.[1]

As a result of my response, GM decided to make a plaster "duplicate copy" or "reproduction" of my 1953 design in the GM Styling Section model shop using flexible rubber mold techniques. Other 1953 national winners — Warren M. Bakken and Arnold L. Joslin, the first and second national junior winners — also had duplicate copies made of their models. These models joined the Futurliner No. 10 exhibit of 1953 Guild models traveling from city to city. Over the years, some Guild national winners did sell their models, but often felt great regret years later.

I also prepared a 9-page set of drawings and sketches to describe how I made my 1953 national winner. Titled "Evolution of the Completed Model," the handout was made to respond to all the requests from potential FBCG contestants who wrote letters to me seeking advice and details on how I designed and constructed my 1953 award winning model.[2] Even the Craftsman's Guild technical staff asked for a detailed explanation. So, I was very busy.

Later, GM sent me an offer for another all-expenses-paid trip to Detroit. This time it was an invitation to interview with personnel from the Styling Section and possibly win a job offer. The GM job proposal was one of many offers I finally received, including others from aircraft companies nationwide. The final decision, however, was to continue my edu-

Four Wykes brothers outside their Rockwell International office building in February 1983. The Rockwell International newsletter *Skywriter* published an article entitled "A Rockwell Engineering Brotherhood," accompanied by this photograph of the four aerospace engineers (left to right) John H., Raymond P., Donald H. and Robert C. Wykes. Raymond and Donald were twins and Guild members. Don won the 1950 Senior Division Third State Award for Massachusetts ($50) and Ray won 1951, 1952 and 1953 Senior Division First State Awards for Massachusetts ($150) and the Regional Awards as well as the 1953 First National Senior Scholarship award (courtesy Raymond P. Wykes).

Raymond P. Wykes holds his refurbished and repaired 1953 national award winning Guild model, Los Angeles, California circa September 1991. Ray completed the reconditioning of his model from December 1990 to September 1991 after his retirement from Rockwell International. The restored 1953 model is currently enclosed in a clear display case and is tied down to a shelf on his entertainment center prepared for the next big earthquake, as is everything else in his apartment. He declined Fisher Body's request to sell his award winning model to them in 1953. Mother Nature tried to destroy it in October 1989, but Ray has held on tightly to his Guild memories and refused to part with what symbolized a significant milestone in his life (courtesy Raymond P. Wykes).

cation and accept a very good job offer from North American Aviation's Los Angeles division. The fact that the Art Center School, the premier automotive styling and design school, was located in the Los Angeles area also made the North American offer very enticing.

The interest in attending the Art Center School and, again competing with some of the most talented design students, was satisfied after a grueling equivalent of 1.5 years full-time as a transportation and product design major. I was gratified that I could successfully compete with the best student designers from all over the world.

Because of the expensive tuition and the exhaustion of my financial resources, I reluctantly returned to my engineering job at North American Aviation. In 1962 I rejoined my engineer brothers John H., Robert C. and Donald H. Wykes, who also worked there. We all became Rockwell International employees in 1973 as the companies merged. All four of us Wykes brothers were honored in the Rockwell International newsletter called the *Skywriter* in February 1983, in an article entitled "A Rockwell Engineering Brotherhood."

I retired in 1990 from Rockwell International after 35 years in the aircraft design and stress analysis profession as an aerospace engineer. My official title at that time was advanced stress/structural analysis specialist, and I was working in the company's research and development unit in El Segundo, California. During my career, I worked on advanced composite structures and stress analysis for aerospace and experimental vehicles, although I had begun my career in a unit concerned with the conceptual design and analysis of hypersonic space vehicles. The first vehicle I worked on in 1956 was an "aerospace plane," which evolved more than 20 years later into a type of space shuttle orbiter. In 1968, I was issued a U.S. Patent (# 3390853) for a "Variable Geometry Re-Entry Vehicle." I was also involved in the X-15 Program, but my favorite project was the Highly Maneuverable Aircraft Technology (HiMat) and the concept of aeroelastic tailoring.

Postscript

John E. Holding, who lived near us on Carberry Street in Medford, Massachusetts, developed an interest in our JAL Club, and competed in many local and international model airplane competitions with us. He came over frequently to visit and "talk shop" while we were both competing in the JAL and FBCG competitions. Eventually, John Holding tied with Wallace Bakken for the 1955 Fourth National Senior Scholarship worth $1,000. This was one of the few times in Guild history when GM awarded duplicate scholarship amounts, strictly based on merit, as neither contestant was the son of a GM employee. Warren M. Bakken, the younger brother of Wallace, was the top 1953 National Junior Scholarship winner of $4,000 the same year I won my national Guild award.

Notes

1. An earthquake epicentered at Northridge in Southern California in October 1989 destroyed most of the contents of Wykes' Los Angeles apartment, including his 1953 award-winning Guild model. From December 1990 to September 1991 he restored and restructured the model, which had fallen five feet to the floor from a shelf along with other memorabilia.

2. "Evolution of the Completed Model" was a nine-page package of drawings and sketches with technical descriptions and notes were prepared by Wykes for the FBCG Technical Department, Fisher Body Central Engineering Offices, Fisher Body Division, Warren, Michigan, circa 1953.

9

The Kid from West Texas Wins Against the Odds!

Charles R. Foreman (1950–52, 1954)

As I flew towards Boston and the 2008 Fisher Body Craftsman's Guild reunion and model car exhibition, all of my old doubts returned. After all of these years, how would my two model cars compare to the national award winners at the show? Would my cars be the brown shoes at a black tie event? I would soon find out.

The FBCG competition is a shining memory of things I accomplished as a teenager. I grew up in the West Texas town of San Angelo and had been interested in cars, motorcycles, and airplanes since I was a child. When I was 6 years old, I remember drawing a picture of my father's 1935 Ford. I avidly read magazine articles about the new postwar models coming out. I was particularly interested in the 1948 Tucker with its advanced styling and rear engine and in the new 1949 Ford and Mercury designs. In the early 1950s, the General Motors concept cars, such as the Buick LeSabre, Buick Wildcat, and Chevrolet Corvette stirred my artistic interests.

The Craftsman's Guild became the perfect outlet for my automobile design interests. I first learned about the Fisher Body design contest from a presentation given at my junior high school in the autumn of 1949. Two young men from GM gave the presentation in the school auditorium to the male portion of the student body. I built and entered model cars in Guild competition 1950, 1951, 1952, and 1954 from the age of 13 to 17. In 1952, I won a Junior Regional Award and an all-expenses paid trip to the FBCG Convention in Detroit.

West Texas is economically centered on farming, ranching and oil. In the 1950s, there was little manufacturing, but certainly no automotive design or manufacturing. My father was a carpenter and knew nothing of building models of this type. Although several boys at my school talked about entering the FBCG competition, I know of no one else in my school or city that built models and entered the competition during that period. Thus, I had no mentor and no help in car design or model building from anyone. I did, however, have support and encouragement from my parents.

Design Process

There were three things I did have in my favor going into the FBCG competition: I was good at drawing and sketching, I had scratch-built some model airplanes, and I had taken a woodworking class in the seventh grade. My shop instructor, Mr. O. D. Adair, taught me the fundamentals of cutting, shaping, and finishing wood.

I generally spent the autumn months designing the Guild models and then started construction of each car in January. I always finished just before the shipping deadline in early June. My father helped me make a sturdy shipping box from ⅜-inch plywood that I used for sending all four models to Detroit.

My main guide for model car making was the instruction booklet that came with the FBCG member packet. It showed three different ways to build a model, but seemed to emphasize the "clay-based model-plaster and mold-plaster model" approach. That seemed time-consuming and risky, so I did not use clay to design any of my four models.

Scale Drawings

I could visualize, in a three-dimensional sense, what I wanted the car to look like. Even at age 12, when I built my first FBCG car in 1950, I was well aware of compound surfaces. I had been drawing cars and other objects since age 6, and I avidly studied pictures of new cars in magazine advertisements. At the time I didn't think about it, but now I see that I really could visualize car shapes in three dimensions. I just assumed that everybody could do that! It just came naturally to me as music ability comes naturally to others. Later in college, descriptive geometry was one of my favorite subjects. I now realize that this ability to visualize objects spatially was a big benefit in my eventual career in aircraft structure design and analysis.

Before starting construction of my Fisher car model, I made several drawings, including initial sketches of the design, a perspective drawing, scale three-view drawings, and three or four section-cut drawings which accurately related to the other views. Details and contours were correctly placed in side, top, front, and rear views. I used section-cut drawings to make accurate templates that were used to check contour during carving. I used templates at approximately two inch spacing to check vertical plane symmetry about the longitudinal centerline of the model. All of this detail drafting took quite a lot of time initially, but greatly aided in sculpting the wood. I then carved the balsa wood until I achieved the desired contours.

Painting

There was considerable improvement in paint quality from my first model in 1950 to my fourth model in 1954. The paint quality of my first two cars was not very good, as they both suffered from surface "orange peel." The green paint on my 1952 Regional Award winner was better. I painted the first three cars using enamel spray cans. For the 1954 model, I used a small spray gun and automobile lacquer paint. This process resulted in a smoother surface with no orange peel. Air supply for the spray gun was the spare tire (filled to about 60 pounds per square inch) from my first car, a 1940 Ford sedan.

Wheels and Tires

Wheels for the first three models were turned on a "jury-rigged" lathe powered by an electric drill. My father had a ¼-inch electric drill that came with a bench mounting bracket.

I used the bracket to attach the drill to a small workbench in my parents' combination workshop/laundry room. I fabricated a tool rest at the end of the drill motor. To make wheels, I rough-cut pieces of balsa wood to an approximate round shape (2 ¾ inches outside diameter) and drilled center holes. I bolted a drill adapter shaft to the balsa and clamped it into the drill chuck. I used tools from my X-Acto set — which I still own today — as lathe tools to shape the wheels. I used cardboard templates to obtain the cross-sectional design shape of the wheel, hubcap, and tire. The last set of tires I used was supplied by the Guild free of charge in 1954.

1950 Entry (Ages 12–13)

This was an aerodynamic "cab-forward" design with a short axle-to-cowl ratio and beautiful compound surfaces. Front and rear fender lines were integrated into the compound door surfaces for a smooth continuous contour, and the front and rear tires were covered

Charles R. Foreman, almost 13 years old, is shown with his first Fisher Body Craftsman's Guild entry, in June 1950. This was a two-door fastback sedan carved from laminated balsa wood and painted Forest Green enamel with gray indicated window areas. The picture was taken in front of his home in San Angelo, Texas, just before the model was shipped to General Motors. The design won a 1950 Junior Division Honorable Mention Award for Texas ($25) (courtesy Charles R. Foreman).

with sheet metal down near the whitewalls. The rear quarter panel "kicked up" starting right at the rear window.

I used laminated balsa wood construction for this first car. Seven 1-inch-thick balsa wood planks were saw-cut to side view profile of the car, leaving about 3⁄16-inch extra material. I used wood glue to bond the planks together. I carved the front bumper, grilles and rear bumpers from balsa and then sealed, primed, and silver painted them. The turned balsa wheels were attached to the body with screws, and the wheel cover area painted silver or aluminum color.

Similar to all of my model cars, I scored grooves into the balsa body for the doors, hood, and trunk. I laid out the locations of the grooves in pencil on the finished balsa bodies. Then I cut small v-grooves in the balsa with an X-Acto knife and finished the grooves by sanding them using fine sandpaper wrapped around a knife blade. For this Guild entry, I was given the Junior Division Honorable Mention Award for Texas of $25.

1951 Entry (ages 13–14)

Similar to the 1950 design, the front and rear fender lines were integrated into the compound door surfaces for a smooth continuous contour. The front and rear tires were covered with sheet metal down to just above the whitewalls. Many individual parts and pieces (metallic and non-metallic) were shaped for window trim, side moldings, headlight trim rings, and the integral front bumper with grille.

Side trim and window trim were fabricated from 1⁄8-inch diameter aluminum purchased at a hobby store. I wanted polished metal trim that I could attach to the car instead of painting-on trim. Split tubing was the solution I would come up with at the time, and I used a fine hacksaw blade to saw the tubing longitudinally. The tubing was held in a vise while sawing, and it was difficult to keep it from bending or folding. I attached the trim to the balsa model with tiny aluminum pegs made from sheet aluminum. I tapped these pegs through holes drilled in the aluminum trim and then filed them flush with the trim. Some unevenness can be seen in the trim on the car. This process was so difficult and time consuming that I reverted back to silver (or aluminum color) painted-on trim for my 1952 car.

I shaped the headlight nacelles, side scoop trim, and taillight surrounds from sheet aluminum. These parts were cut to shape with a hand shear, then final-shaped by filing and sanding. The headlights were painted black.

Bumpers were carved from balsa wood. I carved recesses into the front and rear of the model car body to hold the bumper assemblies. Care was taken so that the bumpers fit into the recesses without gaps. This was a slow "trim and trial fit" process. They looked like they were made from real aluminum stock because of several layers of paint and lots of detail sanding. After carving the balsa bumpers to final shape, I sanded them with 400 grit sandpaper, and then applied a coat of wood sealer. I applied two coats of primer and sanded between each coat. Then, I applied two coats of aluminum paint and sanded before each coat with 400 and then 600 grit sandpaper. All of these coats were applied by brush.

I attached the bumpers, taillight trim, and side-scoop trim with straight sewing pins pushed into the balsa wood model. The bumpers were also glued in place, and the tires were attached with screws. This entry won the 1951 Junior Division First State Award for Texas of $150.

1952 Entry (Ages 14–15)

Similar to the two previous designs, my 1952 version had front and rear fender lines integrated with the compound door surfaces. Front and rear quarter panel sheet metal covered the front tires to below the centerline and rear tires down to whitewalls. The sculp ted bumpers had a nine-tooth grille in front and dual jet-type exhaust ports in the rear. Silver trim or brightwork ran the length of the car model along the rocker panel creating a long, lean body profile. The roof was notched above the driver's side for ease of entry and exit. The headlight lenses were translucent and featured an etched cross-hatched grid or pattern.

For the construction of the body, I followed the methods and techniques recommended in the FBCG booklet for model aircraft-type construction. The base of the model was a ¼-inch balsa wood plank. Ten ¼-inch thick balsa bulkheads were installed laterally at about one-inch spacing along a longitudinal centerline. Three bulkheads at ½-inch spacing were used in the windshield area where the contour changed abruptly. I cut the bulkheads to approximate shape using section drawings I had made. I left some extra material on them for later contour shaping. Bulkheads were notched into the base plank laterally and glued in place. I used solid balsa blocks at the front and rear of the car to achieve sharp compound curves, which could not be accomplished with the thin balsa surface plating. I cut these blocks to rough contour before gluing them to the base. Before applying the surface of double thick veneer, I carved the front and rear blocks to near final shape.

Surface veneer consisted of two layers of 1/32-inch thick balsa wood sheets. I glued these across the bulkheads, or ribs, in small strips to achieve the compound surface contours. Strips were ¼ to 1 inch wide with tighter contours requiring more narrow strips. Surface strips were glued to the bulkheads and end blocks with model airplane glue. I used straight-pins to hold the strips to the bulkheads while the glue dried. As I fitted the surface strips to the bulkheads, I trimmed the bulkhead shapes as required to achieve an aesthetically pleasing contour. I used cardboard templates to check and help maintain symmetry about the longitudinal centerline of the model.

After the first layer of veneer was completed, I sanded the whole surface to remove rough edges before applying the second layer. I always used a sanding block when sanding the final surface. I applied the second layer of balsa sheeting in a similar manner to the first layer. After completing the second layer, I filled any cracks between strips with wood putty. I then carved and sanded the end blocks to final contour to smoothly blend the surfaces.

Bumpers were carved from balsa wood and finished as described above for my 1951 model car. I fabricated the nine vertical "teeth" in the front grille from ⅛ inch-thick balsa sheet, and then attached the balsa wood tires with glue. For my efforts, this design won a Junior Division First State Award for Texas and a Regional Award, including $150 and an all-expenses paid trip to Detroit for the Guild's annual convention.

1952 Regional Trip

My trip to the convention in Detroit in 1952 was, and still is, one of the most memorable experiences in my life. I was 15 years old, and it was my first time to travel alone, having only been out of Texas once before. The round trip from San Angelo to Detroit was my

Rear view of the model entry made by Charles R. Foreman (ages 14–15) which won the 1952 Junior Division First State Award for Texas ($150) and Regional Award (Region 16, Texas, New Mexico, Arizona). For this design, he won an all-expenses paid trip to the Guild convention in Detroit (courtesy Charles R. Foreman).

first train ride. From Fort Worth to Chicago, I traveled on the *Texas Super Chief* and had my own roomette. We left Fort Worth at about 2 P.M. and arrived at about 10 A.M. the next day in Chicago, where a GM representative met me at the train. Looking back now after years of air travel, the train was a very nice, luxurious way to travel.

GM went to great lengths to assure that the Guild convention was an outstanding experience for all of the competitors. The tours and events were well-organized and interesting. They took care of small details, such as providing spending money and suggesting appropriate clothes to bring on the trip.

All of the boys from the Western states gathered in Chicago before heading onward to Detroit. GM had reserved a suite in the Conrad Hilton Hotel for us so that we would have a place to hang out until the train for Detroit left at midnight. Several of us caught the "El" out to Wrigley Field to see a baseball game. There was a television in our hotel room, and that was the first time I had seen that marvel. It was also exciting to be in Chicago that summer since both the Republican and Democratic National Conventions had just been held there.

The next morning, August 19, we were met at the Detroit railroad station by several Cadillac limousines and taken to the Sheraton Cadillac Hotel. We spoke with GM designers, including GM Styling Vice President Harley J. Earl, and toured a Chevrolet forging plant and a Pontiac assembly plant. We took a yacht ride on the Detroit River to the Old Club on Harsen's Island and stayed for a day playing cards, swimming, and relaxing. Later, we had lunch at the Detroit Athletic Club.

The highlight of the trip was, of course, the awards banquet at the great ballroom of the Sheraton with many GM executives and other industry and academic dignitaries. Harley Earl and GM's director of research and development, Charles F. Kettering, were also there. After a great meal and several speeches, the scholarship award winners were announced. Although I did not win a scholarship, the trip was a wonderful experience and quite an adventure for a 15-year-old from West Texas.

1954 Entry (Ages 16–17)

The sports car body style was new to the Guild competition in 1954, and no one knew for sure what a practical American sports car looked like except for the Chevy Corvette. With my 1954 balsa wood design, I think I captured the proportions and essence of the modern sports coupe. It was low to the ground with a compact greenhouse and open wheel wells exposing the tires. It featured a windshield that swooped up to the mid-roof section — almost like a sunroof— and the rear window was a wraparound, fastback shape. The bumpers, grille, and window trim were shaped from soft flat aluminum and highly polished with imbedded headlights and taillights in the intricate bumpers. A thin piece of brightwork running the length of the design, about mid-height, made the metallic green body look long and low. Convex-shaped hubcaps were formed by hand from soft aluminum using a homemade tool and die set consisting of a concave wood form and large convex stove bolt head.

For this entry, I used a different approach for the "chrome" areas. I used the edge of 1⁄32 inch thick aluminum sheet as the chrome trim at mid-body height, from front to rear. Window and side trim were about 1⁄8 inch-wide and were pushed into grooves in the balsa wood body. I had to cut the window trim in a flat pattern so that the exposed edge would follow the curved greenhouse contour.

I used cardboard templates to develop the flat pattern shape of the front and rear bumpers. The templates were fit to the car's contours by "trial and trim" iterations. I transferred the final shapes to aluminum sheet and cut them out using a hand shear. All trim and bumper pieces were filed, sanded, and then polished until smooth, shiny surfaces were obtained. I carefully formed the flat shapes to final bumper shape and pressed the bumpers into cut grooves in the body. Glue was used to permanently attach the bumpers. The vertical bars in the bumpers were 1⁄16 inch-diameter aluminum wire. I curved them to shape and inserted them into holes in the balsa.

Finally, I attached the official Guild rubber tires with screws. The metallic green paint — a popular automobile color in the mid 1950s — was sprayed automobile lacquer. This Guild entry won a 1954 Senior Division Honorable Mention award for the State of Texas and $25.

Post guild

Today, the Guildsmen I remember specifically from the 1952 convention are Stan Mott from Los Angeles and James Powers from Lubbock, Texas. Powers was the Senior Regional Award winner from my region, which included Texas, New Mexico, and Arizona. Stan Mott became a cartoonist whose wild and crazy art appeared in *Road and Track* magazine for

This two seat sports coupe was Charles R. Foreman's 1954 entry in the Fisher Body Craftsman's Guild competition at ages 16–17. He won a 1954 Senior Division Honorable Mention Award for Texas ($25) with this design. The model was finished with metallic green lacquer and had black indicated window areas. Bumpers, grille, and trim for this solid balsa wood model were made from aluminum sheet material. The headlights and taillights were embedded in the intricate front and rear bumpers. The convex aluminum hubcaps were cold-formed (courtesy Charles R. Foreman).

many years and later in *Automobile* magazine. Stan had won regional awards four times but never a national scholarship. Since he was 19 years old, 1954 was the last year he could enter. Stan was an interesting guy who drove a 150 pound go-kart around the world in 1961 to 1964.[1]

After my Guild experience, I really wanted to make a career in automobile styling. This was not to be, however. In the 1950s, I was quite aware of the Los Angeles Art Center School, but it was way beyond my financial means. I graduated in 1962 from Texas Tech University with a degree in Mechanical Engineering. Afterward, I went to work at LTV Aerospace in Dallas, now Vought Aircraft Industries, as an aircraft structural engineer. My specialty was advanced composite materials. I started working with advanced composites in the 1960s when the technology was in its infancy.

I was a project manager in the research and development department at Vought Aircraft. The highlight of my career was working on the design development of the B-2 bomber in the 1980s. I retired from Vought in 1999 but continued to work part-time. My career in the aircraft industry was interesting and rewarding, and I have no regrets now that I did not get into automobile styling and design. In the past, I have owned three Ford Mustangs (1964, 1969 and 1971), three Corvettes (1962 fuel injection, 1966, 1977) and a Pontiac GTO (1964). I have always been fascinated with on and off road motorcycles and still enjoy snow skiing each year in Colorado with my wife, Claudia.

2008 Guild Reunion

I had no contact with Guildsmen through the years until 2006 when I bought John Jacobus' book, *The Fisher Body Craftsman's Guild: An Illustrated History*, and learned of the ongoing Guild activities. I attended the FBCG reunion and exhibition in Boston in August 2008. John encouraged me to bring two of my models to the 2008 Guild Reunion for the Museum of Fine Arts, Boston, exhibition and retrospective.

Charles Foreman, age 72, is shown in this 2010 photograph holding his 1952 Regional Award winning model. The Regional Award (for Texas, New Mexico, and Arizona) was an all-expenses paid trip to the Fisher Body Craftsman's Guild Convention in Detroit. He graduated from engineering school and went to work for LTV Aerospace (now Vought Aircraft Industries). Foreman worked on the development of the B-2 bomber and specialized in advanced composite materials (courtesy Charles R. Foreman).

This was an exciting prospect, but I was quite concerned about showing my cars. I chose to show my 1952 Regional Award winner and my 1954 entry. The models had been stored for many years and had fifty-plus years of grime, nicks, and scratches. I spent about three months restoring them — surprisingly, they looked quite good! However, I was still concerned how their quality would compare to the top winners to be displayed at the museum. The exhibition was very successful, and my fears were unfounded. Although the national winners were of obviously higher quality, my cars received compliments and praise from both museum patrons and fellow Guild members.

All four of my model cars still have the original paint. Recently, I have done minor restoration on the cars, including cleaning and polishing the body paint surfaces. I repainted the silver bumpers on the 1950, 1951, and 1952 cars and polished the aluminum trim and hubcaps on the 1954 car. I also replaced small pieces of trim on the 1951 and 1952 cars. The model aircraft construction technique with 1/16 inch thick balsa wood skin may seem fragile, but my 58-year-old 1952 model is still in good condition.

I had no further accomplishments in the field of industrial design since my career was in engineering. However, the Boston FBCG Reunion and museum exhibition reinforced my feeling that the Guild experience was one of the major highlights in my life. It was a great pleasure to meet and talk with Guildsmen in Boston who had become professional automobile stylists and designers.

Several months before the August event, I told John Jacobus that I remembered Stan Mott from 1952 and had tried to find what had happened to him since his retirement from the auto magazine pages. I had previously contacted *Road and Track* and was told that they had lost contact with him. John thought a report on Stan would be of interest to the FBCG reunion attendees. Ron Will, a 1961 First National Senior Scholarship winner, and I were able to track him down. Stan is retired and living in Germany. Stan is a fine artist and humorist, and samples of his work can be found on his Web site.[2] Who can ever forget his Cyclops vehicle?

Conclusion

Sixty years later, I can see how my Fisher Body Craftsman's Guild experience benefited my future career in at least two ways. First, the experience of building the model cars to meet shipping schedules helped develop my organizing, planning, and scheduling skills. Second, it increased my confidence in my ability to take on and complete challenging tasks. Winning the Regional Award and attending the national convention showed me that I could and should strive for bigger goals such as a college education and career advancements.

Would the kid from West Texas have had the creativity to be a successful auto stylist? I will never know, although I have always maintained a strong interest in automobile design. Over the years, I have read all I can in books and magazines about auto design and designers as well as attended classic and custom car shows and visited automobile museums. Automobiles must be in my genes!

Notes

1. William Glen Davis, "Around the World by Kart!" *Mechanix Illustrated*, May 1962.
2. Some works of the artist and humorist Stanley Mott, Guildsman, can be found online at http://sbiii.com/cyclops/cyclops.html.

10

Guild Defined "Road to the Future" for Designer

Adrian A. Bruno (1954–1955)

I believe we are born with certain innate skills, abilities and talents, and — at ages 15 and 16 — the Fisher Body Craftsman's Guild helped me discover mine. For this teenager from Rochester, New York, there was never any question that my road to the future was clearly defined. After a highly successful, decades-long career in engineering, product design and marketing, the Guild was never wrong.

At the time of black and white television — Bob Hope, Lucille Ball, Jackie Gleason — and the infancy of rock 'n roll music, high schools in New York split students and course work into either college preparatory or vocational technical tracks. Since I was interested in science and engineering, and had fairly good grades, I took the college-bound path. I would be the first in our family to attend college and paying for it was going to be a major issue.

In the summer of 1953, at age 14, I first saw a magazine advertisement for General Motors' Fisher Body Craftsman's Guild contest. Since I was always sketching cars, this seemed to be right up my alley, and it would give me a chance to win a college scholarship. I sent in the enrollment form and soon received a package in the mail with contest rules, how-to booklet, body-style dimensional specifications, and a set of molded model rubber tires.

I had never done any model building before the Guild, but had been exposed to my father's workshop and was aware of local hot rod-creations with their hand-rubbed lacquer paint jobs. From my earlier drawing activities, I already had several refined car designs from which to choose on my drawing board. All I had to do was pick one and build it. Fortunately, I had already developed an interest in woodworking through a school shop class and had built some small projects such as corner shelving, napkin holders, and magazine racks. So, while craftsmanship was not a completely foreign idea, I understood the Guild's complexity was much greater and the expectations were much higher.

1954 Entry (Ages 14–15)

By the fall of 1953, I had decided to build a two-door hardtop for the 1954 contest and avoid the complexities of detailing an interior. I started by building a clay model from one of my concept drawings to fully develop all the body surfaces. Carefully mirroring left and right sides was very time consuming but crucial. It was essential to achieving high scores in workmanship from the Guild judges.

Clay Model Symmetry

Overall, I followed the Guild literature on techniques for clay modeling and building plaster molds for casting plaster replicas. The issue of symmetry was a primary concern for me and drove my decision to cast from the clay master model rather than transfer the clay design to wood or some other medium with templates. I worked the right side body contours for days until they suited my eye from every perspective and captured the highlights produced from various light directions. Then, I proceeded to mirror the left side using cardboard templates traced from the first side. These were generally spaced every inch and sometimes closer where shapes were complex. Over and over, at each location, I pushed the template lightly into the clay and carved away high spots until each section through the car was perfectly symmetrical. Connecting and smoothing the spaces between sections produced a perfectly symmetrical final clay master. While this was extremely tedious and time consuming, it was the key element for the success that followed.

Plaster Model

Following processes described in the Guild's literature, I built a mold in sections around the clay model, removed and reassembled the sections, and then sealed the inner surfaces with shellac. After coating the sealed surfaces lightly with Vaseline as a release agent, I was ready to cast the plaster working model. The first pour worked perfectly, and I had an exact replica of my clay model. I carved wheel well pockets in the underside for the front wheel axle and the skirted rear wheels. After trimming and smoothing mold flash areas, filling small air bubbles and light sanding, the body of the model was ready for sealing and priming. I was very pleased with how well everything had progressed on the first try.

Brightwork

All metal work such as hubcaps, bumpers and window trim were fashioned from aluminum, buffed to a chrome-like finish and glued to the car body. Headlight and taillight lenses were made from scraps of clear or red acrylic, respectively.

All the window trim was formed from $\frac{1}{32}$-inch thick by $\frac{1}{16}$-inch wide soft aluminum ribbon stock that came on small spools obtained from the hobby shop. The medium gray indicated window surfaces were recessed about $\frac{1}{32}$ inches from the roof and pillars. The window trim was simply butted up to the recess edges and surface, and then glued to the painted body. I used rubber cement adhesive for some parts; for others, I used Elmer's glue sparingly and held parts in place with masking tape until the adhesive dried. Taillight trim pieces were cut from $\frac{1}{32}$-inch thick soft aluminum sheet scraps formed to fit the body and surface glued in place.

Bumpers

The bumpers were formed from ³⁄₈-inch, ⅛-inch and ¹⁄₃₂-inch thick aluminum sheet scraps found in my dad's workshop. I made a cardboard template for the rear bumper, carefully trimming and fitting the mating edge to the body. The outer edge shape of the bumper was sketched by eye on the cardboard and trimmed with scissors. The template was used to lay out the profile on the aluminum work piece. The aluminum was cut and hand filed to a precise fit with the body while the outer edges were filed to shape. The wraparound bumper parts were cut from ¹⁄₃₂-inch thick soft aluminum sheet and easily formed to the body shape and secured with pins and glue. The front bumper was thicker, but easier, as it was essentially straight at the mating surface. The wheel/side trim pieces were cut, hammered, hand filed and tediously hand formed to shape. All brightwork was sanded with progressively finer paper and buffed to a chrome-like finish with jewelers' rouge using small 1 inch diameter buffing wheels in a Dremel tool. The heavier bumper and side trim pieces were secured with a few steel "dowel pins" fashioned from finishing nails, drilled into the mating surfaces, and then surface glued just like the window trims.

Tires and Wheel Covers

Because the front and rear tires were only partially exposed below the sheet metal, all four tires had to be cropped at the appropriate vertical height. The problem was how to reliably mount the remaining sections of rubber wheel/tire. A hole was drilled into the flat cut tire surface and a wood screw turned in with about ½-inch protruding.

A cardboard template was made to lay out wheel locations and wheel locating holes were drilled into the bottom of the model with slightly recessed rectangular pockets carved around them. The holes were filled with wet plaster and the wheel with protruding screw head inserted and leveled with a flat board. When the plaster dried, the wheels were precisely located, flat and very secure.

The wheel covers were turned on a small lathe from scrap aluminum rod, buffed to a bright finish, cut to the proper height to fit the remaining sections of the front and rear tires, and then glued in place with rubber contact cement.

Paint Job

After many days of agonizing over the endless color options, I selected a two-tone color scheme with a medium yellow body and black top. The glass areas were painted a satin gloss medium gray. The model body surfaces were sealed with shellac followed by light fine-grit sanding. Subsequent primer and color coats were all stock automotive lacquer colors of the time, obtained from a local auto paint store. With three different color paints, a lot of masking was required, but this was accomplished easily using tape and newspaper trimmed to precision with an X-Acto knife.

Dad brought home an airbrush from the graphics art department at work along with the user's manual and instruction guide. We had a small compressor in our workshop for an air supply. Neither of us had experience with the airbrush, so I experimented with scraps of material to find out how much to thin the paint and how to handle the brush for a proper paint job.

This 2-door hardtop model entry designed and built by Adrian A. Bruno (ages 14–15) of Rochester, NY, won the 1954 Junior Division Third State Award for New York ($50). The "working model" was cast from plaster of Paris using a clay master model. An airbrush was used to paint the model a two-tone medium yellow lacquer and black lacquer. A satin gloss medium gray indicated the window areas. The bumpers, window trim and hubcaps are highly polished aluminum. Clear and red acrylic plastic were used for headlights and taillights. The model took seven months to build (November 1953 to June 1954). The lessons learned on this first attempt were crucial to the success of his second entry in 1955. As this 2010 photograph shows, the 1954 model has been untouched and shows the effects of 56 years in various home display locations (courtesy Adrian A. Bruno).

The superb level of control available with the airbrush minimized overspray issues and fine wet sanding between coats made for glass smooth results. I airbrushed two coats of primer and about eight coats of lacquer, wet sanding between each and hand rubbing the final coat with compound to a high luster.

The entire design and build process for my 1954 Guild model took hundreds of hours over a seven month period from November 1953 to June 1954. During this time, while my friends were out cruising, I was spending most evenings holed up like a hermit in my dad's small basement workshop after doing my homework, working on my car and listening to 1950s music on the radio.

Finally, it was time to ship the 1954 model to Detroit. All during the construction I had a nagging concern about shipping my fragile plaster masterpiece. Of course, I pictured every accident scenario as an absolute certainty. The engineering and construction of the shipping crate would be all that stood between a possible dream outcome and a wasted seven months of tedious effort. I designed and built a sturdy box of ¾ inch thick solid pine board with screws and glue. I suspended the model just off the wheels on a raised base in the bottom of the box. The base had two heavy metal pins protruding upward that inserted into matching holes drilled deep into the plaster car body. The car was held down on the pins by 2-inch wide foam and cloth covered posts hanging down from to the top of the box to the hood and trunk areas. I shipped it and hoped for the best.

In June of 1954, I was notified by letter that I had won the Junior Division Third State Award for New York and $50. I was informed that a Guild Signet Ring and Certificate of Design and Craftsmanship would follow. Of course, I was elated to have won something but also disappointed that I had not placed higher. In my heart, I knew I could do much better and immediately started work on a new entry for the 1955 contest year.

1955 Entry (Ages 15–16)

This time my focus shifted. Building my first car had answered most of the uncertainties regarding fabrication technique and shipment so my confidence was much higher. I was free to focus primarily on design and craftsmanship. For the 1955 car, I used the same proven methods from the previous year but decided to kick up the degree of difficulty and build an open sports car complete with detailed interior.

For me, the objective this time was to create an integrated, elegant design package that was unique and pleasing to me from every angle but not necessarily ground-breaking or exotic. This car design had much more complex contours, and I paid much more attention to detail in the surface development and color choice. I wanted the model's scale and proportions to emulate viewing a life-sized automobile from a distance. This would require meticulous attention to the relative scale of every element including door, deck and trunk gaps. This was achieved mostly by eye, but I double checked my drawings by comparing them to photos from magazines and car brochures of fine European sports cars that I thought had reasonably good proportions.

Perfect Styling and Design

At the time I started work on the second car in 1954, the Corvette was the only U.S. sports car, so most references were European, such as the Mercedes 300SL Gullwing, Jaguar XK 120 and 140, and various models from Aston Martin, Alfa Romeo, Ferrari, Maserati, and Lancia. All of these cars had full wheel openings, but I stuck with the American trend of the time and skirted the rear wheels. In retrospect, I wish I had gone with open rear wheels to widen the rear stance for better appearance and handling.

Interior Design

The most difficult part for the 1955 car was the interior detail leather work and learning to form and trim a clear wraparound windshield, all of which were new to me. These items took much longer than I had planned, and I was feeling the dreaded deadline squeeze.

The seats, dash, door panels and transmission tunnel were covered in caramel color leather from my mother's old purse. I remember it was extremely soft and stretched easily over the carved wood seat bottom and back forms, as well as the dashboard curves. After stretching, it was tacked in place on the back side of the seat forms which were then glued in place. I used rubber cement to glue the leather to all other surfaces.

The steering wheel was a hoop cut from clear acrylic sheet. It seemed to take forever to file it to a round, circular shape as well as to notch the tiny finger grooves in the back side. The acrylic steering wheel was stained amber with a wood stain, and the cross-spokes were fashioned from copper sheet scrap and then buffed before being attached to the steering column with a silver colored pin. The circular speedometer and tachometer areas were just blank cut-outs in the dash panel.

Clear Plastic Windshield

I read the Guild's literature but finally came up with my own method to fabricate the windshield. First, I carefully shaped a male wood form, or wood pattern, to represent the rake angle and curvature of the inner windshield surface. This was covered with thin felt

Adrian A. Bruno's 1955 entry, which won the Second National Junior Scholarship Award ($3,000). The original construction began in late 1954 with a clay model based on selected concept drawings when he was 15 years old. The working model was cast in plaster of Paris using molds taken from the clay master. After the sealer and primer coats, about eight color coats of "Bittersweet" lacquer (a red-orange automotive lacquer) were applied with an airbrush, wet sanded between coats and then hand rubbed and compounded to a high luster. All metal work was fashioned from aluminum, buffed to a chrome-like finish and glued to the car body. A clear acetate plastic windshield was heated and formed over a wooden pattern. Light lenses were made of clear or red acrylic and interior soft surfaces were covered in caramel colored leather (courtesy Adrian A. Bruno).

cloth and temporarily mounted to the dashboard surface of the model car with a little glue. The wood pattern was about ½ inch taller than the desired final plastic windshield shape. I used the leading edge of the wood pattern as a guide to carve a shallow windshield "seating groove" in the body. I made a template for the windshield from thin cardboard, taping it over the wood pattern and carefully trimming and fitting the lower edge until it fit perfectly into the windshield groove. I used the template to trace the shape on the plastic acrylic sheet. After cutting and edge conditioning, the acrylic windshield blank was dipped in boiling water and shaped over the male wood pattern which had, by this time, been removed from the model's dashboard area. I used a piece of foam-backed felt and finger pressure to conform the acrylic to the felt-covered male wood form. It required at least four tries over several days to get it right. This loss of time caused me to make short-cuts in other parts of the interior finishing process so as not to miss the shipping deadline. The top and bottom windshield bright trim pieces were fabricated and surface glued as previously described.

Minimal Brightwork

I chose to go with a minimum of brightwork, or metal trim, for a clean look that would let the body lines, contours, reflections and shadows speak for themselves. I did make one concession to the unconventional by exposing chromed exhaust headers in a scooped-out area just behind the front wheels. All the 1955 car trim was polished aluminum attached with adhesive using the same methods for fabrication described for the 1954 car. The grille and front bumper were much more intricate for the 1955 car, but otherwise the trim parts were fewer in number and considerably refined in design and proportion.

The Last Details

The front tires were fully exposed with open wheel wells, and the rear tires were covered by fender skirts. As with the first model car, these wheel covers were turned on a small lathe from scrap aluminum rod. After the rear tires and their covers were cropped to the proper height, all four aluminum wheel cover parts were buffed to a bright finish and then glued in place. The front wheel covers were topped with buffed copper knock-off wheel hubs, the shape of which was repeated in the buffed copper steering wheel cross spokes.

For the exterior color I chose a rich-looking, red-orange automotive lacquer called "Bittersweet." As with the first car, I airbrushed two coats of primer and about eight coats of color lacquer, wet sanding between each and hand rubbing the final coat with compound to a glass-like, mirror finish.

June Deadline

When the June deadline arrived for shipment, the interior detail was complete enough to meet the contest interior requirements, but certainly not up to the standards set by my own exterior. I had run out of time — there was no option but to ship and hope for the best. To this day I feel that a little more work on the interior might have made a difference in the final outcome. On June 24, 1955, I was helping my dad repair a washing machine in the basement when a Western Union telegram arrived. Mom called us upstairs and handed me the telegram, which said:

> YOU WIN FIRST STATE AND REGIONAL AWARDS. CHECK FOR FIRST STATE AWARD FOLLOWS. RUSH YOUR RING SIZE. GUILD IS ALSO GIVING REGIONAL WINNER BLUE SPORTS JACKETS TO BE WORN AT CONVENTION. PLEASE RUSH FOLLOWING INFORMATION—YOUR AGE, WEIGHT, HEIGHT, CHEST AND WAIST MEASUREMENTS, SUIT SIZE. WE ARE SENDING STORY TO LOCAL PRESS. HOPE TO ARRANGE GUILD TROPHY PRESENTATION AT YOUR SCHOOL THIS FALL. J P WINES SECRETARY FISHER BODY CRAFTSMANS GUILD.

I let out a loud war whoop and leapt straight up in the air, planting my dirty hands on the white ceiling of the living room. Those hand prints remained undisturbed for many years as a memento of that exciting moment in the life of a 16-year-old boy. I was going to Detroit! The telegram was followed by formal letters, including a $150 check for the state prize and information on train travel arrangements to Detroit. There were newspaper articles with photos of me and the model car and as well as publicity for the upcoming national convention in early August. There would even be a 30 minute coast-to-coast radio hookup from the regional winners' banquet ceremonies in Detroit, where the names of nine national scholarship winners would be announced, as two seniors tied for the fourth national scholarship award (Wallace E. Bakken of Grand Forks, North Dakota, and John E. Holding of Medford, Maine).

This photograph of Adrian A. Bruno and his Second National Junior Scholarship award winning model appeared in the October 24, 1955, edition of the *Rochester Democrat and Chronicle* newspaper accompanied by an article titled "Curiosity Pays for East High Senior, Netting $3000 College Scholarship." He was age 16 when this photograph was taken. This kind of publicity certainly inspired other latent young talent in the Rochester area to try their own hand at the Fisher Body Craftsman's Guild contest (courtesy Dennis Floss, Democrat and Chronicle Media Group).

1955 Guild Convention

At 16, I had never traveled much by train or car outside of New York so the trip to Detroit was an adventure. My relatives were at the train station to see me off, carrying signs of congratulations and encouragement. General Motors did an absolutely wonderful job of hosting us for four days with excellent accommodations, interesting factory tours, sightseeing, and recreation. Of course, the highlights for us were the tours of the GM design studios, meeting professional designers, and seeing actual concept vehicles at the proving grounds. We had a most enjoyable experience and later received a very extensive photo album from Fisher Body to preserve our memories. But before that, the main event loomed for 40 very nervous young men.

The awards banquet was held at the Fisher Body General Office Auditorium at the new GM Tech Center in Warren, Michigan. The banquet auditorium was full with hundreds of attendees seated at large round tables on the main floor. The 40 regional winners were seated in a long elevated row of tables, next to a long podium row full of GM executives, with our models lined up in front of us. The award winners were announced.

I remember the lump in my throat as the $1,000 Fourth National Junior Scholarship prize was awarded to Carl C. Ulmschneider, Jr., of Bellevue, Washington. Then, the $2,000 Third National Junior Scholarship award went to Anthony Caracciolo of Trenton, New Jersey. My own chances seemed to fall with each succeeding announcement. The conciliatory thoughts began in my head: just being invited here was reward enough, the quality of the work represented was very high, that management's decisions were final, and everyone there was already a winner. All very true but I needed that scholarship! When my own name was announced for Second National Junior Scholarship winner I could have left my hand prints on the ceiling of that auditorium. Afterward, Thomas F. Greene of Medina, Washington, took the First National Junior Scholarship prize and $4,000.

After my announcement everything was a blur, but upon returning home after all the accolades and extensive press coverage, I had time to reflect on the whole journey. Of the many thousands of contest entries, I was one of only 20 in my age division selected for the national competition, and then, one of only four to be awarded a scholarship. I was proud of my accomplishment and appreciative of the $3,000 scholarship I so desperately needed in order to attend engineering school. But then the natural "What if?" questions kicked in. What if I had finished the interior better, would that have made the difference for the first place award? I wished I could have started building another model for the next year, but the rules did not allow a National Scholarship winner to compete again.

Little did I know that 53 years later I would get a chance to finish that interior the "right way" and to continue the Guild experience at a special exhibition held August 2008 at the Museum of Fine Arts, Boston. I completely restored my 1955 model for the MFA event, and in the process, the blank instrument panel cut-outs in the dash panel were enhanced with bright trim rings, a radio console and a small silver switch plate on the drive tunnel.

I guess I would have to agree with the theory that some people are just born with certain innate abilities. How else can you explain this group of Guildsmen showing such creativity, drive and high skill levels at such a young age? For me, it all seemed relatively easy; there was no question that the future road I would follow was clearly defined. Obviously, winning such a competition also brought social recognition, pride and financial reward. However, the primary benefit was the enormous boost of confidence that encouraged and accelerated my further growth.

In retrospect, I realize how blessed I was to have discovered and verified what was to become my life work at such an early age through the Guild. Following those years, I went on to engineering school, earning a bachelor's degree in mechanical engineering, and entered the workforce. Work was interrupted by a tour of duty as a commissioned officer aboard a U.S. Navy destroyer during the turbulent Cuban missile crisis under President John Kennedy. On return to civilian life, I earned a master's degree in business and began a very rewarding 50 year — and still counting — career in new product development and marketing. Currently, I am a senior executive at Weber Stephen Products Company, home to the famous backyard

Adrian A. Bruno in August 2008 at the FBCG model car exhibition at the Museum of Fine Arts, Boston. A native of Rochester, New York, Bruno is the former vice president of R&D and currently a senior executive at Weber Stephen Products Company. With over 47 years in design, management, engineering and marketing, his background includes experience at corporations such as Corning, Amana, and Middleby. He obtained degrees from Bucknell College and Syracuse University and was a commissioned officer with destroyer duties in the U.S. Navy. After 53 years of storage in various locations the model was in very bad condition with cracked and peeling paint, loose or missing parts and a badly rotted leather interior. After restoration the model was exhibited in August of 2008 at the Museum of Fine Arts, Boston (courtesy Adrian A. Bruno).

Weber Grill. Here, the simple charcoal kettle has evolved and expanded into sophisticated gas and electric outdoor cooking appliances as well as outdoor kitchen centers.

Whether it is cars, appliances, toys or furniture, the fundamental principles employed in the Guild experience still apply in the continuously evolving modern business world. I still relish the stimulating creative process that has resulted in many exciting new products and 40 U.S. patents in diverse fields, including glass ceramics, cooking products, mechanical products and ski accessories. Using today's powerful computer-aided design and analysis tools, we can now accomplish in days what took months to do in 1955. But the underlying creative thinking process has not changed — the new tools simply allow modern designers and engineers to do much more, better, faster.

11
From the Farm to GM Design Staff

Charles H. Stewart (1951–1955)

As a self-taught artist, I found many outlets for my creativity while growing up on, of all places, a tobacco farm in central North Carolina — an unlikely environment for a budding artist. Because of the General Motors Fisher Body Craftsman's Guild, I began my auto design career in 1951 at age 15.

I had spent my childhood making toys and gadgets and constantly sketching cars and airplanes. Although I had no formal art classes, I became the "local artist" at my school and church, as I made many posters, signs, school play backdrop paintings and even a church baptistery background painting. We had no television, but the world I saw in magazines and heard about on the radio captured my imagination.

I had always been a sickly and asthmatic child who was allergic to just about everything on the farm. My parents soon knew I would likely pursue a career away from home. Fortunately for me, they understood that I would be an artist of some sort, and they didn't expect me to spend the rest of my life farming. Although the world of art or industrial design was far removed from anything in their experience, they accepted the fact that I was determined to be creative and didn't discourage me. So when I came across a Fisher Body Craftsman's Guild ad in *Popular Mechanics* with its enticement to design and build a model car that could win a college scholarship, I immediately sent in the entry form. I was eager to win that money!

My first model in 1951 involved a lot of trial and error as I struggled to design and build that one-twelfth-scale, dark blue Cadillac-looking sedan. I modeled it in clay and made a five-piece plaster mold of the finished clay. This method was suggested in the *Guildsman*, a newsletter that began arriving from Fisher Body soon after I enrolled. I greased the mold as directed, poured plaster into it, and then trimmed, sanded, and painted the resulting model to the best of my ability. This was my first deadline-driven design project and the beginning of a lifetime of design deadlines. It was shipped from the local Railway Express Office at the last minute of the June 30, 1951, deadline.

That first year was also my last year to be eligible for the Junior Division for 12 to 15 year olds, and I surprised myself by winning the Junior Division First State Award for North Carolina ($150). I looked forward to trying again the next year in the Senior Division.

In 1952, my senior division entry — a bronze sedan with cream color roof— was a more ambitious undertaking. I made some polished aluminum trim parts, instead of just using metallic paint to simulate chrome parts as I did in 1951. It was another plaster model, but the design was more original with a floating roof supported by three canted thin rods on each side, instead of the usual rear vertical sail panels. The roof was supported in the front by a central pillar structure at the windshield. The paint job was much better than my 1951 entry. However, I was competing with more experienced boys now in the Senior Division who were 16 to 19 years old, and I only won the Senior Division Third State Award for North Carolina ($50).

Nevertheless, I was determined to build a better model in my senior year of high school for the 1952–1953 competition. By using the school's workshop, I made polished aluminum bumpers, grille, and window trim. It was another "future Cadillac" made of plaster with a metallic green body and a cream-colored roof. Looking back, I think styling took a back seat to improved craftsmanship, helping me earn a Senior Division Second State award for North Carolina ($100).

Upon graduation from high school in 1953, I had two top goals. First, I wanted to enter the University of North Carolina at Chapel Hill, where I had earned a scholarship and would start that autumn. Second, to design and build a FBCG model that would win a National Scholarship. I had been sketching ideas for a new 2-door fastback sports car with neat proportions. The sports car was a new Guild competition body-style beginning in 1954.

After three Guild competition entries, I had never seen another Guild model, only photos of winning models in the *Guildsman* and in a few newspaper articles showing the local state winners. In June 1953, the local newspaper reported that 15-year-old Arnold Joslin from nearby Raleigh had won the Junior Division First State award for North Carolina and the Regional Award for North Carolina, South Carolina, and Georgia as well as the Second National Junior Scholarship award worth $3,000. I wanted to meet him in person and see his model, a black sedan. We looked up the Joslin family name in the phone book, called, and were graciously invited to visit. It was an eye-opening experience for me to see the high level of workmanship and finish that he had achieved in his Junior Division model. I resolved right then that my 1954 black, two-seater, fastback model would have comparable national-level workmanship and finish.

Even though I did not have a well-equipped workshop like the one Arnold and his dad shared in their basement, I arranged with my local high school to use its shop after school hours to cut, grind, hand file and polish my aluminum parts. I noticed Arnold had made really nice window moldings by using modified HO gauge model brass railroad track, having them chrome-plated and inlaying them in fine grooves filed into his wood model. The idea of filing off one flange of the H-section rail, resulting in a T section that could then be carefully shaped and inserted in grooves, had been detailed in earlier editions of the *Guildsman*. This window trim approach worked well for me on my new model, which was carved from poplar wood.

I developed the design in clay, but made cardboard templates to guide me in carving the wooden model. The design featured a cantilever roof, bumper-to-bumper side trim and fender skirts in the rear with partially covered front wheels. I used polished aluminum for the grille and bumpers and tried hard to have a flawless paint finish! To achieve a high-gloss, lustrous black lacquer paint finish like Arnold's car, I used a new continuous-action hand sprayer with more wet-sanded primer coats. I was rewarded with the best finish I'd achieved yet.

Today, after more than 56 years, I can still feel the thrill of opening that telegram from General Motors on a hot and steamy North Carolina afternoon in July 1954. The telegram informed me that my model had won Senior Division First State Award for North Carolina

This entry, made by Charles H. Stewart (age 18) of Lillington, North Carolina, won the 1954 Senior Division First State Award for NC ($150) and Regional Award (North Carolina, South Carolina, and Georgia). This 2-seater, fastback sports car was sculpted in clay and transferred to poplar using templates. The window trim was chrome-plated brass HO model train track inserted into grooves, and other bright work on the model was polished aluminum. It was finished with a black paint job using a household hand "bug" sprayer and employed white indicated window areas. His design featured a cantilever roof line, long axle-to-cowl ratio, bumper-to-bumper side trim, and quarter panels that covered the tires down past the hub caps. 1954 was the first year the Guild allowed the sports car body style (courtesy Charles H. Stewart).

($150) and a Regional Award. This was heady stuff for a farm boy who had never been out of the state. I couldn't really imagine how fabulous that trip to the Guild convention in Detroit would be. The trip and the festivities in Michigan, including being introduced on a nationally televised broadcast of the awards banquet held at the Detroit Masonic Temple, was the stuff of dreams. And although my workmanship didn't measure up nationally, my styling score was high, and I was encouraged to try one more time in 1955.

My commitment to building better 1954 and 1955 models involved my dropping out of college in the spring semesters in order to concentrate totally on the models — hoping for a national award. As I mentioned before, my parents were amazingly supportive, and to this day I am in awe of my mother actually turning over our dining room for my use as a workroom! It had large, almost floor-to-ceiling windows that faced north and afforded good light for all my various projects.

My last model was another 2-door, fastback, sports car with the body painted with Chevrolet "Gypsy Red" lacquer and a black lacquer louvered top. I developed the design in clay, but made cardboard templates from the clay model to guide me in carving the poplar wood. It sported "Lucite," or clear plastic, bullet-like headlights that I turned and polished on a school lathe. The front and rear tire profiles were more exposed than in 1954 with sculpted wheel well openings. It featured polished aluminum wraparound bumpers and grille as well as some chrome-plated trim elements for the tops of the rear quarter panels. I typed a list of conceptualized automotive features I ascribed to this model and glued it to the bottom of the model for the Guild judges to read.

This 1955 model entry, again, won the Senior Division First State Award for North Carolina ($150) and a Regional Award. Again, I sat in suspense during the Guild's scholarship awards banquet held that year in the Fisher Body Auditorium at General Motors' new Technical Center in Warren, Michigan. I watched, as I had the year before as, one by one, other boys were called to the podium to receive their coveted National Awards. Again, I didn't make the cut!

Charles Stewart styled another 2-door fastback sports car in 1955 at age 19. This design won the 1955 Senior Division First State Award for NC ($150) and Regional Award (North Carolina, South Carolina and Georgia). A clay model design was developed and transferred to poplar using templates. The paint finish was a production Chevrolet "Gypsy Red" lacquer for the lower and black lacquer for the roof. The model featured Lucite or clear plastic bullet-like headlights, front and rear shaped wheel well openings, polished aluminum wraparound bumpers with grille, and some chrome-plated trim on top of the rear quarter panels (courtesy Charles H. Stewart).

The next day the Guild's Technical Director, Norman E. May, told me that my model had ranked sixth in the national scoring overall with high styling scores. I still hadn't reached the perfection of workmanship achieved by my competitors, but I was rewarded in many ways by being at two Guild conventions and meeting some great industry people. Many Guild officials and GM designers I met at those conventions encouraged me to pursue my dream of becoming an automobile designer.

One designer in particular, Bob Veryzer, who served as a Guild styling judge in 1955, acquainted me with his alma mater, Pratt Institute in New York. I knew little about Pratt at the time and had thought the only school for a serious auto design student was the Art Center School (ACS) in Los Angeles. I had applied to ACS for entry in the autumn, although

I really didn't know how I could afford it. I had been accepted, given a class schedule and even set up with a part-time job to help with expenses. At the Guild convention in the summer of 1955, Veryzer encouraged me to apply to Pratt, which was a lot closer to home than California. He and other GM Styling executives also told me GM was subsidizing the industrial design program at Pratt, and I could expect some tuition help if I did well my first year. My scholarship would probably increase year by year as I progressed toward a bachelor's degree in industrial design with an eye toward working at GM Styling.

Charles H. Stewart began his auto design career at age 15 with the Craftsman's Guild and competed 5 times (1951 to 1955). In addition to the Regional Awards he received in 1954 and 1955, He won several state awards for North Carolina (1951 First State Junior, 1952 Third State Senior, and 1953 Second State Senior). He dropped out of college during the spring semesters of 1954 and 1955 to focus on his Guild scholarship aspirations. Stewart studied industrial design at Pratt Institute in New York City and was hired by GM Styling in June 1959. In 1960 and 1961 he was appointed a FBCG Design judge to help score the multitude of Guild model entries. One of his Oldsmobile Toronado proposals, a sketch with bulging fore and aft wheel well housings, became a prime image of a Guild high school recruiting poster. He worked in the Chevrolet, Oldsmobile and Cadillac Exterior Studios designing new GM products and retired in 1997 after 38 years of service (courtesy Charles H. Stewart).

Immediately upon returning home from the 1955 Guild Convention, I applied for entrance to Pratt Institute. Although it was late in the summer, I received the entrance exams, and completed Pratt's talent test. I was accepted just in time to enter in September and immediately cancelled my enrollment at ACS. To this day, I feel that my Fisher Body Craftsman's Guild participation had an influence on my quick acceptance at both ACS and Pratt.

The extra effort I put into my last two models did pay off for me monetarily. I desperately needed money for college, especially to get started at an expensive school like Pratt. So when GM offered to buy my 1954 model and then my 1955 model for around $500 each to display in the GM "Motoramas," I didn't hesitate to sell them. Five hundred dollars went a long way toward school expenses back then. With financial help from my father, supplemented with my savings from summer jobs and various loans, I was able to enroll in Pratt Institute more or less by the "skin of my teeth." As it worked out, GM did help me with tuition throughout my four years at Pratt and covered my full tuition in my senior year.

The years spent in New York at Pratt Institute constituted an incredible culture shock for a farm kid like me. After I got used to the New Yorkers and their zest for life, I began to enjoy the classes, welcomed the challenges presented by Pratt's teaching techniques, and absorbed the Bauhaus School of Design philosophy. New York City was an industrial design student's dream city, and just being there inspired my artistic creativity every day.

The next culture shock for me was the move from New York City to Warren, Michigan, to work for General Motors Styling staff. The GM Technical Center was new and ultra-modern, the epicenter of automotive technology and design, when I arrived there in 1959. GM's automotive design aesthetic dominated the automobile market in those days. Over the years, I worked with many former Guildsmen, like myself, styling and designing new products for General Motors. Some of my Guild co-workers were Design Executive Chuck Jordan (1947 National Senior Scholarship winner), Chevrolet Exterior Studio Assistant Chief Designer Robert Cadaret (1950 National Senior Scholarship winner), fellow Oldsmobile Exterior Studio Designer William Charles Mason (1954, 1955 and 1956 Regional Award winner, and 1957 First State Maryland Senior Division winner) and Preliminary Design Studio Assistant Chief Designer Terry Henline (1957 National Styling Scholarship winner, 1958 National Senior Scholarship winner).

In 1960 and 1961 I was a Guild judge, in 1962 I helped design show cars for the 1964–65 World's Fair, and in 1964 Charles Mason and I began working together in the Oldsmobile Exterior Studios, which lasted for several years. One of my production Olds Toronado proposals, a sketch with bulging fore and aft wheel well housings, became the prime image of a 1967 Guild high school recruiting poster. I retired from GM Design (formerly GM Styling staff) in 1997, but still reside in a small northern suburb of Detroit with family and friends. I still communicate with some of my fellow Guildsmen.

The Fisher Body Craftsman's Guild definitely paved my route to Pratt, and later, to working at General Motors as an automobile designer. The over 38 years I spent at GM, mostly in the Chevrolet, Oldsmobile and Cadillac studios, were challenging and rewarding, much like my teen years spent competing in the Fisher Body Craftsman's Guild.

12

From Ford Country to GM Interior Design to DeTomaso Mangusta Devotee

George E. Anderson (1954–1956)

Growing up in "Ford Country" with family ties to Henry Ford and the Rouge plant, I was possessed by car design at a very early age, having decided when I was seven years old that I would become either an automobile designer or a cowboy.

Living in Detroit was a definite advantage for a young car enthusiast like myself because we were surrounded by the inescapable presence of many car companies. In my community or nearby we had the Ford Rotunda, the GM Building with Chrysler, Packard, Nash headquarters and plants easily within the area. The plants were good places to spot the new cars before they hit the showrooms. We were also just a couple miles from the then world famous Livernois Avenue lined with used car lots, new car showrooms and those big searchlights that lit up the evening sky in the fall beckoning new car buyers.

I got the model building bug from my father, Emil, an electrical engineer and model railroader since the 1930s, who had even worked on a train layout for the New York World's Fair in 1939. He was quite interested in model building, in general, and built a PT boat and a Japanese Zero model airplane for me. My father would occasionally take me downtown to his model railroad club at the old Union Station, around town to visit many wonderful hobby shops and to the home train layouts of his friends around the city. These visits have left lasting memories with me.

We also did auto enthusiast activities together such as the midget car races at the Motor City Speedway on Eight Mile or the air show at the Michigan State Fairgrounds. Another event we liked was the introduction of all the new 1949 Ford products at the Dearborn test track.

We had an excellent shop class or "Manual Training" as it was called at Edgar Guest Elementary School in Detroit, where the instructor, Mr. Perrine, taught us to do nice, clean, precise three view mechanical drawings of our woodworking projects. This skill served me well in the Fisher Body Craftsman's Guild program. My older brother, Al, built a model sailboat for his class project with the goal of entering it in the annual *Detroit News* Regatta on Belle Isle, an island out in the Detroit River. This turned out to be a magnificently beautiful model, and an inspiration to me, with a smooth white hull, translucent white sails, and a varnished wood deck with all the brass fittings gleaming in the sun.

My brother's stunning model sailboat inspired me to put my efforts into building a model car for entry in the Fisher Body Craftsman's Guild model car contest. Again, living in Detroit, I had been exposed to the Craftsman's Guild models over the years because the winners from every state and region were put on display annually in the lobby of the GM Building on West Grand Boulevard across from the Fisher Building. I was able to ride my bike down there and study the models up close as they were lined up in their glass showcases with the builder's names, regions, and awards. This was very inspiring and overwhelming, at the same time, because the quality of the models, particularly the seniors, was extremely high and they looked totally professionally made. When I finally got over my apprehension, I applied to the Guild and received the plans and then a set of tires. I began to lay out my design with my drafting tools and was careful to follow all the dimensions given on the Guild specification sheets.

1954 Entry, Age 15

I chose to design a 2-door sport coupe with a glass canopy or greenhouse rather than a sedan, now that the 1953–54 rules had been revised to allow this body style. I wanted to include a continental spare tire mount on the rear because I didn't like the tacked on appearance of the ones that were being done at the time and thought that I could do a better job of integrating it into the design of the car. I felt that I was vindicated by the introduction of the Pontiac "Bonneville Special" show car at the 1954 Motorama that used a similar integrated spare tire design. I had tilted my tire cover slightly forward at the top rather that straight vertical position like the Pontiac "Bonneville Special" show car.

Construction of my model was started in a manner that was similar to my brother's sailboat model by gluing pine boards together and carving the block of wood to shape. The carving went pretty well with the exception of fighting the end grain of the wood to get a smooth shape, particularly on the very rounded front end. The rear fenders also presented a challenge to get them both the same shape. The production of the spare tire cover was simplified because of my father's lathe in his little machine shop in our basement. I was also able to produce very credible turned and polished solid aluminum wheel covers that dressed up my model considerably. The wheels were not required to turn so the tops were cut off to fit into the fender openings and fastened with wood screws to hold them in place. The wheel covers were secured with "Pliobond" glue (an industrial adhesive) and press fit into the rubber tires. Since this was the era when every high-class car had whitewall tires, I painted the sidewalls white with a brush and pin striping enamel.

The rear fenders were carved with a deep v-groove shape, almost like a double fin with the valley getting more shallow and washed out as it came forward to meet the backlight. This was a real challenge to get the transition on both sides looking the same but when finally accomplished, the v-shape provided a pleasing contrast to the soft, rounded, sculptural forms that blended into the circular shape of the spare tire cover.

The front-end theme was a wide horizontal shape in contrast to the vertical emphasis of the rear design elements of the model. The front bumper started high at the extreme outer corners of the hood and swooped down smoothly to form a soft v-shape in the middle that protected the center grill bars. Grooves in the aluminum surface simulated headlights that were hidden behind doors in the outboard upper tips of the bumper. The front fenders trailed back from the bumper tips in a high horizontal peak line that turned sharply downward to become a vertical air scoop at the front edge of the rear fenders.

The passenger compartment upper (e.g., canopy or greenhouse) was conceived as an all glass greenhouse with normal drop down type side glass and a wraparound windshield, but the rear glass wrapped from the deck lid all the way up overhead to form a clear glass sunroof and terminated at the top of the windshield. The glass pieces would be all outlined with flush chrome trim to set them off and anchor all of the edges of the glass.

I planned to paint my Fisher Body model a two-tone turquoise and white, which was a very popular color combination in 1953. As I started the project, I spent a lot of time sanding and priming with spray cans of Sears Hot Rod primer and sanding again and painting and sanding, and painting some more, to eliminate any traces of the wood grain and to get a perfectly smooth finish. I was gratified when my aunt exclaimed that it looked like it had been molded from plastic rather than having been carved from wood. At this point, the car model was a solid white color similar to my brother's sailboat model. I decided the shapes would show up better in a dark color and found a new dark metallic plum color paint that seemed to fill the bill and looked good with a dark gray upper canopy color.

All this time I had been treating the model as a sculptural shape without much concern that it would need substantial chrome trim to compete with the models that I had seen at the GM building in previous years. Then I came across an article in the paper saying that 3M had developed a chrome tape that I thought might be the answer to my current dilemma, since I wanted a flush look and the deadline for entries was drawing near. This new product was not readily available in stores, so I would have to go over to the 3M offices on the east side of Detroit. It took most of a day, but I managed to accomplish this using several different streetcar and bus lines to get there and back.

The 3M material was a pretty good solution to the problem but my technique left something to be desired. The model suffered from my lack of experience and I was running out of time. I had to live with it. I very carefully boxed up the model in the prescribed shipping crate and placed it in my *Detroit News* delivery bag and rode my bicycle down to the designated Fisher Body receiving dock and was able to just beat the deadline for model entries. Needless to say, my shortcuts hurt the quality of the model and I did not qualify for any awards in my first year of participation.

1955 Entry, Age 16

I was determined to improve my model and to do better in the 1955 contest the following year. The 3M material was quite stiff and did not conform well to compound surfaces, so I decided not to use the tape for compound surfaces and proceeded to cut off the front and rear bumpers of the 1954 model entry and carve them again, but this time, from solid aluminum. This required a lot of time and effort, but made a major improvement in the appearance of the model. I sawed the new front bumper from a solid chunk of half-inch thick aluminum and then laboriously filed it down to the finished shape with much "trial and error" to get the proper fit and finish. The rear bumpers were cut from the same aluminum stock and were even more finicky because they had to incorporate a red plastic lens on the top as well as the exhaust tips protruding through the bottom.

I re-sprayed the paint in the same dark metallic plum color with dark gray to simulate an all glass canopy or greenhouse. I did not have any fancy paint spraying facilities, and just used spray cans of Tempo Touch-Up Metallic Plum for the body and Dark Gray for the upper. I hesitated to redo all the windows with polished aluminum trim strips seen on

This model, on display at the 2004 Guild Reunion held in Detroit, was made by GM Designer George E. Anderson when he was 16 years old. The design earned a 1955 Senior Division Honorable Mention State Award for Michigan ($25). This 2-door sports coupe was carved from pine and featured an integrated continental spare tire mount and a glass canopy or greenhouse. Also shown in the background (L to R) are the rear views of the Guild models made by Michael A. Pietruska (1965 Naational Scholarship), Richard Pietruska (1963 National Scholarship) and Ronald C. Pellman (1960 National Scholarship) (courtesy Harry E. Schoepf).

the top award winning models. I was torn here because I thought that the thickness of the metal gave a heavy look and I preferred the flush look of the 3M material. I then found out, by experimentation, that the 3M material worked better if it was cut from wide pieces using a flat pattern that fit the contours of the model, so I employed this flat pattern technique as much as possible, with much better results than my previous entry.

My revised model was entered in the contest in 1955 and, even though the basic design had not changed, the detail revisions and the metal parts that I had made were quite a considerable improvement and the model received a Senior Division Honorable Mention Award for Michigan ($25). While it was not a scholarship award, I considered it to be a good result for an entry in the highly competitive Michigan Region.

1956 Entry, Age 17

I entered the contest again the following year (1956) and should have designed a new model that incorporated everything that I had learned from previous years entries, but this was not to be. Upon graduation from high school, I was looking for a job to make some money for college and had applied for a couple jobs in downtown Detroit doing artwork. Well, I got both jobs, so now during the day, I was doing the catalog illustrations for an electrical company, and in the evenings I would do the design and lettering for an embroidery company a few blocks away. This left little time for the design and construction of a new model so all I could do at this point was clean up and polish the existing model and enter it again. I sent in that same old model, which was getting pretty tired and dated by this time and, as could be expected, the exercise resulted in no further awards from the Guild.

About this time I put together a portfolio of my artwork and applied for entrance to the Art Center School in Los Angeles, that had the reputation for being the best school in the country for car design. I included photos of my Fisher Body Craftsman's Guild model car and I am certain that building the model and entering the contest played a major role in their consideration and acceptance of my application. The determination and self-discipline required to complete a model and receive an award for its design and craftsmanship spoke for itself. Once at the school, I was soon able to receive a scholarship sponsored by Ford Motor Company through my efforts in the class assignments, and that scholarship allowed me to finish my education at the school.

These were interesting times in Los Angeles, with car shows held every weekend in restaurant parking lots and various golf courses around the city. The larger car shows were held at the Pan Pacific Auditorium or the Long Beach Arena. Sports car races such as the annual Los Angeles Times Grand Prix were held at Pomona and Riverside raceways and there were the power boat races on the inland lakes. One of the class projects involved taking photos at the Los Angeles International Airport, and at that time, you could walk right out on the tarmac among the DC 6's and Lockheed Constellations and shoot pictures to your heart's content, a far cry from today's elaborate security precautions.

In January of 1960 there were six of us graduating from the original class of 36 freshmen that had started together. A seventh classmate stayed at Art Center to work on his master's degree. Of the six graduating, two were native Californians who decided to stay on the West coast, another was Canadian and went to an aircraft company in Toronto, and three of us were hired by GM while Ford did not hire anyone that semester. Since we were all enamored with GM's radical new rear engined, air cooled Corvair, it seemed like GM would be a more exciting place to work than Ford, so I became the black sheep of our Ford family and accepted the offer from General Motors to work in their interior department.

At GM, the designers had originally done the exteriors and interiors in the same department but when they opened the new GM Tech Center in Warren in the mid-fifties, the Styling Section featured five specific interior design studios, one for each of the various divisions. Steve McDaniels was in charge of the interior design studios and George Pollard was his assistant. It was George Pollard who came to Art Center with Charlie Jordan and interviewed the graduating seniors. I had included a couple of interior illustrations in my graduation portfolio, and I was interviewed by the GM Design representatives and was offered the opportunity to fill an opening in the interior design studios. I accepted the offer and this led to a fascinating and rewarding 43 year career at GM design in automotive interior design. My other two classmates did not remain at GM long term, as one was released after the ninety-day trial period and the other switched from automotive to the Graphics Department, eventually leaving to run his family business.

I was extremely fortunate that the very first sketch that I did in a production studio was accepted and went into production. It was just a heater face on the Corvair, but was my first project and it went into hundreds of thousands of vehicles for several years thereafter. My second project was a Corvair show car for the Detroit Auto Show in 1960 and it was so well received at the show that it also went into production. I continued to enjoy the

George E. Anderson, career GM designer, with his 1955 Guild model entry he displayed at the MFA Boston event August 9–10, 2008. He attended the Art Center School in Los Angeles with help from a Ford Motor Company scholarship and graduated in 1960. He was hired by GM Styling at the GM Tech Center and enjoyed a 43 year career in the area of automotive interior design, retiring in 2003. He worked on show cars for special events like the World's Fair (1965) and the GM Pavilion at Epcot Center in Orlando, Florida. Of particular note are the interior projects he worked on involving the Buick Riviera ***Silver Arrow*** and the ***Pasha***, a pearlescent yellow Oldsmobile 98 convertible. Outside of GM, he had several sports racing car projects such as a Cadillac engined SCD III in 1965 and Jerry Hansen's Wolverine in 1966 (courtesy Elaine M. Anderson).

constant challenges of the interior design department because of the safety, comfort, and perceived quality issues that were increasingly difficult to satisfy as government regulation and competition from Europe and Asia ramped up year by year. I always enjoyed working on special show vehicles and we would sometimes do as many as half a dozen of these special show car interiors during a year, with extra efforts for the New York World's Fair, the opening of GM's Epcot Pavilion in Orlando, Florida, and other high profile events. One of the interiors that I designed was for the Buick Riviera "Silver Arrow" that is now at the Sloan Museum in Flint. Another project was the "Pasha," a pearlescent yellow Oldsmobile 98 Convertible done for the New York World's Fair in 1965.

While working at GM we were able to view the Fisher Body Craftsman's Guild entries every year and I was always impressed at the sheer number of entries annually and the daunting task of checking every one for dimensional accuracy and compliance as well as for their design and craftsmanship qualities. Looking back on the tremendous effort expended every year to judge the numerous entries, we should not be surprised that the decision was eventually made to discontinue the contest.

Outside of my career at GM, I was also fortunate to be able to aid in the design and construction of a couple of successful sports racing car projects in the mid–1960s, the Cadillac engined SCD III in 1965 followed by Jerry Hansen's Wolverine Chevrolet in 1966. The Wolverine has now been restored and recently competed at the Goodwood Revival in England. I currently own a 1969 DeTomaso Mangusta that was designed by Giorgio Giugiaro and I have always felt is one of the most beautiful cars ever built. I have taken home several awards for the car at the Meadow Brook Concours d'Elegance, the Eyes on Design car show, the Italian Happening, and the Concours Italian Style, with appearances at the Cranbrook Concours d'Elegance, Ypsilanti Orphan Car Show, GM Employee Classic Car Show, Saturn Employee Car Show, the Franklin Village Jim Beaudette Classic Car Show, and the Woodward Dream Cruise.

Since my retirement from GM in 2003, I have kept up my interest in model railroading and railroads in general, and recently co-authored a book describing railroads in the copper mining area of Michigan's Upper Peninsula. I am currently building a model railroad layout depicting the copper mines, mills, smelters, and other industries of the Keweenaw Peninsula. I have also continued drawing and painting over the years, with one of my watercolor paintings even gracing the cover of a national model railroad magazine. I would like to do more of this type of artwork in the coming years. My wife, Elaine, and I now spend the winter months in the beautiful Sonoran desert near Tucson, Arizona, thus fulfilling the cowboy part of my early boyhood dreams.

13

Shaping a Model and Forming a Life

William D. Scott (1957–1959)

When I was 12 years old, I saw a newsletter from General Motors' Allison Division in Indianapolis, where my grandfather worked. Although this GM plant made jet engines and transmissions, there was an article about the young men in Detroit who designed cars for the GM Styling Section. It included photos of the stylists working on sketches and clay model automobiles. This left a strong impression and planted the idea that I could design automobiles for a living. Having the Fisher Body Craftsman's Guild as reinforcement, that goal never wavered and set the direction for the rest of my life.

From my earliest days in rural Indiana, I liked to draw cars, airplanes, boats, houses and churches, but my keenest interest was cars. I collected photos of the new Detroit models from newspapers, magazines and brochures. I loved watching for the new model introductions in the fall. I remember that I especially loved Pontiacs but maintained a high interest in many different brands. The mid–1950s Ford Crown Victoria with its sunroof stands out in my memory and probably was a strong influence for my first Guild model. I wanted to draw cars and design things, hobbies that led others to see me as the "oddball" since my peers were interested in tractors, farming and basketball.

I saw an ad in *Boys' Life* magazine for the Guild's 1957 competition just as I was about to enter my sophomore year in high school. I sent for the entry package, which included a booklet on how to design model automobiles, a subscription to the *Guildsman* newsletter, a blueprint plan of required dimensions and scale rubber wheels. My parents helped me set up a work table in my bedroom, and I spent all of my time working at it when not in school.

1957 Model Entry

My first Guild entry at ages 15–16 was a 2-door convertible that was a hybrid, a cross between a Chevy and a Ford. It had rather long, but low, tail fins and a recess, or notch, in the rear deck lid. Carved from basswood, it was painted a coral color using acrylic enamel and employed aluminum bumpers and chrome-plated brass trim. The interior was painted coral with charcoal gray accents — a color scheme adopted from the 1955 Chevrolet Bel Air. My model design featured hooded headlights and shaped wheel-well openings — all representative of the "futuristic" style of many Ford products in the 1950s.

The scale, proportions, believability of design and careful detailing made the model

The model made by Bill Scott (age 15) of Acton, Indiana, which won the 1957 Junior Division First State Award for Indiana ($150) and Regional Award (Indiana, Illinois). This two-door convertible was carved from basswood and the exterior painted a coral color with acrylic enamel. The interior was painted a coral color with charcoal accents. The aluminum bumpers were shaped from flat stock aluminum and other parts of the model employed chrome-plated brass trim. The steering wheel rim and wheel covers were turned on a lathe. The side sweep spear, the shape of the tail fins starting at the small vent at the belt line and the hooded headlights were characteristic of mid–1950s Ford designs. The width of the car is accentuated by the strong horizontals of the light bar, wraparound grille and bumpers. This model shows a heavy influence of the Ford Crown Victoria, a car which had an innovative translucent roof panel (courtesy William D. Scott).

look like a real automobile. All three of my Guild models were distinctive this way. This required some innate ability and some drafting skills, but mostly experience. I did well in geometry and drew house and church plans from a very early age, requiring me to measure and lay out things in an accurate way for a workable plan. Using my drafting skills, I laid out the 1957 model design accurately with a scale drawing of the side, front, rear, and top views and then mapped it out on a block of bass wood.

From my previous model building experiences that included plastic model car kits, stick and tissue airplanes and 4-H Club projects, I knew that my design would have to be limited to the extent that I could make parts to complete the model. For example, the bumpers were made from flat stock aluminum that I could polish, headlights were made from clear plastic dowel rods that were rounded to a convex shape, scored on the face and cut thin like wafers to apply to the model, and the grille texture was cut from bright screen wire and attached over a black painted surface. All of my Guild entries were convertibles, so the full interiors were best handled similar to building a plastic model car. The seats and console were built on a removable floor pan, the inner door trim panels were applied to the sides of the interior walls of the body, and the instrument panel was installed separately from the top.

My father, a skilled cabinetmaker who could do almost everything with his mechanical aptitude and woodworking abilities, taught me a lot about woodworking and building a model. Dad also had a good friend who was an engineer with machinist equipment set up in his garage. With their help and supervision, I made the bright metal wheel covers and a steering wheel rim on a metal lathe. As for my mother, she lent her oven. I learned about how to form a plastic windshield over a wooden form from the *Guildsman* newsletter and practiced in her oven. Mom also contributed by allowing me to use her Electrolux vacuum with a spray attachment to spray the many coats of paint required for a suitable finish, not to mention her putting up with the endless mess of woodchips and dust during the long build process.

As my first finished Guild model was being carefully packed away in a wooden box for shipment to Detroit, it gave me a feeling of pride and accomplishment. But there was also the fear of facing the competition and the fear of failure. After all, it represented nearly a year's worth of work and a total dedication of effort. So, I was elated to learn by telegram that I had won the Junior Division First State Award for Indiana and Regional Award with $150 cash prize. This meant that I would travel to Detroit for the Guild convention that July. This was a really big event for a 15-year-old kid from a rural farming community. It meant taking an overnight train to Detroit alone, being hosted by General Motors, and attending a four day adventure with other kids my age along with the annual Guild awards banquet.

We were treated like royalty during many excursions and shown many aspects of the auto industry, from design studios and concept cars to a Lansing plant tour and the Bloomfield Hills Country Club, all the while accompanied by real working designers. We toured the new, ultra-modern GM Tech Center, the center of world automotive technology and innovation. And, of course, we had the opportunity to see the level of competition with all of the cars displayed in the Fisher Body Auditorium. It was an awesome, exhilarating experience that at the same time was also overwhelming! I then realized that my first convertible model car design looked too much like a Ford design and would not achieve top honors.

1958 Model Entry

My second Guild entry in 1958 at ages 16–17 was a white, 2-door convertible with red interior which took on much more of the GM-look or what I call the "GM genre." This model was carved from basswood, employed chrome-plated brass trim and was painted with

Close-up interior view of Bill Scott's 1958 model entry which won the Senior Division First State Award for Indiana ($150) when he was age 16. Building a closed coupe may have been easier, but for Bill, the additional details along with build complexity added substantially to the overall challenge and to the value of the model itself (courtesy William D. Scott).

acrylic enamel paint. The forms of the body were more jet aircraft fuselage shapes.The taillight bumper elements certainly were of the "afterburner" type and the grille was a clean-air intake. There were references to power in the bumper "dagmars" and taillight detailing. The design used vertically aligned headlights, tight circular wheel-well openings, and had clean aerodynamic body lines. In sum, it looked like a future GM product.

While integrating these elements into the design, I also attempted to have a pleasing and exciting design that hopefully would be considered "advanced," or a bit more daring than what you were currently seeing on the street. This was my first year competing in the senior division, so the competition was expected to be more difficult both in terms of design and craftsmanship. I also paid more attention to the interior details and the overall quality and craftsmanship required to be competitive. Similar to the 1957 model, the scale, proportions, believability of design and details of the 1958 model made it look like a real automobile. It won a Senior Division, First State Award for Indiana and $150 cash. My 1958 Guild entry would be a building block model, a stepping stone, on the way to the next challenge.

Since I did not win a regional award for the coveted all-expenses paid trip that year, my Dad and I drove to Detroit that summer to look at the competition and attend the awards banquet as a Guild alumnus. This gave me another opportunity to size up the competition and gave my Dad a chance to actually see and talk to GM design executives to help allay some of his concerns about the viability of a career in car design.

The model entry made by Bill Scott of Acton which won the 1958 Senior Division First State Award for Indiana ($150). The stacked twin headlights embedded in the fender forms bracket the simple air intake front grille inspired by and continuing the jet aircraft theme. The full round wheel well openings cut into the fuselage body with an efficient, functional and minimal approach (courtesy William D. Scott).

My parents had survived the nation's Great Depression; they were hard-working, but not college educated. Thankfully, they were willing to look into and take seriously my desire for a career in auto design. I had already contacted the famous Art Center School (ACS) in Los Angeles for information about their transportation design program. We drove to Chicago that fall to meet with two ACS teachers, the well-known former professional auto designers George Jergenson and Strother MacMinn. After this meeting, my parents were far more receptive to the idea of an auto designer career for their son. The question was how to pay for it.

1959 Model Entry

My third and final model, in 1959 at ages 17–18, was a bright red 2-door convertible with red interior accents. This model was carved from basswood with acrylic enamel paint and chrome-plated brass trim. This design was an all out effort to achieve national winning status. It had to have outstanding design, be advanced in styling and display impeccable craftsmanship.

I started as usual by sketching many design ideas on paper and really thinking about what constituted a truly "advanced design." It really could not look like what one currently saw on the street, but it couldn't be so outrageous or so bizarre as to not be believed. Do cars have to have fender forms, a hood and a trunk, or can they all be in one form? Do headlights, grilles and taillights have to be on the front and rear surfaces or can they be deeply imbedded within the overall car form? These and many more questions had to be explored. I had to think outside the box while observing current trends.

I also had to think about the end goal: was this going to be a real car and judged on efficient packaging or was this a design contest to be judged on design and craftsmanship?

Bill Scott's bright acrylic enamel red 2-door convertible with red interior was carved from basswood and employed chrome-plated brass trim. Deeply recessed or embedded headlights, grille and taillights were "hooded"—a styling accent that was popular among the California custom cars at the time. A unibody form with deep recessed head and taillights gave the plan view a dramatic silhouette. The "fared in" headrests helped integrate the interior with the exterior, making a single complete design statement. The simple, clean forms had been well crafted and executed (courtesy William D. Scott).

Deeply embedded headlights, grilles and taillights would intrude on engine and trunk space, but this model had no operating hood or trunk opening and no engine or luggage, so "advanced" design over function in this case seemed to be the approach to take. As with the other entries, scale and proportion, design credibility and detailed craftsmanship were top priorities so that my 1959 model entry looked like a real automobile.

Quality of finish and craftsmanship also had to be top notch. I couldn't be afraid to hold back on details because they might be too intricate or difficult to execute. The deeply recessed or embedded headlights, grilles and taillights were to be "hooded"—a styling accent that was popular among custom cars in California at the time. My father used thin veneer in his cabinet making skills and so it seemed this would be the best way to achieve this customized look. It would have to be done without a trace on the slightly curved painted surfaces, both the tops and the bottoms, and would require up to eight inlays total.

The taillights were deeply curved and chromed with a mirror-like surface. I had to learn how to bend a flat piece of brass smoothly over a curved form, or mandrel, without any break or fold lines. When inspected by the Guild judges, the reflections of light off these mirror surfaces had to be smooth. Also, a slight under-curvature of these brass parts would allow them to be squeezed or snapped into place, and held firm with tension, on the rear of the model.

The windshield was a challenge. Not only did it wrap around, but it also wrapped over at the top and had a recessed area in the center. I carved a wooden form, or pattern, for the final shape of the windshield and then made

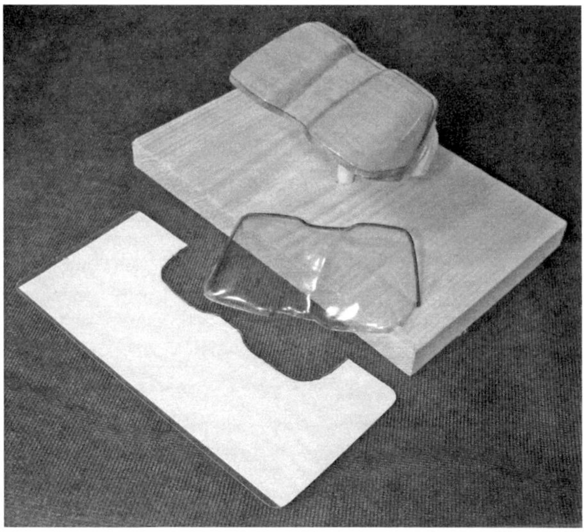

The wood pattern and template used to shape the clear plastic windshield for Bill Scott's 1959 Fourth National Senior Scholarship ($2000) award winner. The carved wooden pattern was fitted into the body cavity leaving a thin space for the windshield material. It was then mounted on a secure platform so that a soft cloth could be stretched completely around the pattern as the heated softened plastic came out of the oven (courtesy William D. Scott).

Close up interior view of Bill Scott's bright red 1959 model. The carpeted floor, lower console and seat bottoms were inserted from below. Door panels were applied to the sides. All of the sweeping lines had to match up as the instrument panel was inserted from the top, clenching the base of the windshield and matching up with the console. Note "Drive by wire" steering controls on the left side of the driver's seat, rather than a traditional steering wheel (courtesy William D. Scott).

many attempts in my Mother's oven to heat and form the shape in plastic without the surface bubbling. This clear plastic material also tended to have a memory as it cooled. The mold had to be modified to compensate for the contraction of this hot, pliable and rubbery material while it cooled.

Placing a chrome-like molding around the edge of this very contoured windshield was also a tough assignment. I found a very small silver tube likely used to make jewelry. I actually cut along one side of this tiny tube so that it could be slipped over the edge of the thin clear plastic and around all of the curves of the windshield. These struggles and many more all had to be overcome to demonstrate top quality. The simplest design forms had to be exquisitely executed, as this was where craftsmanship played an important role in the final product.

The instrument panel also had a dramatic shape with a recessed area on both sides, one for the instruments and the other for design relief. The lower surface on each side swept into a center console. This had to be made up in layers of wood, sandwiching them together as a unit and fit precisely into the cowl with just enough of a gap to clinch the base of the windshield and hold it in place. Here I learned that what fit before being painted didn't necessarily fit afterward. Some rework had to be done for the final fit.

This 1959 design won the Senior Division First State Award for Indiana, Regional Award and Fourth National Scholarship award ($2,000) and another rewarding trip to Detroit for the Guild convention. I enjoyed the opportunity to speak with actual working car designers, as by this time I was clearly bent on the path of college and a career in automotive design. Now I could ask meaningful questions about what such a career was like and what I needed to do to reach that goal. On this trip we toured interesting sites in and around the Detroit area and GM facilities. At the GM Proving Ground in Milford, Michigan, I remember the surprise of seeing a car flash by above the bus as we were on the high speed test track, slowly making our way around the high-banked turn on the lower flat lane.

Bill Scott (age 18) of Acton, Indiana, is shown in a white tuxedo holding his 1959 award winning Guild model. His Fourth National Senior Scholarship award as a high school senior was a major milestone towards financially securing an opportunity to attend college (courtesy William D. Scott).

The two trips I had to Detroit were wonderful opportunities to experience time with other boys with similar interests and to view the world of auto design. They were instrumental in forming a goal, a clear target to work toward, and providing a focus for achievement. Of course, the main thing was winning the scholarship, which made it possible to attend ACS.

I attended one year at the Purdue University Extension in Indianapolis to get some of my liberal arts academics out

The 1959 Fourth National Senior Scholarship award with the Fisher Body Craftsman Guild model was the accumulation of challenges met and skills learned, all of which helped Bill Scott in both his professional and personal life. He holds the same winning model in a recent photograph. The discipline it took for him to make this winning Fisher Body Craftsman Guild model prepared him for what lay ahead in life and a satisfying automotive design career. He retired from GM Design at the Tech Center in Warren, Michigan as an executive, director of design (courtesy William D. Scott).

of the way. At age 19 and driving my first car, a 1956 Chevy convertible, I left the country life in Acton, Indiana, to attend industrial design classes at ACS. I majored in transportation design and graduated in June 1963.

After graduation, I began my career with General Motors Styling, later renamed GM Design, as an interior designer at the Tech Center in Warren, Michigan. I worked in a number of divisional design studios, becoming an assistant chief designer and later, in 1971, a chief designer for Chevrolet and, in 1981, for Pontiac. In 1991, I was promoted to the director of design for the Oldsmobile Studios, both interior and exterior, and then the Saturn Studios. Eventually, I became the director of interior design for all divisions and for design quality for GM Design. I retired in July 2001.

A fascination with cars and a strong will to pursue an idea, combined with a determination to complete goals, have been great strengths in many stages of my life. They helped me while competing in the Fisher Body Craftsman Guild, studying and graduating from Art Center School, and over the next thirty-eight years with General Motors Design. I had a great career beginning as a designer and retiring as an executive level director of design. I am now able to look back over a satisfying career in automobile design and feel that I have made a contribution to the industry. I credit much of this to my early experience with the Fisher Body Craftsman's Guild.

Top: 14-year-old Philip J. Rauth from York, Nebraska, carved this 1947 entry from Honduran mahogany. The front engine, 4-door sedan was painted two shades of dark blue Duco lacquer (the same paint his brother Vincent had used for his 1946 national award winner, except he reversed the upper and lower colors). The model featured plexiglass window inserts painted on the inside with aluminum color paint, brightwork shaped from aluminum and wheels/tires homemade from mahogany. Phil Rauth spent more than 220 hours designing and building this model (courtesy the Rev. Philip J. Rauth). ***Bottom:*** Roger D. Teter's 1949 model, made from a clay master model using plaster of Paris, was painted with blue automotive lacquer and had white indicated windshield and window areas. The design featured wraparound bumpers and rectangular headlights. The wooden bumpers and grille were painted silver, the wheels/tires were hand-carved and the polished wire window trim was held in place by multiple soldered pins (courtesy Roger D. Teter).

Top: This 2-door hardtop sedan constructed in 1951 by 14-year-old Gordon D. Williams was carved from balsa wood. The undercoat was Carlotta Brown and the body color was a 1951 production Oldsmobile color called Palm Green. The brightwork consisted of chrome-plated brass bumpers, grille and delicate window trim. The turned maple wheels/tires had wide whitewalls stylish at the time (Kaile Hart Photography). ***Middle:*** Wilfred C. Keagy (age 18) cast this 2-door passenger sedan in 1952 from plaster of Paris. It featured carved walnut bumpers painted with Chrome-a-Lac and polished aluminum window frames (courtesy Wilfred C. Keagy). ***Bottom:*** Inspired by his trip to Detroit the previous year, William A. Keyser, Jr., responded with this two-door in 1952. The model, carved from poplar wood, had chrome plated brass brightwork, turned maple wheels/tires, shaped opaque Plexiglas window inserts and a sprayed and hand rubbed lacquer finish. It took 5 to 6 months to design and develop, involving two preliminary clay model studies (Geoff Tesch Photography).

Top: Charles R. Foreman's 1952 entry was painted with turquoise enamel and had black indicated window areas. The model was fabricated using model airplane type construction, with balsa wood bulkheads and a thin balsa skin (courtesy Charles R. Foreman). *Middle:* Warren M. Bakken's previous models were constructed of balsa wood, but for 1953 he tried a plaster model. The bumpers and most of the trim were chrome-plated brass. The window trim was stainless steel half-round dental wire with dozens of small brads soldered on the back to act as fasteners. The wheels/tires were turned on a home-made lathe from solid aluminum. The 15-pound model is still in original condition (courtesy Warren M. Bakken). *Bottom:* Wilfred C. Keagy (age 19) from Lafayetteville, Pennsylvania, carved this 2-door sedan from poplar in 1953. He used clear acrylic plastic inserts for the window areas (painted light gray on the inside surface) and chrome-plated brass for trim (courtesy Wilfred C. Keagy).

Top: In 1953 Arnold L. Joslin (age 15) of Raleigh, North Carolina, designed this 2-door sedan using clay for a reference, but the model was carved from clear white pine. Grooves were cut to accept trim pieces and the bumpers were sculpted from aluminum. The wheels were made from plaster and had aluminum wheel covers (courtesy Arnold L. Joslin). ***Middle:*** Raymond P. Wykes' refurbished 1953 model. The model had been heavily damaged after falling in an earthquake in October 1989. In restoring the car, Wykes repaired and revised some of the chrome and detail parts and repainted it a two-toned blue similar to his 1968 Mercury Cougar. The original model was painted a Cadillac black body color with a Ford canary yellow color for the roof. The restored model is 95 percent the same as the original (courtesy Raymond P. Wykes). ***Bottom:*** Wallace E. Bakken cast a clay master model in plaster for this 1955 design. Half-round stainless steel wire was used for window trim, and other brightwork like the bumpers and grille were chrome-plated brass (courtesy Warren M. Bakken).

Top: Adrian A. Bruno's 1955 entry began in late 1954 with a clay model concept drawing and when he was 15 years old. The working model was cast in plaster of Paris using molds taken from the clay master. After the sealer and primer coats, about eight color coats of "Bittersweet" automotive lacquer were applied with an airbrush, wet sanded between coats and then hand rubbed and compounded to a high luster. All metal work was fashioned from aluminum and buffed to a chrome-like finish. A clear acetate windshield was heated and formed over a wooden pattern. Light lenses were made of clear or red acrylic and interior soft surfaces were covered in caramel colored leather (courtesy Adrian A. Bruno).

Middle. George E. Anderson's 1955 award winning model on display at the MFA, Boston. It featured hand-sculpted aluminum front and rear bumpers, a lathe turned spare tire cover and polished aluminum wheel covers, and employed 3M chrome tape for trim (courtesy Ronald J. Will). ***Bottom:*** This Bolero red and Dover white sports car model was built by Herman I. Rauth (age 18) of York, Nebraska, in 1956. Carved from poplar, the model featured a clear plastic windshield with intricate aluminum trim, highly polished aluminum parts (bumpers, grille, light bezels and air vents) to simulate chrome, hand-formed, spun aluminum wheel hubs and double striped whitewalls (courtesy Herman I. Rauth).

This 1958 entry by Walter O'Neill (ages 19–20) of Springfield, New Jersey, was painted with Duco automotive lacquer. Its "moon" hubcaps were hand-shaped Plexiglas discs laminated to polished aluminum discs. Walter developed a base primer coat consisting of talcum powder mixed with Tester's model airplane dope. The design had crisp clean lines, enclosed front wheels and a perfectly scaled "Club de Mer" decal depicting the model's name located just forward of the side trim (courtesy Walter F. O'Neill).

The fastback sports coupe designed and built by Thomas L. De Fazio (ages 17–18) of Philadelphia, Pennsylvania, for the 1958 contest was carved from basswood and painted in a custom blended, medium metallic blue Duco lacquer with light gray indicated window areas. A hand-operated insecticide sprayer was used to apply the lacquer paints. The design featured headlights recessed into the fenders, taillights recessed into the body below the rear bumpers, a full width front bumper and a pair of split rear bumpers, fully exposed front wheels/tires and partially exposed rear wheels/tires with highly polished aluminum wheel covers (courtesy Thomas L. De Fazio).

George Herzog (age 15–16) of Falls Church, Virginia, built this asymmetric station wagon in 1958. The coral and ivory color scheme was suggested by his mother to appeal to the female consumer. George used the same sculpted lines on both sides except in reverse directions. The front and rear sculpted shapes of the model were also asymmetric. It was constructed of plastic wood and plastic metal hand-molded on a balsa wood armature. A non-functional convertible top feature was explained on a card, as allowed by Guild rules (courtesy George Herzog).

The deeply recessed rear areas of Bill Scott's 1959 model provided interest in the design, but created a challenge to build. Thin wood veneer completely surrounds the chromed curved metal background for the taillights. Following painting, the curved background was squeezed together and sprung into place, trapped inside the veneer. The taillights and bumper were installed separately. In front, deeply recessed headlight pockets set up a strong central theme which runs through the vehicle. The recess in the windshield carries this feeling into the interior, picking up with the finished panel between the seats, flowing between the headrests and ending between the twin rear recessed pods. The car design had no steering wheel. The car was controlled with two levers, one in the door and one in the console. Steering was accomplished via fore/aft movement in the right-hand and left-hand controls (courtesy William D. Scott).

Frederick J. (Bud) Magaldi (ages 15–16) used white pine, aluminum, injection molded plastic and acetate in this 1960 entry. It featured a front end design inspired by the tiger shark, fully exposed wheels/tires with highly sculpted wheel well openings and mild fins with integral high mounted taillights that stopped short of the rear bumper. Other features included a handle bar instead of a steering wheel, aluminum seats covered with leather, polished aluminum front and rear bumpers and wheel covers made from round injection molded checkers game pieces painted silver (courtesy Frederick J. "Bud" Magaldi).

The 1961 entry by Felix R. Collazo (age 17) was made from balsa wood with a hand brushed silver canopy. The hubcaps were cut from an artist's aluminum paint palette (courtesy Felix R. Collazo).

Opposite, bottom: This 1961 entry is Roger Schneider's favorite among his models. The design was aerodynamic with good proportions, a low profile and a more sophisticated look than previous models. It featured razor-sharp fins made of balsa wood and a molded black plexiglass bubble roof (courtesy Roger L. Schneider).

Above: This model designed and built by Ron Will (age 18) in 1961 was carved from poplar and coated with a fiberglass resin wood sealer to enhance the paint. It was painted in "Twilight Mist" lacquer (a 1961 Oldsmobile 98 production color). The exotic, asymmetrical design featured sweeping parabolic curves centered around a single driver and was inspired by the Virgil M. Exner, Sr., asymmetric show car the "XNR." Some of special features envisioned included a horizontal steering control bar for better occupant crash energy management, a rotary engine and shock absorbent bumpers. The model featured a jet aircraft–type plexiglass windshield, chrome-plated brass trim and thin whitewall tire lines drawn with a compass (courtesy Ronald J. Will).

Samuel T. Kjellman (age 16) entered this richly featured model in 1961, carving the body from a block of laminated poplar and using acrylic lacquer over primer and wood sealer. The dual clear acrylic windshields were stretched by hand over a wooden pattern, after being heated in an oven. The seats are carved in soft foam and covered in supple stitched, white glove leather. The trim is formed copper and plated with bright nickel. The floor was covered in black velvet to simulate carpeting (courtesy Samuel T. Kjellman).

In building this symmetric, two-passenger convertible with a wedge-shaped top view in 1962, 19-year-old Tom Covert carved the body from kiln-dried poplar and sealed it with a polyester wood sealer. Sprayed in "Wedgewood Mist," it was trimmed with 25 small rhodium-plated brass parts for a brilliant appearance. The instrument cluster trim was made of stainless steel. The acetate windshield was actually an open-ended canopy which swept up and over the occupants. About 30 to 35 acetate windshields were heated and shaped over a wood pattern and then graded to find the "perfect" competitive part. Among the special features envisioned were heated seats with a vibrating lumbar section (courtesy Thomas L. Covert).

A three-quarter front view of Paul Bonfilio's (ages 17–18) 1962 entry. Visible are the dual sealed beam headlights supported by a horizontal bar suspended under the integrated front bumper. Also evident is the simple shaping of the flat surfaces that define the design (courtesy Paul Bonfilio).

Samuel T. Kjellman (age 17) in 1962 created this highly sculpted 2-door sports car which truly transcended the mental constraints of the rectangular carved block approach. His highly fluid design bears influences of fender lines of classic cars and features external running lights and body contoured grille bars inspired by late '40s designs. The trim parts were formed and fitted from polished soft aluminum and some were chrome-plated. The cylindrical parts were turned by lathe (courtesy Samuel T. Kjellman).

Jeff Jones (age 13–14) produced this single seat experimental sports car with wing-shaped front and rear fenders, open wheels/tires, flip-up headlights and a single stick control (acceleration, braking and steering) in 1963. This was an Open Competition model carved from sugar pine and spray painted a burgundy candy apple color with a gold trim theme. The body color was clear coated. The model had a clear plexiglass canopy, pearlescent white vinyl upholstery and body colored bumpers. The conical brass hubcaps were cut from cabinet drawer pulls (courtesy Jeffrey A. Jones).

Opposite, top: Geza A. Loczi (ages 19–20) entered this Open Category, 2-door sports car with 1 and 2 seating in 1964, while attending the Art Center School in Los Angeles. The model was built in the tradition of the Art Center School's famous model-making program. Designed with a diamond-shaped tire tread, the front and rear wheels/tires had front and rear motorcycle-type suspension systems for "at speed" operation, while the two outrigger wheels/tires were retracted. Carefully conceived and designed in clay, this was a fiberglass lay-up model with an upper and lower half. The body was painted a silver color and the wheels were sand-blasted brass with a clear coat. There were brass and black bumper details. The architecture of the car led to an exciting and dramatic design for Loczi's sixth Guild model entry.

Opposite, middle and bottom: Geza A. Loczi (ages 20–21) produced this futuristic and beautifully sculpted 2-door hardtop, Open Category model in 1965, building it in fiberglass in the best tradition of the Art Center School. The graceful top view had a wide front end and a tapered rear end. The result was a wide track width front axle and a narrower track width rear axle. From the side view, the design flowed from a forward point, up and over the passengers' heads, and down to a graceful tapered rear end. This sculpture required warped and twisted surfaces which had to be carefully controlled with "highlight" studies. The chrome-plated bumper bars housed a theoretical crash avoidance radar system. The original half-round shiny metallic window trim was lost during a Fisher Body exhibit which toured the country in 1965 (courtesy Geza A. Loczi).

1964
sixth model

1965
seventh model

1965
seventh model

This sports coupe, crafted in 1965 by Frederick J. (Bud) Magaldi (ages 20–21) from Charlestown, Massachusetts, featured a V-shaped front end with polished aluminum bumper, a sweeping wrap-around V-shape windshield, a "swing-up" canopy, and dual under-slung vertical grilles. Construction techniques and materials included a master clay model, a multi-piece plaster mold, fiberglass and polyester resin lay-up for the finished body of the model, aluminum, highly polished stainless steel wire and metallic Firefrost Maroon lacquer, a 1965 Cadillac paint color. This model was inspired by the Jaguar XKE, in that it has a long hood, a very personal greenhouse look, and a short rear overhang (courtesy Frederick J. "Bud" Magaldi).

Robert W. Lawhn (age 20) of Houston, Texas, created this rear engine, wedge-shaped, aerodynamic Open Category sports car design in 1967. The front end was tapered, with a narrower front track compared to the rear. The model was carved from white pine, covered with fiberglass resin. Bumpers and grille were chrome-plated brass, window moldings were stainless steel wire and wheel covers were cut Ford chromed dashboard knobs. Occupant entry was by way of a canopy which included the top and window glass surface. The designer envisioned radar enhanced rear vision system (courtesy Robert W. Lawhn).

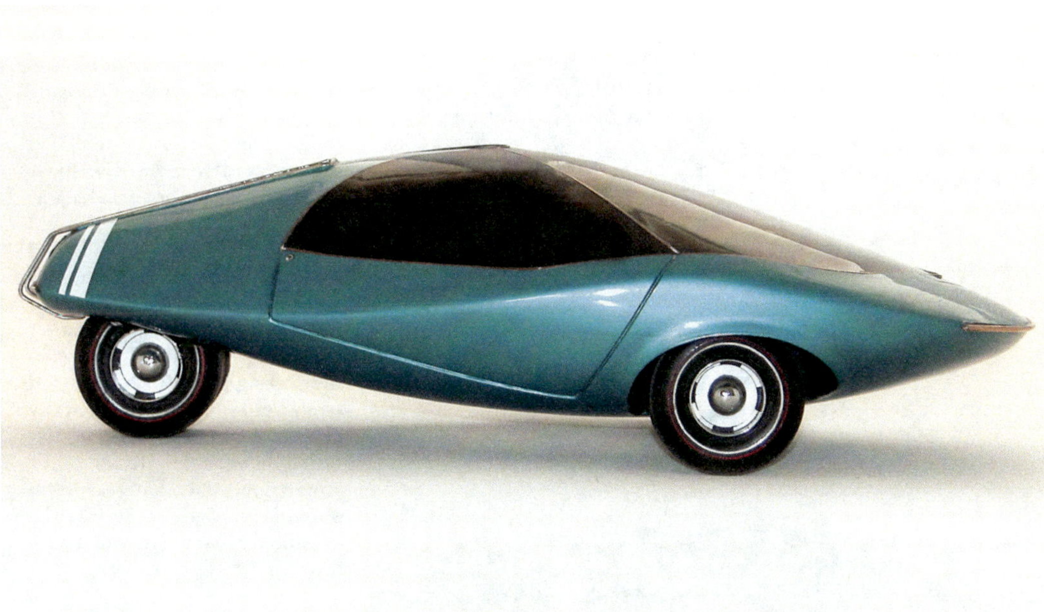

Terry P. Graboski (ages 18–19) entered this unique design in 1967. The teardrop-shaped, two-door sports coupe was designed for the Open Competition with a rear engine and a large louvered rear deck for cooling. Carved from balsa wood and sealed with hard fiberglass resin, the model was finished with a metallic blue automotive lacquer, black indicated window areas and chrome-plated brass bumpers. Window trim was made from inlaid sterling silver wire filed flat. Awarded 99 out of 100 points for design originality, the design featured a tapered rear end with a narrow track of rear steered wheels, concealed headlights and "Baby Moon" wheel covers. It took Terry 6 months to design and construct the award winning model (courtesy Terry P. Graboski).

Tom W. Graboski (age 20) produced this design for an urban electric commuter car, similar in size and proportion to the current Smart ForTwo, in 1968. This Open Competition model had a trapezoid-shaped wheelbase (narrower in the rear), was carved from closed-cell, rigid polyurethane foam, and had inlaid silver wire filed flat for window trim. Among other things, it featured a smaller diameter set of wheels/tires, custom designed and fabricated from fiberglass. The "Rally Wheels" had fine "Firestone" lettering on the sidewalls. Note the scored door, hood and rear hatch opening grooves as well as the outside rear view mirrors (courtesy Tom W. Graboski).

Spencer L. Mackay's (age 18) 1968 entry, a 2-door sports coupe, was designed for the Open Competition. A clay master model was used and transferred via sections and templates to closed-cell, rigid polyurethane foam — a faster new method being promoted by the Guild. The model was surfaced with a hard thin layer of Liquitex modeling paste and then coated with Bondo polyester body filler. Brightwork was highly polished aluminum. The large V-shaped windshield accommodated forward and side vision and was split down the middle with a structural support member. A bold piece of aluminum brightwork dominated the centerline of the model from the cowl area to the rear end. Note the complementary green wheel covers and tire sidewall accents plus the side marker lights in the front view (courtesy Spencer L. Mackay).

14

A Story of Almost

Walter F. O'Neill, Jr. (1956–1959)

It was the 1950s, an exciting time for automobile design, and my eyes were on automobiles that were sleek and stylish. Cars were being transformed from dull, ugly and utilitarian sedans into pieces of rolling sculpture. Each autumn, it seemed as if the whole country held its breath waiting for the new models to be unveiled, and I was one of the anxious lot. I visited local Springfield, New Jersey, dealerships with my dad looking at the latest samples of what was soon to hit the road.

When the pink and charcoal gray 1955 Chevrolet Bel Air convertible was unveiled – Wow! What a departure from the past offerings. It had a two-tone paint job, a wraparound windshield, ample chrome trim and the first hint of finned rear fenders. More importantly, it had pizzazz, as in, "What a neat bucket." At the same time, Detroit introduced the first American two-passenger sports cars, the Corvette and the Thunderbird. All of these exciting new concepts generated my interest and passion in automotive design, leading me eventually to the Fisher Body Craftsman's Guild.

In 1955, I discovered the Fisher Body Craftsman's Guild quite by accident. My woodshop teacher, Mr. Herbert Boblin of Regional High School in Springfield, New Jersey, maintained a filing cabinet to store plans and dimensions for student projects. He encouraged perusal of this information. There I found a 1952 brochure produced by the Guild to generate student interest in building 1:12 scale model cars for its annual competition. They were offering substantial cash awards and college scholarships. I had been heavily involved in building models of airplanes, boats and trains and thought some were pretty good — why not cars?

I was 17 years old at the time and not quite sure I had the talent to actually design a scale automobile. I soon began drawing cars all the time on any available surface. My figures seemed square and boxy – not at all what I was dreaming about. Then I met Eric Norton, a previous Regional High School graduate who was several years older and had entered an FBCG model in 1950, winning a Senior Division Honorable Mention award for New Jersey. He was now at Pratt Institute in New York City studying Industrial Design and would later go on to have a successful design career at General Motors.

He reviewed my drawings and made the critical observation that there should be no straight lines in automotive styling, that every line contains some form of gentle curve. What a revelation! Of course, my designs were full of ruler-drawn straight lines. I took his advice, and from then on drew only in freehand and in three dimensions. I doodled anywhere at any time on any paper pad and book cover I could find. I was consumed with cars!

Mr. Boblin showed me how to glue together a wood block from four pieces of poplar, which he said was a good choice for carving. He also lent me several of his personal hand tools, as I had no significant carving tools other than an X-Acto tool set. The art of carving with gouges, chisels and spokeshaves came more easily to me than drawing. I also seemed to excel in my mechanical drawing class, and there spent my extra time drawing cars to 1:12 scale. My teacher, Mr. Albert Rothbard, called my designs too radical. He once said, "Who-ever heard of a wraparound windshield or fins on a car that size?" He was a good teacher, but not very forward thinking.

1956 Model Entry (Ages 17–18)

So, my first entry in 1956 was a two-passenger open roadster with a wraparound windshield, low fins and a metallic paint scheme of Glen Lake Blue and Colonial White. It was attractive with flowing lines, but the finished details in the cockpit were not up to par. Mr. Louis Fredericks, my metal shop teacher, allowed me to use one of the school lathes to turn the aluminum wheel covers, steering column and steering wheel for this Guild entry. While I was able to complete it on time, my results only won a Senior Division Honorable Mention award for New Jersey. I swore the next year my car would be better.

I graduated from high school and went immediately into the Army to fulfill the military obligation men had at the time. I elected to serve for six months of active duty and was honorably discharged by December 1956 with a 7½-year active reserves obligation. I was released from the Army in time to build a second model in 1957, and I continued to doodle and keep a scrapbook. I planned to build a fastback coupe design to avoid the challenges involved with an open cockpit roadster.

1957 Model Entry (Ages 18–19)

Using another block of poplar, I discovered a new technique of bending fine steel wires and inserting them into the wood by cutting a fine groove and gluing them in place. After this, a layer of plastic wood was applied over the entire length of wire and then sanded uniformly. After the paint process, a light scuffing with fine steel wool would allow the wires to reappear again. This would accentuate the design elements of the model and would serve to define windows from body panels. It was tedious work but provided a nice crisp line. The 1957 model had hubcaps turned on a lathe, made from bar stock. The whitewalls were cut from heavy paper. In order to cut perfect circles, I mounted a piece of razor blade in a small compass. The "brightwork" was made from aluminum tubing and scrap aluminum stock, then highly polished on a homemade buffing wheel.

One of the continuing challenges was to duplicate the left and right sides. I simply drew a center line and transferred measurements with a divider. I had worked exclusively "by eye" without grids or templates. Another challenge was painting — Mom's vacuum cleaner spray attachment worked okay, but I needed a better system. The finished model with gold body and concave front wheel well accents won a Senior Division Second State Award for New Jersey ($100). I think it looked a little too much like a rival Chrysler product.

I wanted so badly to go to the FBCG Convention, talk with other Guild members and meet actual GM stylists. For that reason, I was determined to make my third model my

The 1957 Guild entry made by Walter F. O'Neill (ages 18–19) which won a Senior Division Second State Award for New Jersey ($100). This 2-door fastback coupe design was carved from a block of poplar and painted a metallic gold with black front wheel well accents. The indicated window areas were painted black. The hubcaps were turned on a lathe and made from a bar stock. The front wheels were fully exposed and the rear wheels were covered. Fine steel wire was used for bright work. Walter was pleased with the overall design; in retrospect, the fins seemed a little outrageous, but not so in 1957! He often felt this model resembled a Chrysler product and not what the "GM eye" was looking for (courtesy Walter F. O'Neill).

best effort. After observing previous winners in the Guild literature, I decided in 1958 to adopt a design that would be more pleasing to the GM eye, and it turned out to have a strong Pontiac resemblance.

1958 Model Entry (Ages 19–20)

This time around, I used a new method for applying a base coat. Before painting, I mixed talcum powder with Testor's airplane "dope," applying it with a brush and then carefully sanding it. It provided a good surface for multiple coats of finish. I also purchased a simple spray gun outfit from Sears Roebuck and Co., and the painting was much improved.

My design was simpler with flowing lines, enclosed front wheel wells, black windows and silver Duco automotive paint with black accents — another coupe. One final detail was apparently very important: I applied a very tiny decal from a plastic model kit of a GM concept car known as the Club de Mer. I applied this to an area that suggested this was the model's name. This decal was so small you almost needed a magnifying glass to see it.

I no longer had access to my wood shop or metal shop, so I had to be creative with items like hubcaps. I used clear ¼-inch Plexiglas shaped by hand and backed up with a polished aluminum disc to resemble chrome. They looked like the "moon hubcaps" that were popular at that time.

My 1958 model still was not judged high enough to win the trip to the convention, earning a Senior Division Second State award for New Jersey. However, it did receive high

"Springfield Youth Receives Prize in Nation Wide Auto Contest" was the headline that accompanied this photograph in the *Newark Evening News* on August 21, 1958. The article described Walter O'Neill's prize winning model and the $1,000 National Styling Scholarship. The news article made him feel like a celebrity at age 20. Only Walter knew that the information was inaccurate because the winning entry was still in Detroit. The photograph shows the previous year's black and gold model (1957 entry) that won the Senior Division Second State Award for New Jersey ($100) (courtesy Ms. Donna M. Reed, Media General Operations, *Newark Evening News*).

The emphasis of this 1959 two-door sports coupe, painted glossy black with silver canopies and made by Walter F. O'Neill (ages 20–21), was to move away from conventional thinking. He was rewarded with a Senior Division State Award for New Jersey. It was quite advanced for its time with dual aircraft cockpit canopies and long flowing lines, and the design envisioned would have had translucent fins that would glow softly at night when the running lights were activated. As conceived, there would be no separation from taillight to fin. When the brake light was lit, the entire fin would glow red. In addition, there were "feelers" or antennae protruding from the headlights and taillights that would alert the driver to the proximity of another object (something like a "curb feeler" of the '50s). There was also a translucent horizontal bar in the grille, from headlight to headlight, which would serve as a running light similar to that of the much later Mercury Sable (courtesy Walter F. O'Neill).

marks for styling and went on to win one of the ten National Styling Scholarships and a $1,000 award. I guess the judges were really impressed with that tiny Club de Mer label.

The phone rang early in the morning at our home in New Jersey right after the 1958 Guild Awards Banquet in Detroit. We were all asleep. The Western Union operator, on the other end, announced that I was the recipient of a National Styling Award and had won a $1,000 scholarship. I said, "Thank You!" and then hung up. My father asked me, "Who was that?" I replied, "My model won something," and I went back to bed.

Soon after, my photograph appeared in the newspaper with my previous year's model since my 1958 model had not yet been returned from Detroit. Even with the wrong car, I was a celebrity. After reading the newspaper article, a tool company sent two sets of carving tools, one to me and one to my shop teacher.

After the Guild Convention in Detroit, General Motors' Linden Plant hosted a luncheon, which I attended, at the Robert Treat Hotel in Newark, New Jersey, for the local Guild contestants. After lunch, a GM representative approached me and mentioned the uniqueness of the decal design.

After I had been declared a styling winner, I was sure that I wanted to become an automotive designer. My dad and I drove to Detroit in the family's 1958 Buick Century hardtop convertible unannounced and met with a GM executive at styling headquarters. At this time, I was attending Rutgers University studying engineering. I told him I wanted to be a designer and perhaps enroll in the General Motors Institute (GMI). He told me simply to finish college, get my degree, and if I still wanted to be a designer to keep in touch. It

Walter F. O'Neill at home reminisces about his three surviving models 54 years after his first Guild entry in 1956. He is surrounded by his 1957 State winner (metallic gold and black) in the left foreground on table, his 1958 Styling Scholarship winner (on top of the shipping box) and his 1959 black and silver State award winner (on table, right side foreground in front of the tall National Styling Trophy). Unfortunately, no photographs exist of his first Guild model built at ages 17–18, a 2-passenger open roadster that won a 1956 Senior Division Honorable Mention award for New Jersey. The 2-seat open roadster, carved from poplar, had a wraparound windshield, low fins and a metallic paint scheme consisting of Glen Lake Blue and Colonial White (courtesy Walter F. O'Neill).

was not a very encouraging visit — talk about letting the air out of your sails! I didn't realize then that the 1958 Guild competition models were still on display in Detroit, and it was never mentioned during the interview. It would have been helpful to see the other entries. As we drove straight home, the dream of becoming an auto designer slowly began to slip away.

1959 Model Entry (Ages 20–21)

I was now contemplating my next model. A Guild newsletter showed canopies and windshields made by some vacuum process, but I did not have a clue how to copy such designs. So, in 1959, I stayed with a solid top coupe design, a glossy black body and silver dual roof canopies. It was my fourth model and it would be my last. I am sure it was the "best," as it was very advanced for its time. I included dual aircraft type canopies and long translucent fins designed to glow softly when the headlights were on. The model had "feelers" that projected from the headlights and taillights to alert the driver of the proximity to another object. A translucent bar between the headlights in the grille served as a running light — a design similarly used by the auto industry in the Mercury Sable.

It won another Senior Division State Award for New Jersey, but still no trip to the Detroit convention. What a bummer! I worried that the "Special Features" card attached to the bottom of the model and used to describe special details had somehow became detached from my model.

Soon my focus turned to real cars. As a 17-year-old, my first car was a 1952 Ford Victoria, 2-door hardtop (tan color with a blue top). This was the top-of-the-line Ford product. My second car was a 1956 Ford Victoria, two-door hardtop, painted mandarin orange and colonial white, with a continental kit. To my way of thinking, this was absolutely the hottest car you could own.

No longer eligible for the Guild competition, I needed a new project. I began a five-year restoration of a 1931 Model A Ford Roadster from the ground up and still have that vehicle more than 50 years later. I have been an active member in the Antique Automobile Club of America and numerous other car clubs, having restored and owned many interesting cars. This is what happens when you are car crazy! Unrelated to my talent and passion for cars, I became an insurance agent and had a successful career with a major insurance carrier. I have been retired for more than 17 years.

It wasn't until 2005, when I read John Jacobus' book, *The Fisher Body Craftsman Guild, An Illustrated History*, that I realized the significance of my National Styling Scholarship. I traveled to Wayne, New Jersey, in April 2009 and saw some of the Guild models made by fellow Guildsmen (John Jacobus and Ronald Steinhilber) at a model car show. The models on display included First State, Regional and Styling Scholarship winners and satisfied my need for a first-hand look at these works of art. Fifty years after building my models, I realized how close I had come to choosing the correct path to becoming an auto designer.

If only I had won that trip to the FBCG Convention, the outcome of this story might have been entirely different. My devoted wife of fifty-one years, Mary Ellen, has suggested that I read Robert Frost's poem "The Road Not Taken" and "get over it."

15

Balancing School, Styling and Model Building

Thomas L. De Fazio (1953, 1956, 1958–1960)

I grew up in South Philadelphia in what remains a pleasant and well-tended Italian-American working class section of the city. Most fathers were steadily employed as craftsmen, workmen, or in the building trades. My father was a maintenance machinist and toolmaker, responsible for maintaining a mid-city factory's production machinery. Our families bought goods and automobiles that were made in America.

Growing up in the city, an older friend introduced me to the design and styling features of cars and of the shapes of racing cars. His knowledge and opinions fell on a fertile mind, and I soon began sketching all kinds of automobiles. An engineer, a friend of the family, loaned me his copy of Raymond Loewy's book, *Never Leave Well Enough Alone.* I saw the Raymond Loewy Studio's work on the street as the Studebaker Starlight coupe and on the Pennsylvania Railroad. I saw other impressive designs on the road, like the 1938 Lincoln Zephyr coupe, the elegant 1941 Continental, and the handsome 1949 Cadillac Fleetwood sedan. I was hooked on automobile design.

Then, in 1952, a General Motors representative visited the George C. Thomas Junior High School, where I was an 11-year-old student, to make a presentation about the Fisher Body Craftsman's Guild (FBCG). It was a revelation — GM would offer prizes to little kids and teenagers for styling cars. I submitted my enrollment card and joined the Guild. Then I did nothing. But my father had learned of the FBCG and its prizes, and he bribed me to prepare an entry. I've forgotten the small *quid pro quo*, but I entered and collected. My entry was sad. It was a poorly finished, two-door sedan constructed from balsa, two-tone light gray and blue. It won nothing, and that was appropriate.

At the beginning of the summer in 1955, at age 15 just before starting the 11th grade, I gave the FBCG another try. No bribe was needed this time. I wanted to be an engineer. College loomed, and money was needed. The FBCG had expanded its specifications to accept sports coupes for judging. As it turned out, I focused my Guild attention in 1956, 1958 and 1960 trying to perfect, with varying degrees of success, two-door sports coupe designs. I had been struck and inspired by the simple elegance and beauty of a recent 1955 National Award winning roadster made by Adrian Bruno and featured in the Guild's newsletter. My skills had improved significantly working with and finishing wood and aluminum, and I was eager to compete again.

1956 Entry (Ages 15–16)

This was a very low profile, sleek design with a highly-raked windshield. The windshield and side windows were plane-curved rather than compound-curved. To show off the body styling, brightwork was kept to a minimum with no windshield or window trim and the use of simple "corner" bumper guards rather than a full-width bumper. The front and rear corner "L" shaped bumper guards were shaped and polished aluminum. A bright metal screen for the front grille was set back within a rectangular air intake duct. The circular headlights were recessed into the body structure, and rear taillights were sculpted into the corners of the rear quarter panels. The tires, with wide whitewalls, were fully exposed with open wheel wells. The lathe-turned aluminum wheel covers were highly polished.

I prepared a model, but there were flaws in my design and shortcuts to my work. I worked from sketches and drawings without a clay model. I had a life outside the basement and away from the workbench, and I did not devote the time to my entry that typical national winners did. At the suggestion of a friend of my father's, a cabinetmaker who made replicas of antique furniture, I worked in solid basswood. My father was occasionally called in to work on Saturdays, and I would accompany him. There, I formed and finished my aluminum bumpers with a vise, files and buffing wheel, and I turned my aluminum hubcaps on one of my father's employer's lathes. I liked what I produced, though my workmanship was not nearly up to FBCG national standards. I made the deadline and submitted my model, which turned out good enough to win a Junior Division First State Award for Pennsylvania and a Regional Award that included $150 and an all-expenses paid trip to the annual Guild convention in Detroit.

This Guild model entry made by Thomas L. De Fazio (ages 15–16) of Philadelphia earned a 1956 Junior Division First State Award for Pennsylvania ($150) and Regional Award (Pennsylvania and Maryland). Carved from basswood, this two-door sports coupe had a medium blue-green body and a light blue-gray top. The windshield and window areas were indicated by dark gray paint. The model was designed without benefit of clay, but relied on drawings and sketches. The design featured sculpted front and rear bumper guards made from highly polished aluminum, recessed circular headlights, lathe-turned aluminum wheel covers, fully exposed front and rear tires and whitewalls (courtesy Thomas L. De Fazio and image enhanced by Ronald J. Will).

1956 Guild Convention

Detroit was a busy, working city in 1956. GM was flush and knew how to show the regional winners a good time. The extensive GM Technical Center and the proving ground designed by architect Eero Saarinen at Warren were new. GM displayed their prototype gas turbine cars, the Firebird I and Firebird II. We watched a driver take an Oldsmobile sedan around a test track, make a mild sideswipe against a typical steel highway barrier, and—unexpectedly to all—wholly climb over it! A GM executive showed his immaculate Duesenberg Model J roadster; another showed his recent custom-built Cadillac roadster. Regrettably, I couldn't capitalize on the interest generated by the Detroit trip with a follow-up 1957 model, due to my senior year in high school and my preparations for attending the Massachusetts Institute of Technology (MIT).

1958 Entry (Ages 17–18)

It would be hopeless for me to work on my model during my MIT freshman year. As I often worked summers to help pay for college, I designed and built the model car for the 1958 Guild competition during the summer evenings of 1957. Except for an occasional date, a ride on one of my friend's two motorcycles, or some railroad steam locomotive photography, I had little recreational life that summer. With limited time available for model making, I again had to use all my shortcuts as well as my improving skills.

This low-profile, sleek design appeared to be proportionally longer and narrower than my 1956 model even though its overall dimensions were the same. There was a gently curved beltline side view with a hard edge, running from front to rear with its highest point at the

The fastback sports coupe designed by Thomas L. De Fazio (ages 17–18) of Philadelphia, Pennsylvania, which won the 1958 Senior Division First State Award for Pennsylvania ($150) and Regional Award. The model was carved from basswood and painted a custom blended, medium metallic blue Duco lacquer with light gray indicated window areas. A hand-operated insecticide sprayer was used to apply the lacquer paints. In anticipation of the rigorous academic school year ahead, the model was designed and built by Mr. De Fazio during the summer of 1957 (about 9 months in advance of the June 1958 deadline). The design featured headlights recessed into the front fenders, taillights recessed into the body below the rear bumpers, a full width front bumper and a split pair of rear bumpers, fully exposed front wheels/tires and partially exposed rear wheels/tires with highly polished aluminum wheel covers (courtesy Thomas L. De Fazio).

windshield and sloping downward to the rear. Near the rear, this design line reversed direction slightly. There was a very thin roofline and minimal roof structure so as to not distract from the styling theme. Front tires were fully exposed, and rear wheels were partially exposed to the wheel centerline. The side panels were modestly contoured and compound surfaces were also modest. The windshield and windows were plane-curved rather than compound-curved and were trimmed with nickel silver, representing chrome trim. The headlights were recessed into the front fenders. The taillights were recessed into the body below the back bumpers. The highly polished aluminum wheel covers were set off by the wide whitewall tires. The front bumper was a fair shaped, full-width horizontal aluminum stock, with no guards. The rear bumpers were a split pair that protected the rear fenders and bore a license plate between them. The model was sprayed with an insecticide sprayer charged with a custom-blended metallic blue Duco DuPont lacquer.

I finished the model by September 1957 and started at MIT. The school year was taxing, but I met all the stiff requirements to continue my city scholarship. My model was submitted in June 1958. I earned a Senior Division First State Award for Pennsylvania and a Regional Award, meaning I would again go to Detroit. My round-trip ticket to ride the Pennsylvania Railroad's *Red Arrow* soon arrived.

1959 Re-entry (Age 18–19)

There was still no hope I could style a car and represent it as a finished scale model during an MIT school year. Again, the work had to be done during a summer. I read the rules carefully: nothing barred re-entry of a model. I did so in 1959, and was lucky to earn a 1959 Senior Division Second State Award of $100.

1960 Entry (Ages 19–20)

In the late spring 1959, I styled a car that I thought to be quite elegant, one that truly made me proud. I did not work that summer and built a model car instead. I relied on my developed skills, on the knowledge gleaned from two trips to the Detroit Guild convention, and on the shortcuts I had developed over the years of model-making. I had just about 10 weeks that summer to prepare an entry for the 1960 contest. I did so and met my own deadline. My model for 1960 was ready to ship when I returned to MIT for my junior year in September 1959. The model presented well, and it was very well finished. I had national level hopes.

Throughout my FBCG career I do not recall ever having had access to a score sheet, and I never designed my car models in a calculating way. I had made many sketches over the years, and I had developed the sketches I liked that I thought held some promise. The flaws of my 1956 and 1958 cars were clear to me by 1959. The 1956 car was more slab-sided than I had expected it to be, a consequence of working primarily with side-view sketches and without preliminary clay modeling. Its highest point was at the windshield rather than where the driver's head might be, an ergonomic error. And finally, front wheel bounce room was insufficient. The last was a common flaw among FBCG contest entries. My 1958 model car repeated two of the same flaws, but each to a lesser extent. And my 1958 car was less slab-sided.

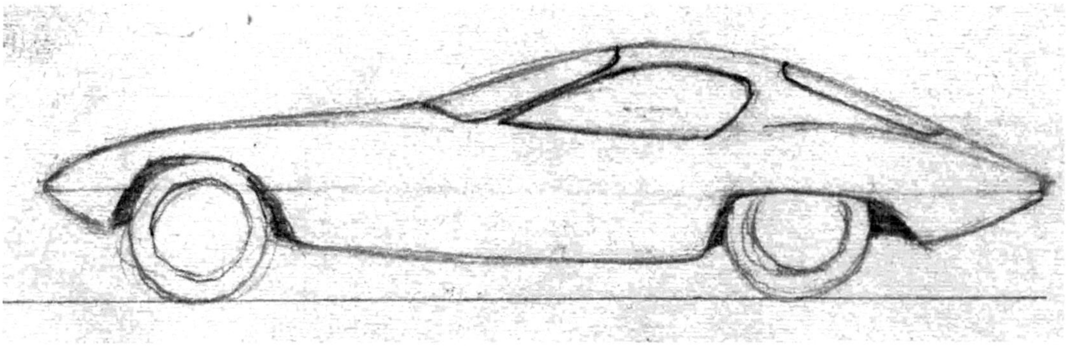

A pencil sketch made by Thomas L. De Fazio during the June 1959 period. It closely represents the completed model that he produced over the summer of 1959 and entered in the Guild competition in June 1960. The completed model differed in that it had a slightly increased front overhang and simpler treatment of the wheel well openings. Producing this 1960 model began by establishing the wheel/tire and body center lines, establishing an accurate side-view outline, and marking many of the model's design lines onto a laminated block of basswood. The wheel well openings were bored with a belly brace and bit, and the outline of the side view profile was cut using a band saw. Most of the subsequent carving and shaping was done with hand tools. This design captured the 1960 Senior Division Second State Award for Pennsylvania ($100) and a National Styling Scholarship ($1,000) (courtesy Thomas L. De Fazio).

All these flaws were corrected in my new design for the 1960 Guild contest. Surely, it helped that the wheels and tires supplied for the 1960 contest year were a bit smaller than those of years past. Again, I was on a very tight schedule, and again, there was no clay model. But I had also learned how to plan and accommodate for three-dimensional shapes in a series of paper sketches.

I carved my model from a single, well-squared basswood block that had been glued together from many longitudinally disposed boards. A first step was a drawing instrument (square, straight edge, dividers, scale) layout of axle center-lines on the sides and bottom, layout of wheel depths on the bottom, layout of an accurate axial centerline all around, and layout of a lateral profile on both sides. The wheels were to be fully exposed. Wheel wells were carefully bored to marked depth manually, using a sharp bit in a belly brace.

The elevation profile, or side view, was carefully cut on a large band saw. Next, still more instrument layout work was required based on the front and rear view design, depicting where the curved sides were to intersect the roof line. The plan view, or top view, shapes of the front and back of the model were marked. Two holes were drilled and steel pins were driven along the bottom centerline to give means for holding the model in a bench vise. This work went fast, and the block quickly took a pleasing shape.

The car's profile now had shape, both top and bottom. The only remaining flats were the car sides, so no more band saw work could be done. The top view gross shapes were manually and carefully cut, then carved away. Then, carving and shaping was done using a plane, chisels, knives, rasp and files, and, finally, sandpaper. Shaping addressed the sides, hood, trunk, and wheel wells. The car had upper and lower compound curved body surfaces, and an upper-lower demarcation pencil line was always maintained, as was a complete axial centerline. Other pencil lines were used, where appropriate, to delimit or guide shaping. Partial templates were used in places to try to maintain symmetry. This was slow work, and it was critical. When shaping was done to my satisfaction, the modest single tail fin was fabricated, glued to the trunk-area centerline, and shaped.

Thomas L. De Fazio (ages 19–20) won the 1960 Senior Division Second State award for Pennsylvania ($100) and a National Styling Scholarship ($1,000) with this Guild entry. Carved from laminated basswood, this two-door sports coupe is painted a custom blended, dark metallic blue "Duco" lacquer with light gray indicated window areas. Paint was applied with a vibratory spray paint device. Because of his academic demands, Mr. De Fazio completed the design and construction of this model by September 1959 (about 9 months in advance of the deadline of June 1960). The design featured fully exposed front and rear wheels/tires, lathe cut aluminum wheel covers and polished aluminum bright work (courtesy Thomas L. De Fazio).

Grooves had to be cut for trim inserts and to indicate functional openings for parts of the body such as doors, trunk, hood and retractable headlights. The trim lines and indicated opening lines were laid out using soft pencil with the aid of templates and a scale. Thin aluminum trim, inset, would denote the window edges. These wider grooves for the trim were scored first by knife and then by a piece of a fine hacksaw, a tool with perhaps a dozen teeth, ground laterally flat.

Thomas L. De Fazio around 1960 (age 20) while a junior in college, with a diamond-shaped FBCG pin on the left lapel. He entered the Guild competition four times (1953, 1956, 1958 and 1960) and won first State and Regional awards in 1956 and 1958. He won the 1960 Senior Division Second State Award for Pennsylvania ($100) and a National Styling Scholarship ($1,000) award (courtesy Thomas L. De Fazio).

The front and rear grille openings were then carefully drilled, carved and smoothed. Finally, the wood was sealed, sanded, sealed again, and sanded again, with the grooves maintained as needed throughout sealing, priming, and painting.

Painting was next. DuPont automotive lacquer was applied using a small Sears Roebuck and Co. vibrating sprayer. A gray lacquer primer was sprayed on in several coats, [with] sanding between the coats. Wheel wells and the grille openings were brush-painted with a dull black model railroad lacquer and then masked off when dry. Windows were sprayed a dark gray, in several coats, sanding in between coats and, finally, masked off.

Color was next. The color chosen was a rather dark metallic blue that was blended to my order by the local Duco distributor. Color was sprayed, sanded, and re-sprayed until a good coat was established. The finish was polished using a fine polishing compound.

Next, trim was prepared and mounted. The FBCG regularly arranged with a Connecticut non-ferrous supply house, Bridgeport Brass, to make available to Guildsmen a package of soft aluminum

Thomas L. De Fazio at the Museum of Fine Arts in Boston in August 2008, with his wife, Sally Alice Ruth De Fazio, seated in the background. He is a retired mechanical engineer and occasional consultant. He holds an industrial arts diploma from South Philadelphia High School and multiple degrees from the Massachusetts Institute of Technology. He has taught engineering courses at the Florida Institute of Technology, had a long tenure at Draper Laboratory, and most recently, did innovative work in mobile robotics with iRobot Corporation (courtesy Ronald J. Will).

rolled plate, bar and round stock, for a modest price. I used the thinnest plate to cut, form, shape and polish trim pieces to delineate the indicated window areas between the windows' dark gray and the body color coat. These trim pieces were tapped into place. The next-thicker size aluminum plate was shaped and polished to represent small door handles and the modest wraparound fore and aft bumpers, and then mounted with integral pins. Grille inserts, fore and aft, used fitted aluminum screening pieces painted dull black, with high points then polished. Tail and turn lamps were represented by bits of shaped plastic cut from red reflectors. Front parking lamps, within the grille opening, were turned from aluminum and filled with clear plastic. Bumpers were pressed in last.

Simple conical hubcaps were turned from aluminum round stock on a tool room lathe in my father's employer's shop. I already knew how to grind lathe tools and use an engine lathe from both my father's training and my high school training. A hubcap was shaped using the compound feed, and a fine, small-diameter ring recess was cut into its surface. The hubcap was cut off, and the next one was turned. The hubcaps were polished on a buffing wheel, and the fine recessed rings were filled with body-color paint. The official Guild wheels were mounted next. Each was attached to the wooden body using several through, flush wood screws so wheels would not turn. Wheels were removed and thin white circumferential stripes were painted on them manually. The wheels were remounted, and the hubcaps were pressed into place. The model was complete.

It was beautiful to my eye, and it fully represented my design concept. On a day with a hazy sun, I took the car out and exposed Kodachrome images. The car was fit to a shipping box, where it was stored until time to be sent off. The heavy and well-closed plywood box buffered changes in humidity that might affect the model's wood. It was late summer 1959, and my 1960 model entry was already complete.

My junior year went quite well. I accepted a summer job in 1960 with an engineering consulting house in Cambridge, Massachusetts, that used all my newly-developing engineering skills. I went to work each day awaiting the telegram. It never came, but a letter came instead. I had won the Senior Division Second State Award for Pennsylvania ($100) and a National Styling Scholarship worth $1,000. Apparently, my model was not the only senior national-level model entered from Pennsylvania. Another senior contestant's model had outscored mine at the state and regional level. And yet, my scoring for styling was sufficient to qualify me for one of 10 FBCG Styling Scholarships.

Months later, and into my senior year at MIT, I received another letter from GM, one soliciting my interest in employment in a GM styling studio. I thought about it, but declined. I enjoyed using my newly developing engineering and analytical skills. I was apprehensive about a star system profession, and I was apprehensive of a stylist's apparent stress level that I had witnessed in 1958 at Detroit. I am not a wholly confident person, and it seemed to me that a styling studio was not a place for someone with modest confidence. I felt that the styling of my 1960 car was equal or superior to all but a few within the FBCG contests, such as Adrian Bruno, the 1955 National Junior Scholarship winner, and E. Arthur Russell, the 1957 National Senior Scholarship winner. But could I create such a well-designed car, to a specification, on demand, year after year? Perhaps I could, perhaps not.

The Guild gave much to very many. Some FBCG alumni went on to successful creative careers in the automobile industry, in many cases working as stylists. Others embarked on careers in other industries involving esthetic design and style.

The Guild gave me much as well. It developed my modeling skills and showed me how someone can successfully expand their own skill set independently of formal training. It taught me as a young man to plan an extensive project and to complete it on deadline. It lured me into enjoyable hard work with a clear goal. It gave me valuable insight into the automobile industry and its creative styling functions. And it covered more than a third of my undergraduate tuition, which was no small return for my efforts.

I am a retired mechanical engineer and occasional consultant. I am interested in American and British steam railroading, precision portable timekeepers, and the history of mechanical technologies. I am a lapsed motorcyclist, and an occasional hiker, bird watcher, bicyclist and windsurfer.

16

Forgotten Dreams Remembered

Felix R. Collazo (1959–1961)

I first heard about the Fisher Body Craftsman's Guild when I went with a friend to the 1958 General Motors Motorama show at New York City's Waldorf Astoria Hotel just a few blocks from my high school. The automaker was showing off its 1959 models, but I was more fascinated with the Guild models they had on display. I could not take my eyes off of them.

The models stirred my 14-year-old imagination. I had been drawing and designing cars since I can remember. In school, when I was bored with a class, I would draw cars and at home I would do the same thing while watching television. At the High School of Art and Design, I was majoring in architecture and industrial design. In reality, the program had more to do with drafting, but I loved it.

I was part of the "in crowd." There were lots of girls, weekend dances, trips to the beach on school half days or parties at various kids' homes while their parents were working. It was a "macho" environment, so we had to be street smart. Yet deep down inside we were really sensitive kids; after all, we were art students. Our crowd was a mixture of Puerto Ricans, Italians, Blacks, Irish, Jews, Germans and more.

I never mentioned my participation in the Guild to anyone, except to my architecture and design teacher, Dr. Erwin T. Muller. He was not very supportive and basically told me not to waste my time or effort on building models. I never knew why, but I never brought up the subject again.

When I got home from school, I would put on my "geek" persona and work on my model for hours. We lived in a sixth floor apartment in the High Bridge part of the Bronx, close to Yankee Stadium. It had three bedrooms, so my room was my workshop. My mother would have to call and beg me to come eat dinner. It was like living in two different worlds: hip in school and geek at home.

My work station was a flat drawing board on top of a nightstand next to my bed, so when I went to sleep, the last thing I saw was my model. When I awoke, it was the first

***Opposite*: Felix Roberto Collazo's 1960 Junior Division Honorable Mention Award ($25) winning model was made from three pieces of laminated balsa wood glued together and painted with drug store spray cans. Headlights were white clothes buttons and the grille was made from lead pencil wood with chrome tape on top. Rear taillights were made from red hair comb ends and the windshield was cut from clear plastic and formed by hand. The hubcaps were fashioned from "Quick" chocolate can tops (slightly rusted in these recent images). A 1959 junior division entry he had made at age 15 was a black balsa wood fastback with a teardrop roofline and employed metal hair rollers for the grille and aluminum amusement park souvenirs for hubcaps (courtesy Felix R. Collazo).**

Rear view of the 1961 black and silver Senior Division Guild model entry made by Felix R. Collazo of High Bridge, Bronx, New York City (courtesy Felix R. Collazo).

sight of the day. I carved my models mostly from laminated balsa wood using a ½-inch wide chisel and razor blades. I tried clay modeling, but it never worked out for me. I used to paint my models in my bedroom using spray paint cans from the local five and dime store. My mother would have a fit because the fumes would quickly engulf the small apartment. I could not open the windows because that would let in the dust and ruin the paint job. Sometimes, I would paint in the hallway on the steps, and pray the people in the other apartments would not come out. My father was a factory worker at a cardboard box company, and even though he could not give me technical support, he always made to exact specifications the hard cardboard boxes to mail my models. To his credit, all three of my Guild model entries arrived intact.

I had not been into model making before, so when I decided to enter the 1958–59 Guild competition, it opened a new world to me. The Guild sent me the instructions and specification sheets. I bought three blocks of balsa wood at a hobby shop downtown that were glued together. My first model was a black fastback with a teardrop roof line that sloped toward the back. I used my sister's hair rollers for a grille, clear marbles for headlights, and red hair comb ends for taillights. For hubcaps, I used the back sides of a star-like amusement park souvenir coins — they were made of aluminum and the perfect size. It did not win anything, but it was a learning experience all the way.

For my second model in 1959–1960 Guild competition, I choose a two-door gold and black convertible. It took a lot of work and attention to detail to complete this model. Since I did not have the tools to make parts, I had to improvise and use anything at hand — grilles from my sister's metal hair rollers, hubcaps made from "Quick" chocolate can tops, taillights using red hair comb ends, side trim from old metal watch bands and headlights made from white buttons.

The interior of my convertible was just as detailed and innovative. For the seats, I cut a soup can in half and shaped it with my hands. Then I used a piece of my father's therapeutic back patch, which he used for back pain, for the seat covers. It was red felt with a sticky substance on the back side which worked great. To this day, I don't think he ever noticed the difference in the size of his patch. I was constantly scavenging, especially in my sister's

Felix Roberto Collazo, age 16, seated in a new 1961 Austin-Healey sports car at the GM Motorama at the Waldorf Astoria Hotel in Manhattan, New York City, around September 1960. Attending with a classmate from the High School of Art and Design where he was a senior, he saw a select number of winning Fisher Body Craftsman's Guild models. He had just won a 1960 Junior Division Honorable Mention Award for New York as a Guildsman. His favorite car while in high school was the 1960 Chevrolet Impala (courtesy Felix R. Collazo).

room. We were a modest family, and money was used for basic needs, so I had to use my ingenuity for my models.

We mailed in my convertible using one of my dad's cardboard boxes. It looked great, and I had high expectations. To my surprise and elation, I received a letter from the Guild stating that my model was judged among the best and had earned a Senior Division Honorable Mention for New York along with a $25 check. To a kid from Puerto Rico living in the Bronx, this was like winning the top award. I was excited, believing I might have a chance of winning one of the styling scholarships. I listened to the radio broadcast the night of the Fisher Body Craftsman's Guild awards banquet, but they did not call my name.

My third model for the 1960–61 competition was a black and silver, sleek-looking two-door hardtop. I was a 16-year-old high school senior with a lot of activities going on.

Felix Roberto Collazo shown in his Army dress uniform. After completing high school, with a major in architecture and industrial design, he joined the Navy and became a draftsman at the Norfolk Naval Base in Virginia. Later, he was stationed in San Juan, Puerto Rico, where he was a tugboat seaman. While working in hotels, sales, and for the Puerto Rico government's Department of Health, he completed his bachelor of arts degree (with help from the GI Bill), joined the Army Reserve and went on active duty again. Working as a water craft operator, retention NCO, career advisor and finally personnel sergeant, he retired in January 2002, at the rank of sergeant first class with 28 years of military service between the Navy and the Army (courtesy Felix R. Collazo).

I chose to make a "solid top," as opposed to a convertible, thinking this would be easier. The paint job gave me a lot of trouble. When I went to paint the silver on the top, it ran into the black paint of the body. I freaked out and started crying. My mother came to my rescue. She took the model and soaked it in paint remover in the bath tub, removing all the paint down to the wood sealer. I will always be grateful for the support and encouragement she gave me to continue. I did just that, but I was short on time and scared of making the same mistake again. This time, I painted the top of the model silver with a small paint brush, but in the end you could see the brush marks. The hubcaps were unique, because they were cut from an artist's aluminum paint pallet and with holes from a paper punch. I mailed it in hoping the design would make up for the paint, but I did not win anything.

I graduated from high school in 1961, and though I had excellent grades in my art classes, my academic grades were not good enough for a scholarship. My parents could not afford to pay for college, and I just wanted a change in my life. I joined the Navy for four years, got married to my wonderful wife, Zory, of 46 years, and eventually had four kids—Shirley, Shelciy, Roberto and Ricardo. I took my GI Bill and earned my college degree in my early 30s. I joined the Army Reserves and went active duty again until my retirement in January 2002.

Even though my life did not follow the path of automotive design and art, it was always my passion. I am a designer and artist at heart. The years I competed in the Guild were some of the most creative and exciting days of my youth. They taught me perseverance, ingenuity and how to see a job through. The models were stored in a cardboard box in a closet for 45 years in Puerto Rico. I had taken them to our home there after my father died and we moved back to the island. Recently, I brought them back to our home in South Carolina. Time has taken a toll on them, but they hold a special place in my home and in my heart, serving as a reminder of that time and place.

Years later, I realized the enormous odds and level of competition I was up against. My region, which included New York and Connecticut, composed one of the largest and strongest regions. Many winning competitors have told about their access to technical, professional and economic resources that I and many other participants did not have. This is not to take away from their glory. But those of us who competed and poured our hearts and souls into a model, yet won few or no awards, see it from a different perspective. Whatever we won was worth ten-fold considering the odds and level of competition at the time. I will always be proud to have been a Fisher Body Craftsman's Guild member and treasure those days of my youth—and for learning to look at everything from combs to buttons with a model artist's eye.

17

From Soap Box Cars to Subaru's Executive Suites

Ronald J. Will (1959–1961)

Dennis and I were terrible at model planes. They were supposed to fly with rubber bands wound up tight to drive the plastic propellers. But, when launched in the neighboring prison field, the planes stalled, hooked or sliced and ended up with their noses buried in the mud. This was the beginning for a person who would climb the ranks to be the manager of product planning and design at one of the major automobile manufacturers.

I was a scrawny 12-year-old growing up in Michigan City, Indiana, on the southern tip of Lake Michigan in 1954. We lived with the South Shore Railroad tracks in our front yard and the Indiana State Prison in our back yard. It wasn't the greatest neighborhood. My best friend Dennis Mitchell (our own Dennis the Menace) lived a few blocks away, and we both liked building things: balsa model planes, boats or anything that we could afford from the local hobby shop on Franklin Street.

Dennis's father worked for General Motors, and one day he brought home a newsletter called the *Guildsman* featuring the company's Fisher Body Craftsman's Guild program. We were both amazed at the shiny, sleek and futuristic looking model cars entered in the competition. We were even more amazed when we discovered that these model cars were built by boys not much older than we were. With our pathetic model building skills, we never even considered entering such a difficult contest. But, we did think we could build Soap Box Derby racers. The *Guildsman* was put in a drawer and forgotten.

Although I finally mastered the model airplane building craft with a giant sail plane model, I succeeded early on at ages 14 and 15 with the All-American Soap Box Derby. This is how I honed my self-confidence, skills and abilities needed for the Craftsman's Guild. My 1957 white derby car, which I built in our dark, cold attic, was sleek and constructed in the shape of a model airplane fuselage and used model airplane bulkhead and stringer construction.

After the 1957 derby races, we moved to a new house nearby in Hobart, Indiana, with a real garage where I could work. I penciled my next derby designs on sheets of butcher paper and studied the pictures of the other aerodynamic derby cars that had won the nationals in Akron, Ohio. My 1958 derby car showed great improvement and was brush-painted black with pink trim. I won heat after heat as my gravity powered car slipped silently down Michigan Avenue until that final race arrived. I was scared to death until I crossed the finish line a winner. This one got me to the finals in Akron, where I won several heats, but not a national scholarship. I had tasted competition and savored close victory.

Soon after, a General Motors representative showed up at our school during the autumn of 1958 to promote the Fisher Body Craftsman's Guild. Now I was ready. I could do this. After the presentation, I discovered that another boy from Hobart, Brian Skogler, had already entered a car in the Guild and had won a 1958 Junior Division Regional Award. We quickly became good friends and are still friends today. Eventually there were eight of us, including myself, all building cars for the Guild (Brian, Dave Richmond, Patrick Saturday, Tom Covert, Wayne D'Mura, Bill Potter and Larry Pecorella.) We designed cars in study halls and even at football games. Our fathers drove us to Guildsmen's homes farther away where we compared ideas, designs and construction tips. We fed off each other and raised the quality level of all our models. Each of us eventually won prizes ranging from an honorable mention to the highest national scholarship.

Grandfather's Carving Tools

For my first Guild model, I glued together some 1-inch poplar boards to create the solid block that was eventually to be my model car. Poplar wood was supposed to be good for carving, and I soon dug in with my old X-Acto blades from my model airplane building days. They really didn't work as well on the hard poplar as they did on the soft model airplane balsa wood. When my dad saw I was struggling, he said, "Come with me." He took me out in the garage to a polished wooden cabinet that had sat on the garage workbench for years. I had never seen inside because it was always locked.

Now I was finally going to see what was in the mystery cabinet. As the door slowly opened, I could see row after row of wood carving tools with beautiful polished wooden handles hanging not only in the back of the cabinet but also on the door. Each blade had a different size and shape from small curves to wedges and flats. My dad took one of them down from its hanger. It was as sharp as a razor blade. He then explained that my grandfather had been a master carver at a chair factory in the early 1900s. These were his tools that he used to carve the intricate bas-relief designs into the backs of chairs and custom tables. I learned that as a young apprentice carver, my grandfather also had to make the beautiful cabinet that now held all of his tools. My dad said that if I took good care of them and used them wisely, someday they would be mine. They were perfect. Granddad's sharp blades carved that laminated poplar block like it was soft balsa. Today the polished wooden cabinet sits on my workbench in my garage.

Derby Paint Craft

As part of my local 1958 derby win, I was sent to the nationals in Akron to represent my city. Traditionally, the city would repaint the winner's derby car with the sponsor's logo. We did the repainting at our local Enyeart Chevrolet dealer's auto body shop. Here I learned the magic of how to get a glass-like finish. Many coats of thick brown primer were applied and then carefully wet sanded with 400 grit sandpaper before applying the black lacquer finish. The lacquer was allowed to dry and "shrink down" before being sanded with 600 grit wet sandpaper and buffed with polishing compound. After Blue Coral waxing, I couldn't believe it was my car. It was a gleaming black beauty that showed my reflection like a mirror. Of course, I carefully re-applied the pink trim by hand with a fine brush.

1959 Entry (Age 16)

After my experience with the Derby, I was ready for a big win with my first Fisher Body model. The Guild's *Designing and Building a Model Car* booklet showed a sculpted clay model as the first step in the process. I'm glad I took their advice, because design ideas in your head do not always work out 3-dimensionally with a scale model. Making these mistakes while carving wood would have been very discouraging. Gradually, using my fingers, broken hack saw blades and a table knife, the lump of clay took the form of a sleek 2-door sports coupe. I made thin cardboard templates from all the clay shapes I wanted to transfer to the final, durable wooden model. Made from laminated poplar wood and coated with wood sealer, this design featured a cantilevered roof line with wraparound rear windows, but no window trim. I used 1958 Chevrolet light Frost Blue metallic lacquer paint for the body with white-indicated windows and white rear wheel well opening accents. The aluminum wheel covers were a very simple design spun and polished on a high school shop lathe. In May of 1959, I sent in my gleaming two-tone blue and white sports coupe to the competition headquarters in Warren, Michigan, in a sturdy wooden box. I even nailed heavy steel bands around the box to further protect my miniature automotive creation.

I waited patiently to hear about my top prize, and finally, the all important Guild letter arrived. I ripped it open to find that I had only won a Senior Division Honorable Mention of $25. I was totally deflated. What happened? I had spent more than $25 just to build my model. Clearly this was not going to be as easy as I thought. I pored over and over my score sheet filled out by the Guild judges. It showed fairly good points for originality and design but just average points for craftsmanship. I decided I needed some expert advice.

Just like in the Soap Box Derby, I realized I needed to know more about "how to win" and went straight to the pros with my Guild questions. No one would be better than the

Ron Will's first attempt at the Guild competition in 1959 at age 16 won a Senior Division Honorable Mention Award for Indiana ($25). This 2-tone, two-door sports car with open wheels/tires was designed in clay and transferred to poplar. The finished wood body was covered with wood sealer and painted a light Frost Blue metallic lacquer (a Chevrolet production color) with white indicated window areas. The design featured a cantilever roof line, wrap around rear window, white rear wheel well accents and aluminum wheel covers. Note the absence of metallic window trim. Door cuts were thin black tape and the taillights were also red tape. Ron later learned it would take door, trunk and hood opening grooves hand-cut into the body surface, with more realistic and sophisticated details, to win the highest craftsmanship scores (courtesy Ronald J. Will).

previous year's Guild winners. I prepared a 3-page questionnaire and sent it to that year's top national Guild winners. Most of them were very nice and sent back hand written detailed information on how they built their cars and what they thought made them winners.

1960 Entry, Age 17

My 1960 design was another 2-door sports coupe painted a deep 1959 Chevrolet Harbor Blue metallic lacquer with white indicated window areas.

For this, my second Guild model, I followed the advice from the Guildsmen I contacted like the gospel. I strove for perfectly inlaid window trim, highly polished brightwork without any signs of scratches and a deep, lustrous paint finish. The styling of the model was clean and flowing with elliptical top view shapes. The clay model allowed me to experiment with shapes that I could not envision on paper. Like the first model, it was a carved from laminated poplar and coated with wood sealer.

Unfortunately, it ended up with an odd combination of brass window trim and chrome-plated brass bumpers. This model was meant to have chrome trim, but the chrome plating businesses I contacted could not figure out how to plate such tiny parts like toothpick diameter door handles and fragile 1⁄16-inch width window trim. The first plater actually destroyed the smaller parts trying to plate them. It was painful to have to make them again and present them on the model as polished brass parts. Running out of time, I also had to re-use the old hubcaps from my first model. Now I had a mixture of brass, aluminum and chrome plated trim on the models, which I am sure cost me points in the competition. I was very disappointed when I sent this model off to Detroit. I knew I could have done better and was expecting another honorable mention as recognition for my efforts.

Shazam — to my surprise, it worked! The model car that was such a disappointment won the Senior Division First State Award for Indiana and the Regional Award. I was headed to Detroit after capturing 82 percent of some 450 available points. The Guild divided up the United States into 20 regions with a junior and senior winner from each region, 40 in all. These 40 boys would receive an all-expenses paid trip to Detroit for four days for a grand tour culminating in the big awards ceremony at the Fisher Auditorium.

TABLE 2
Score Sheet for Ron Will, Model #6427, 1960 Senior Division First State Indiana and Regional Award (Region 11, State of Indiana)

Craftsmanship	*Possible Points*	*Points Awarded*	*Earned Percentage*
Scale Fidelity	50	50.0	
Workmanship	80	56.5	
Painting and Finishing	70	50.0	
Sub-total	200	156.5	78.25%
Design			
Originality of Design	100	81.0	
Artistic Merit	80	68.5	
Practicality of Design	70	62.5	
Sub-Total	250	212.0	84.80%
Total	450	368.5	81.88%

Reference: Original Score Sheet for Ronald J. Will. *How Your Model Scored in the FBCG Competition*, Year 1960, Model #6427, Senior Division, Region 11, Indiana.

This two-door sports coupe designed and built by Ron Will (age 17) of Hobart, IN, won the 1960 Senior Division First State Award for Indiana ($150) and Regional Award. Carved from laminated poplar and coated with a wood sealer, this model was painted a Harbor Blue metallic lacquer color (a 1959 Chevrolet production color) with white indicated window areas. He employed a combination of inlaid brass window trim, highly polished aluminum wheel covers and chrome-plated trim parts. The styling of the model was round and fluid with an elliptical roof and sleek fender lines. The Regional Award allowed Ron to attend the Guild Convention in Detroit (courtesy Ronald J. Will).

After arriving in Detroit, I finally got to see other regional models. Even after building two models, I was still in awe. They were like miniature pieces of jewelry. When comparing them with my model, I knew I was not going to be in the top winner's circle that year. I was right. I didn't win, but now I knew what league I was in and what it would take to win. I talked to the winners and pestered the judges about what I had to do better. Basically, they said a winning model had to shine like a diamond among a table full of Guild models, with a gleaming and flawless paint job, bright chrome-plating and stunningly original styling.

1961 Entry, Age 18

It was 1961 and my last year at home before I would head off to college. I knew I couldn't build a model away from home. It was now or never. Going back to my 1959 Guild winner's survey, I knew the design originality was the key to success, and I would have to gamble on something totally unique to win. At the Chicago Auto Show that winter, Virgil M. Exner, a senior vice president of styling at Chrysler, showcased an asymmetrical sports car called the "XNR." That was it. That was my inspiration.

My asymmetrical design featured sweeping bold parabolic curves around a single occupant, namely the driver. The sides had deeply sculptured speed forms. The model was designed from the judge's perspective, from the top view, and had front and rear asymmetric design lines. None of the quarter panels matched. The model featured a fighter jet cockpit style Plexiglas windshield, and finally, all matching chrome-plated brass trim, and included a unique "floating center" wheel cover design. I spray painted the model Twilight Mist, a new space age looking, metallic lavender lacquer color. For good luck, that was the same color as my father's 1961 Oldsmobile 98.

This 1961 Guild entry, a 2-door sports convertible, was again carved from poplar and coated with wood sealer. But that was not good enough. Both of my previous models showed their wood lamination cracks and exhibited minor sink marks from the wood. I needed a perfect paint score if I wanted to be in the winner's circle. I was told that coating the model with fiberglass resin would seal all this out and eliminate any flaws. Although I had no clue how to use fiberglass resin, I bought a can, mixed in the hardener and painted the thick, sticky, gooey layer all over my model. When it hardened I was in shock! The surfaces were rock-hard and impossible to smooth out with normal sheets of sandpaper. I thought I was in big trouble.

I ended up using ultra coarse 36 grit sandpaper, and even metal files, to smooth down the lumpy surfaces. After that I was able to do the final smoothing with finer and finer sandpapers. It was tough to carve or cut tiny details into this hard resin material, but the work was worth it.

I was happy to find that the resin-coated surface had a big advantage. I was able to use tiny curved riffler files to carefully cut accurate detailed door lines to simulate the door and hood opening cuts. Doing this in wood was nearly impossible. After spray-painting the model, the fine grooves were filled with black India ink to give them a realistic appearance.

Having twice seen the winning models in Detroit, once with my dad and once as a regional winner, I knew that design originality was the most important factor where every detail counted: grille, bumpers, wheel covers, lights, door handles, odometer markings and instrument panel controls. The score sheet awarded the most points (100) for originality of design. I also knew the secret of success in the Guild competition was perfection. If you could see a scratch, a drip of paint or glue or a rough edge anywhere, then so could the Guild judges. Sometimes metal parts didn't fit or work out, even after days of effort. At that point, I had to make the hard decision to throw them away and start over again, learning from my mistakes.

I wanted to paint a thin whitewall on each of the four official Guild rubber tires. To do this, I used a mechanical drawing ink compass with white paint that had been thinned enough to flow from the compass. I made a round aluminum insert for the rubber tire (where the wheel cover would go) and accurately located the center of the tire with a nail punch. This located a fixed point for compass rotation. Because of thick paint globs, drips or over-runs, I repainted the six tires over and over again, cleaning them each time with solvent, until I had four perfect specimens and one spare in case of damage during the model assembly process.

At least a half-dozen Plexiglas windshields were hand-formed in mom's kitchen oven over a male mold dipped in cooking oil. The oil acted as a lubricant and helped avoid surface scratches in the plastic. Some pieces stretched too thin and others were wrinkled. The samples were closely inspected to see which one came closest to being "a perfect part" for the competition. The best windshield part was selected then slowly and meticulously trimmed with fresh sharp blades and filed to fit the tiny channel in the chrome-plated brass window trim.

That perfect windshield, as well as the other perfected parts I made, were wrapped in a soft cloth and stored in a jewelry box below my work bench awaiting final assembly. I never saw the model as one whole assembly until the last day before it was shipped to Detroit.

1961 Special Features (Automotive Design Concepts)

I was particularly concerned with safety because my dad had been severely injured in a head on automobile crash. The steering wheel caused most of his injuries, so I got rid of the steering wheel and designed a pad that could be pulled back into your lap like a seat belt. On the pad top were flush controls and a flush steering knob. There was nothing protruding to injure you and the pad absorbed the impact in a crash. I also made shock absorbent bumpers front and rear. On the doors I had flush fingerprint control locks. The car was to be driven by a rotary engine. Many of these features are common now, but for 1961 these were pretty far out ideas.

I had all these ideas, but this was my high school senior year, and it quickly became a panic with no time to work on the car. I was filling out college applications, cramming for finals and then ... there was the prom. Judy Anderson was a cute blonde art student whom I had asked to the prom months earlier. The prom was scheduled for just a few days before the Guild deadline, and I was frantically behind schedule. I had planned the finishing of the model right down to the last hour when it would be carefully packed in a foam padded box and taken to the railway station for shipping to Detroit. So there it was ... the prom, final tests and the Guild model.

Something had to give. I decided to skip the tests. To do so, I had to confront the tough principal, Mr. Harold Moody. Luckily he believed in the Guild and knew I was good at it. He already had one of my Guild trophies in his office. He said I could finish up the car, then study for a couple of weeks and take my final tests in summer school. What a guy! He also allowed me to walk at graduation, but I didn't get the real diploma until later. The prom was another matter. I couldn't make that up. I went to the prom, but made a horrible weak apology to Judy and then took her home early. We both missed the after-prom parties and the beach parties the next day. Years later Judy still had not forgiven me for that night.

I was working about 20 hours a day completing the model in the last few days. The last big step before final assembly was wet sanding the cured paint to a flat smooth finish with 600 grit paper. After that, the gleaming finish was revealed by carefully polishing the surfaces with fine rubbing compound. If I accidentally sanded or rubbed down through to the brown primer below, I would be a goner. I'd have to go back to ground zero, sand and re-paint the model again. Fortunately, everything went well.

When it was finally assembled, packed and gone, all that was left in the basement were paint fumes, piles of used sandpaper, and tools scattered everywhere. I was physically and mentally exhausted. I think I slept for 12 hours into the next day. Such was the dedication and belief in the Guild by a young teenager.

It seemed to be all worth it because, once again my new wild sports convertible won the first in State and Regional Awards. I was headed back to the national convention. In Detroit, there was a good sign in that all the other regional winners thought I had a winner. But then, they were not the judges, and the model was a very strange design. Had I gone too far with a weird, unorthodox design that the judges might hate? I feared the worst. The

Ron Will at age 18 with 1961 First National Senior Scholarship trophy given to Hobart High School in his honor. Ron made all of his Guild model wheel covers in the high school machine shop classes (courtesy Ronald J. Will).

awards banquet in the Fisher Auditorium was held on the last night of the Guild Convention, and my stomach was churning all day with anxiety. The National Junior Scholarship winners came first, and it seemed like it took forever to get through them. The presentation became a blur. I couldn't concentrate. Finally the National Senior Scholarship awards came around. The 4th place was announced first. I would have been happy with anything at that point. The speaker announced 4th place ... not me ... then 3rd... nothing again ... and then second. Oh no, I blew it. The judges hated it. I went too far. I couldn't possibly be first. Then, there it was, my little Twilight Mist model projected as large as a real car on the banquet

Desert photograph of Ron Will's full-sized, 3-wheel sports car called the "Turbo Phantom." It was fabricated with a lightweight urethane foam and fiberglass body sandwich, and had a tilt-lift canopy, inner roll bar structure and front steering axle. The Turbo Phantom was 7 feet wide in the front and had a low stance (44 inches high). Ron Will left GM to become an independent auto designer and started the Phantom Vehicle Company in Costa Mesa, California. This aerodynamic sports car was the ultimate expression of the Guild spirit — to design and build something new and original. Mr. Will built a 1:12 scale model to begin the project, just like he did for the Guild. The vehicle was built around a turbocharged Honda Goldwing motorcycle engine and other production automobile components. The goal was a lightweight, yet sporty design with good performance and high mileage (courtesy Ronald J. Will).

hall screen, with hundreds of GM executives looking on and applauding. I knew my parents and friends were also listening by radio from home. I sat there and shook my head in disbelief. I was supposed to accept my award at the podium, but I was so stunned that I just sat there. The Guildsman next to me finally nudged me to get up and receive my $5,000 scholarship award.

Conclusion

As it turned out, the Guild was just the first step in a 40 year career as an auto designer. I have thoroughly enjoyed working as a car designer, but the Guild win that night long ago was still the thrill of a lifetime.

After four years of design training from the Illinois Institute of Technology (IIT), where I majored in product design, I began work at GM as a designer. There I worked in the Corvette/Camaro Studios and contributed to the Aerovette and the 25th Anniversary Corvette. After 10 years, I left GM to follow a new dream of designing my own car just as I had done for the Guild 20 years earlier. I became an independent auto designer and went to work on my full-size dream car in Costa Mesa, California. I started with a 1:12 scale model, exactly the same scale as the Guild models, and I wanted my new design to be just as wild and eye catching as my 1961 winning model. It took three years to complete and the result was a two-passenger, 3-wheel, tilt-canopy Honda cycle-powered sports design I called the "Turbo Phantom." It looked like it should be preparing for launch from an Air Force runway. After satisfying my dream, I went back to work as a designer for Subaru in California. I recently retired as manager of product planning and design for Subaru of America, where I created the Outback concept and directed the Forester design.

Looking back, I had some very stiff Guild competition right there in my own back yard, and amazingly, five of my Indiana competitors also became auto designers. Out of my region, seven boys won national scholarships or regional awards; of those, five became auto designers (designated by asterisks). They included:

Brian Skogler * (1957 Honorable Mention, 1958 Regional Award winner, Ford)

Bill Scott * (1959 Fourth National Senior Scholarship winner)

Patrick Saturday (1961 National Styling Scholarship winner)

Allan Flowers * (1962 Fourth National Scholarship winner, GM and Nissan)

Tom Covert * (1962 Second National Scholarship winner, GM)

Geza A. Loczi * (1965 First National scholarship winner, GM and Volvo)

Joseph Wayne D'Mura (1966 Fourth National Scholarship winner).

I remember all these guys and the good times we had designing model cars for the Fisher Body Craftsman's Guild every time I look at my grandfather's tool chest or my three Guild models. The 1961 scholarship model I made is still rock-solid after some 50 years and looks like new. The real grand prize from the Guild was not the scholarship, but the connection to a lifetime of great friends and a career in auto design that a scrawny 12-year-old kid from Michigan City could only have dreamed of.

Ron Will, 1961 First National Senior Scholarship winner, joined many other Guildsmen who displayed their models at the 2008 Guild reunion held at the Museum of Fine Arts, Boston, the weekend of August 9–10, 2008 (courtesy Ronald J. Will).

18

Wheels, Wheels, Wheels: Chronicles of a Mid-Level Guild Winner

Roger L. Schneider (1957–1962)

It was the mid–20th century and an exciting time to be a kid in love with cars. America had become a post-war nation on wheels, and everyone wanted a car that was longer, wider, more colorful and more unique than the one his neighbor was driving. Detroit automakers and the independents were more than willing to provide the style-hungry buyers with everything they wanted—and more. And every September the latest offerings were unveiled at dealerships around the country. I loved to draw cars, collect new car brochures and see the latest models from Detroit.

Growing up in Crystal Lake, Illinois, a small Midwestern town near Chicago, I developed my own ideas about automobile design. My Dad was a teacher and a very practical man who couldn't be bothered by style and status. For him, the car of choice was always a maroon or green Studebaker. At one point, we owned a green 1952 Champion four-door sedan that was a dealer demo. Studebaker launched one of the most beautiful and revolutionary cars (1953 to 1955) designed by Raymond Loewy called the Studebaker Starliner series. One of the most stylish cars in my neighborhood was a 1954 Mercury Sun Valley featuring a transparent panel at the front portion of the roof. Our next-door neighbor—like my Dad—was always loyal to one brand and every other year or so purchased a new Nash. My favorite had to be his 1954 Nash Statesman, a model in the "Airflyte" series styled by Pinin Farina. In 1953 and 1955 we all marveled at the introduction of the new two-seat American "sports cars," the Corvette and the Thunderbird. My favorite car belonged to friends lucky enough to own the flagship car of the 1955 Studebaker lineup, a two-tone gold and olive colored "President Speedster" hardtop coupe with simulated wire wheels and wide whitewall tires.

My big opportunity to personally express my interest in automobile design came, appropriately enough, in September of 1956. I first became aware of the Fisher Body Craftsman's Guild contest in a *Boys' Life* magazine advertisement. Soon after that, I saw an article in the *Chicago Tribune* about an upcoming display of winning 1:12 scale model cars from the Guild's previous contest that would take place at the huge Marshall Field's department store in the Chicago Loop. I convinced my parents to take me into the city to view the futuristic models of cars designed and built by teenagers like myself.

One look at the table of shining 1:12 scale cars on display and I knew this was something I had to do. Being 13 years old, I was eligible to be a junior division contestant. I completed the necessary forms and before long received my official model design specification sheet and first issue of the *Guildsman* newsletter. I eagerly pored through the articles in the newsletter, underlining tips I thought I could use to help me get started on my first car.

Although the *Guildsman* newsletter offered detailed advice on how to begin the design process using clay, I opted to directly carve my first model car in white pine. My Dad and I visited the local lumber yard to pick out some ¾ inch thick wood that we cut and glued together to form a rectangular block — it was a process we would repeat five more times in years ahead. And then I was ready to go, having diligently sketched my plans for a 2-seater convertible.

1957 Entry (Age 13)

My first model entry at age 13 was a cream-colored convertible with blue accents carved from white pine and painted with a brush. I was taking wood shop and mechanical drawing as a junior high student, and I was fortunate enough to get permission from the shop teacher to make building my first car the major project for the spring semester. That provided access to the many of the shop's hand tools and a specific time to work on the car each day. As the weeks went by, my convertible model began to take shape, and I now had to begin solving a lot of small design details. I referred back to the creative and practical suggestions in the Guildsman newsletter on possible materials for bumpers, headlights, seats and a steering wheel. I quickly decided my next car would have a roof— it was a lot of work detailing the interior of a convertible!

Little by little, I was able to find the necessary materials. The single headlights were made of the open ends of thimbles with clear Plexiglas inserted into them. Aluminum tubing from the hardware store worked well for the bumpers. The wood screws that held the front bumper were described in my "Special Features" sheet attached to the bottom of the car this way: "Screws on front bumper are parking lights." A narrow horizontal slot, with rounded corners, was cut into the body just above the tubular bumper to serve as the grille. Viewing the front end of the model with its headlights sticking out and the slotted grille, the design had a strong resemblance to a frog.

Red and clear Plexiglas was cut, shaped and recessed into the rear fenders for a taillight and backup light combination. Clear acetate was cut and shaped into the wraparound windshield. For the steering wheel, I followed a *Guildsman* suggestion to use the top from a fountain pen with the shirt pocket metal clip bent out perpendicular to form a gear shift lever. A plastic cap from an aspirin bottle was cut out to have 3 spokes and was attached to the pen top with a wood screw to complete the steering wheel assembly. For a final styling touch — suggested by my Mother! — sponges were cut to shape the seats, then covered with fake leopard-skin cloth.

One detail that gave me a lot of trouble was how to handle the wheels and hubcaps. I came up with the simple solution of using flat washers that just fit in the recesses in the Guild-supplied rubber tires for the hubcaps. The long woodscrews that held the wheels in place filled in the small hole left by the washer. Since I had smaller wheel well openings for the rear fenders, I ran into trouble in making the wheels fit the shape. The best I could come up with was just cutting off the tops of the wheels and the washers and fitting them into the openings. I vowed that I would solve that problem a different way on my 1958 model.

Finally it was time for the final sanding, sealing the wood and painting. My school didn't have any paint spraying equipment, so the high gloss enamel cream color and blue paint was applied with a brush. I had a narrow blue stripe on the rear fenders, which was tricky to control by hand. I also used the blue for the dashboard and the inside of the door panels, plus inside the wheel wells. One of the last steps was to do my best in cutting thin grooves into the hood and doors to represent the openings using an X-Acto knife. I quickly realized that for future cars, I should put in the grooves *before* painting. The heads of straight pins served to represent push buttons to open the doors. The bumpers and other brightwork were attached, and the interior was installed. So the details were now complete — crude, but complete!

The school year and the model-building year were almost over, and there was just one more construction project that had to be completed: building a small crate to protect my masterpiece in transit to Detroit. With all the work completed, I gently lowered the car onto the dowel rods that would support it inside the shipping crate, installed the lid, tightened the wood screws, and, finally, attached the official Guild shipping labels. Off I went with my parents to our local Railway Express Agency. Eventually, I received my completion letter and the evaluation of my efforts. I wasn't an award winner, but I felt like one for having followed the project through to completion.

1958 Entry (Age 14)

My 1958 entry was a two-tone, two-door hardtop, again carved from white pine, with a medium blue body and white top. White paint was also used in the carved concave areas on both sides of the model, behind the front bumper and the front wheel well openings.

As a reward for completing my 1957 model car, my Dad bought me a Dremel drill tool set which greatly speeded up the process of doing the rough shaping of my 1958 white pine model. I was also able to use the set to create sufficient recessed wheel well areas in the wood so the semi-exposed wheels could fit up into the body. This eliminated the sawed-off wheel look. I got away from the rather bulbous look of the previous year's convertible and created a car with a more angular look. The front and rear side profiles were V-shaped, and the top view had a fairly deep V-shaped recessed area for hood air intake ports. There were also air intake ports along the roof edge at the windshield line.

I adopted the dual headlight designs that were popular in 1958 production cars and fitted the ½-inch diameter ends of four thimbles with clear Plexiglas that were scored with thin grooves to give the appearance of glass. Small nail heads served as parking lights centered above the headlights. For the taillights and backup lights, I cut triangular pieces of red and clear Plexiglas and positioned them to be visible from the rear and the side.

I purchased various shapes of soft aluminum that were offered in the *Guildsman*, using them to create the bumpers, a front grille insert and "moon disk" hubcaps. The aluminum bar for the grille had 13 drilled holes to simulate air intake ports. The Guild-supplied wheels were re-designed in 1958, to be slightly smaller in outside diameter (2¼ inches, down from 2 7/16 inches) and this stroke of luck suited the proportions of my model. I also painted the raised molded rubber sidewall area white to simulate a thin whitewall. Although I still painted the car with a brush, the overall design and finish were greatly improved, and the final entry brought me a Junior Division Honorable Mention Award for Illinois of $25.

This 2-door hardtop was made by Roger L. Schneider of Crystal Lake, Illinois, while in the 8th grade (age 14). With a 2-tone medium blue and white brushed paint job and black indicated window areas, the model won a 1958 Junior Division Honorable Mention State Award for Illinois ($25). Carved from white pine, the model employed the bright rims from thimbles, along with clear Plexiglas disc inserts for headlights, and made use of various aluminum shapes purchased from the Craftsman's Guild Technical Department (courtesy Roger L. Schneider).

1959 Entry (Age 15)

In 1959, I decided to create another two-door hardtop, this time with a bright yellow body and white and black accents. It had a wedge-shaped black canopy, or "greenhouse." The overall design was far more angular than the previous car, and the sides were more sculpted. The window pillars featured dangerous looking fins that angled and pointed to the rear of the car with red running lights at the base. The fins also served as dual radio antennae. I made the hubcaps using flat metal buttons and created delicate spinners out of bent straight pins. Smaller luminescent red buttons made the centers. And, finally, I had a novel idea of headlights being mounted in long tapered tubes that could be removed to easily change headlight bulbs — not a very practical idea, but one that fit well with the angular overall design of the car.

Toward the end of July, an article appeared in our local newspaper announcing two state award winners from Crystal Lake, Illinois, that year. At age 15, I had won a Junior Division Honorable Mention Award for Illinois of $25 and quickly made arrangements to meet the Senior Division First Place state winner. He was Richard Soderberg, a 21-year-old design student at the American Academy of Art in Chicago who lived just two blocks from my house.

It was a great experience because my older neighbor provided me with a lot of great advice. The first was what I probably would have received if I could have spoken with the contest judges in person. He correctly noted that my 1958 and 1959 entries were extremely slab-sided and really looked like what they were — cars carved out of a big block of wood! The sides of the roof were almost flush with the sides of the doors, resulting in a pretty unpleasant look. He suggested I try to work some curved shapes into my next designs and ensure that the roof area was offset from the sides toward the center of the car. Of course he was right. I was just too close to my project to see what was so obvious!

This two-door hardtop, painted a bright yellow with white and black accents, was made by Roger L. Schneider at age 15 and featured a wedge-shaped black canopy. Headlights were mounted on long black nacelles (envisioned to be removable), and metal fins at the rear of the roof panel served as dual antennae. Note the wide whitewalls and hubcaps made from large brushed-metal buttons with red centers. This is the last year that Roger used a brush to paint his cars. This car brought him a 1959 Junior Division Honorable Mention State Award for Illinois ($25) (courtesy Roger L. Schneider).

Richard also offered a great construction tip that would have made things a lot easier for me when I tried to carve the very thin tapered sections at the front of my yellow car. He suggested using balsa wood to create any long, thin straight or curved shapes. Once these were glued into slots cut into the main wood block, they could be blended into the rest of the car body using filler. After the two types of wood were primed and painted, they would all blend into a continuous surface.

His other terrific advice related to the painting and finishing of the car. I still have my pencil notes on lined note paper suggesting that I use "Martin Senour spray automotive touch up lacquer from the local NAPA parts store, many coats of NuSand primer surfacer and Mac's Dynamite Rubbing Compound." He assured me that if I used these products and did lots of wet-sanding between coats and finished off with the rubbing compound and a good paste wax, I'd have a finish just like a real car. I could hardly wait to get started on my 1960 model and put all of his suggestions to work.

1960 Entry (Age 16)

My 1960 entry was a silver blue, two-door fastback coupe carved from a combination of laminated white pine and balsa wood. This was a solid-top construction with a charcoal gray canopy. I focused on getting a more aerodynamic and fluid look with sweeping curves and more sculpted body surfaces. Extra emphasis went into avoiding the flat sides of my two previous cars and giving more offset surface relief between the roof and the sides of the car. Bright aluminum trim was used laterally to accent the fastback canopy shape and highlight the integral roll bar. Aluminum wheel covers were turned on a lathe and plastic spinner hubcaps from a 1:25 scale AMT plastic model kit were applied for a finishing touch. For the paint job, I used the lacquer from automotive touch-up spray bombs.

During the previous summer, I had been building some AMT plastic scale model cars and realized some of those parts could be used to create nice details on my Guild model. As a senior in high school, I had access to the metal shop and for the first time I was able to turn the cylindrical aluminum stock I purchased from the Guild Technical Department on a metal lathe to create really nice looking wheel covers. I drilled large holes in the center of each aluminum wheel cover to create a recess in which to mount AMT scale hubcaps, with 3-point spinners, from a 1957 Mercury model kit. Since the AMT models were 1:25 scale, these were a perfect size to add delicate details. For the front of the car, I turned aluminum stock into Buick-type "Dagmar" bumpers; for the rear, I bent thin flat aluminum stock to form a very delicate flat loop-shape running the full width of the body. The taillights were mounted inside the loop on the visible body surface. A rectangular license plate holder was used to cover the butted ends of the aluminum loop. AMT scale taillights from a Pontiac Bonneville model made perfect taillights.

The most exciting improvement that year was the car's finish. As my mentor had urged me to do, I made a trip to the local NAPA parts store to check out the various spray cans of touch-up lacquer and pick my dream color. I had no idea there could be so many cans of spray paint on one wall display rack—all in different colors! I finally decided on a really great 1958 Chevrolet color called "Silver Blue." For the large roof area, I picked a spray can of a charcoal gray color. Now began the process of sanding, filling and priming as well as spraying light gray primer. Eventually, I was ready to apply my final coats of "Silver Blue" paint and use that legendary "Mac's Dynamite Rubbing Compound." The final results were well worth the many hours of work. Unfortunately, I again fought and lost the battle to create deep, uniform grooves representing the hood, trunk, door, and headlight openings. Although I had cut all the grooves beforehand, I had to apply more coats of spray paint than anticipated to get the surface and color right, and many of the groove shapes filled in almost completely.

Roger L. Schneider's 1960 Guild entry was a 2-door fastback coupe painted silver blue with a charcoal gray canopy. This design left behind the slab-sided look of his previous Guild models. With more aerodynamic and fluid styling, balsa wood was glued into the white pine body of the car to provide the thin, delicate curved fender shapes. The black canopy was accented by a highly polished aluminum roll bar trim. For the first time, he made hubcaps from turned aluminum and added plastic model kit hubcap centers. New also was the use of spray can–applied, auto touch-up lacquer for a far superior finish to that of his previous models. The silver blue lacquer Roger selected for this model was a 1958 Chevrolet production color. This model won the 1960 Junior Division First State Award for Illinois ($150) (courtesy Roger L. Schneider).

With a much-improved overall design, lots of nice details and a paint finish I was really proud of, I could hardly wait to see how the entry would do in competition. Even though I got a late start due to a gym class injury, I thought this effort might be a winner. Finally a letter arrived announcing that I had won the Junior Division First State Award for Illinois and a prize of $150.00. I received a trophy for myself as well as one for my school.

1961 Entry (Age 17)

The year 1961 would be a more challenging one. At age 17, I moved into the senior age division, where the competition would be much more demanding. Overall, that year's car was probably my best entry, both in terms of design and craftsmanship. It was a two-door sports coupe with a very aerodynamic, low profile and good proportions. Painted a deep blue, it featured a molded black Plexiglas canopy and turned aluminum wheel covers with plastic AMT model kit spinner inserts.

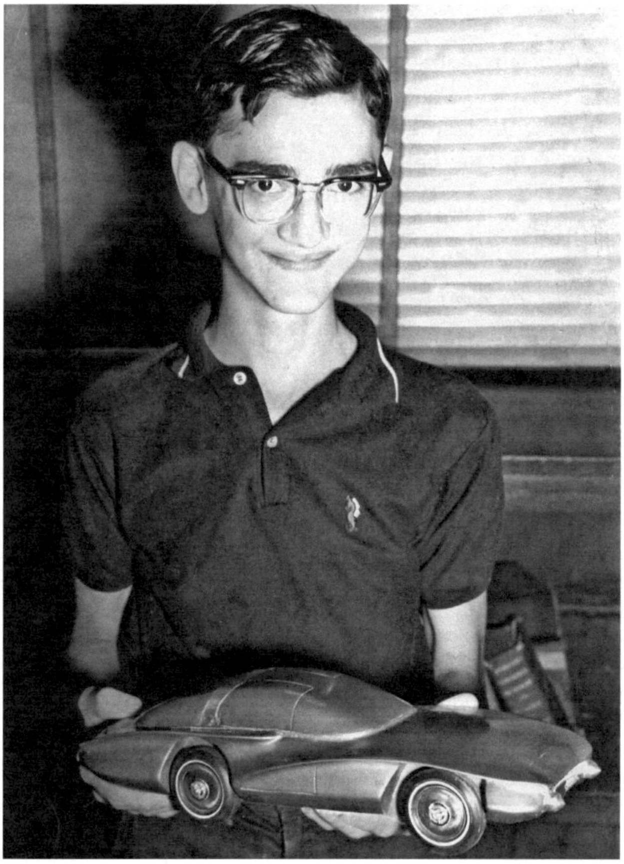

A photograph of Roger L. Schneider (age 16) which appeared in the *Crystal Lake Herald* on August 4, 1960, along with an article announcing that he had just received the 1960 Junior Division First State Award for Illinois ($150) in the Fisher Body Craftsman's Guild model car design contest (courtesy Mr. Dan McCaleb, *Northwest Herald*).

I employed all the ideas I had learned the previous year to create a spectacular looking roof by heating and molding a sheet of thin black Plexiglas into the desired shape. It took quite a few tries using our basement oven, but I finally got the right shape. The amazing gloss of the Plexiglas made the car glow. I described the functioning of this one-piece roof as pivoting at the rear to lift up for access to the interior through the wide doors.

All the construction steps went smoothly, and the car looked fantastic as I lowered it into the shipping crate that had already been used four previous times. I could now only try to wait patiently to see how the judging turned out. The news was a little disappointing. I was the recipient of the Senior Division Third State Award for Illinois of $50. The model car had earned 285 out of 450 total points, or 63.3 percent of available points, and I was very happy with this result considering the much stiffer competition among the older contestants. But, it was still a bit of a letdown from my previous year's first place.

1962 Entry (Age 18)

This would be my senior year in high school, and I wasn't sure there would be enough time to build a winning car. Finally, I decided to give it one more try. It was a very challenging school year, and some evenings and weekends were spent working at the local Walgreens drugstore to earn college money. Spare time for building my model was getting tough to find. As the end of the school year arrived, I realized I should have decided against building another car at this busy time. Although I was able to complete my entry, the painting became very rushed. By the time the car arrived in Detroit, there had been a major paint failure, and the car obviously wasn't the level of quality to be considered competitive.

It was truly a case of what might have been. Sadly, although this final car received my very best ever "Design Category" scores, these were completely negated by the "Craftsmanship Category" scores, which obviously had to be very low. At least I was able to take some measure of pride in having seen this project through to completion one last time.

During the previous spring semester, I had a wonderful opportunity to take a series of Saturday design classes at the Institute of Design at Illinois Institute of Technology in Chicago. That experience convinced me that it was the school I wanted to attend. One of my dreams came true when I received my acceptance letter from the Institute of Design to start as a full time industrial design student the next year. My moderate success in the Fisher Body Craftsman's Guild had encouraged me to start working toward a career in design.

An interesting connection with the Fisher Body Craftsman's Guild contest occurred in my junior year at the institute. A new student from Indiana started in the industrial design program that year as a sophomore. He was none other than Ronald J. Will, the 1961 First National Senior Scholarship winner. Ron and I immediately began a friendship that has continued to this day. While at design school, I also made friends with two junior year students, both Guildsmen from Indiana: Brian Skogler, a 1958 Regional winner, and Allan Lee Flowers, who took the Fourth National Senior Scholarship in 1962. Early in their design careers they had worked as automobile designers for Ford and GM, respectively.

I completed my bachelor's degree in industrial design in 1966 and spent four years designing railroad and mass transit control equipment at Westinghouse Air Brake Company in Pittsburgh, Pennsylvania. During those next four years, I had the opportunity to work with engineers and manufacturing specialists to design a number of brake control systems (locomotives and freight cars). The most exciting projects involved the design of prototype cab control units to be used for mass transit systems nationwide.

Then, I took a daring leap and decided to utilize my photography and graphic design interests, my problem-solving training, and my budding love for the city of Pittsburgh in a massive project. After gathering together a group of other young men, I founded a new cultural magazine, which was called *Renaissance: Pittsburgh's Showcase Magazine*. After a very difficult period of 18 months, I and a few others received an offer to go to work for WQED-TV and continue working on the magazine, now called *Pittsburgh Magazine*. For the next eight years, I was involved in a variety of printing and publication projects for the station, which had become a national production center for the entire Public Broadcasting Service.

Following my years in public broadcasting, I had the opportunity to work initially in advertising, then as the organizer of a print department. Currently, I serve as the plant manager for American Thermoplastic Company, a nationally recognized manufacturer of custom-printed loose-leaf products for businesses and college bookstores around the country. The company is located in suburban Pittsburgh and utilizes the screen-printing process, as well as digital printing on vinyl and other plastic materials, to produce high-quality graphics on custom-made products. In 2010, I celebrated my 32nd year at the company.

Although I never became an automobile stylist or designer, I have enjoyed a very interesting and challenging career in design, publishing and manufacturing. Still, my interest in cars and car design has never diminished. One of my hobbies in my younger days was rebuilding engines and painting my own cars. I have been an avid automotive literature collector for more than two decades, and I have owned and enjoyed several collector automobiles, namely an MGB/GT and a 1965 Porsche 911.

I was fortunate to marry my wife of 40 years, Vicki, who is also an avid automobile enthusiast. We spent many years in the local Porsche Club of America planning rally and show events and working together as regional newsletter editors. Since I sold the 1965 Porsche 911 (one of the earliest 911s registered in America), our lives have involved far less exciting models from Volkswagen, Audi and Subaru — but we still enjoy our cars. Although I don't work on cars much anymore, each year when I hear the sound of the locusts, I still find myself drawn to the car showrooms to see the latest models now created by car designers and stylists worldwide.

Roger L. Schneider (age 65) photographed in the fall of 2009 with his 1960 Junior Division First State Award for Illinois ($150) winning model car in-hand. This was the highest award he received while competing in the Guild 1957 to 1962. Following graduation in 1966 from the Institute of Design at the Illinois Institute of Technology, he worked first as an industrial designer and later as a graphics and print manager (courtesy Roger L. Schneider).

19

From Windshields to Sugar Scoops: A Guildsman Makes the Grade

Thomas L. Covert (1961–1962)

Growing up in Michigan City, Indiana, in the 1940s and 1950s I discovered at an early age that I had a natural affinity for art and creativity. This turned into a lucrative hobby of painting sand dune scenes, harbors, and fisheries, as my hometown was situated right at the southern tip of Lake Michigan. It was no surprise, as I came from a creative and talented family. My dad was a craftsman, carpenter, plumber, and woodcarver, while my mother sewed many of our clothes, enjoyed knitting and made award winning quilts.

Tom Covert (age 17–18) sketching cars at a motel in Warren, Michigan, while on a 1960 trip to view the model cars submitted to the Craftsman's Guild competition. Ron Will's father had driven Larry Pecorella, Tom Covert, Ron Will and Leland Will to Warren, so they could view the field of 1960 Guild model entries (courtesy Ronald J. Will).

Left to right are Larry Pecorella, Tom Covert, Ron Will and Leland Will, who traveled to Warren, Michigan, to view the 1960 Fisher Body Craftsman's Guild model entries submitted from all over the country. Knowing what the competition was about became the first step toward producing a national caliber winning design. Later, they found out that Ron Will had won the 1960 Senior Division First State Award for Indiana ($150) and Regional Award in the competition (courtesy Ronald J. Will).

We were also a musical family, with my Mother playing the violin, my older brother taking up cello, and my younger sister singing and playing the piano. So to be part of this ensemble, I took violin lessons. I thought I must have been very good because my instructor would cry when I played. The truth emerged one day when he suggested that a career in art might be my true calling. So crafting and design became my hobby.

Among my circle of art friends was my high school buddy Ron Will. Although Ron moved a few miles south during his junior year, we kept in close touch and even sketched car designs when our football teams played against each other. Ron was always designing and building something and had been a local winner in Chevrolet's All-American Soap Box Derby Competition. In addition, he entered a 1:12 scale model car of his own design and construction in the 1959 Fisher Body Craftsman Guild Model Car Competition and won a Senior Division Honorable Mention Award for Indiana and $25. I figured that General Motors' Fisher Body design contest was for people with special talents, but Ron said "No, it's for people like you and me that possess perseverance." In 1962 I would learn first-hand exactly what he meant by that word "perseverance" and how my whole family would be tested. I sent for the Guild information and was soon hooked on the idea. I was determined to build a national award winning model car.

I had a chance to drive with Ron and his father to Warren, Michigan, to see the display of 1960 Guild competition entries from across the country. Ron had entered that year, and we wanted to survey the competition. We arrived at Fisher Body Division headquarters on Van Dyke Avenue, which was next to the GM Tech Center, while the design judges were in the process of scoring the models for design and craftsmanship. There was a football field full of models, with rows of tables covered with model cars of every shape, size, and color imaginable. After strolling along the aisles and viewing the cars, we observed that out of several thousand entries there might have been a few dozen that qualified as national contenders. The competition was really stiff at the top. And the best news of all soon came: Ron's 1960 model car had won a Senior Division First State Award for Indiana worth $150.

Pat Saturday, another friend from Michigan City, also became interested in the Guild. So now we had three friends competing against each other. Ron and Pat were determined to build national award winners. I knew I did not have an advantage over Ron because of his past experience and that it would be a tough learning curve for me.

1961 Entry (Ages 17–18)

A clay master model of the 1961 design was made on a wood armature with plasticine clay and transferred to the kiln dried poplar wood using cardboard templates. I was careful to repeatedly check scale fidelity against full-size car parts and accessories in order to achieve a realistic automobile look (rather than toy model look). Maximum and minimum vehicle and body dimensions set by the Guild rules had to be checked and re-checked.

My 1961, two-door sports coupe was inspired by the aircraft industry, as were many production automobiles in the late 1950s and early 1960s. The high ground clearance, denoted as door sills at or near wheel centerlines, and front ramp angle were to look like an aircraft in lift-off mode, and the dual canopies were reminiscent of aircraft of the time. For example, GM's Firebird III featured dual jet aircraft-style canopies. (I always thought dual canopies would be good for feuding couples — they could be separated while driving and unable to speak without using an intercom.) There were concave coves behind the front wheel well openings as well as in front of the rear wheel well openings.

The front end had a fighter jet-type air intake with no grille, and a bar of polished aluminum protected four high intensity driving lamps made of cut-off Plexiglas brand rods shaped and polished with rubbing compound. On both sides of this opening were bullet shaped turn signal and running lamps mounted on a shock absorber-type system for protection. The rear of the model had a split bumper made of polished aluminum with two jelly bean shaped tail lamps made of red and white Plexiglas for the turn, and a backup and running lamps system mounted below the bumpers.

The body of the model was painted with 15 coats of a DuPont cherry red lacquer over primered wood, and the dual canopies were made glossy black. This was done with compressed air cans and a mixing bottle for the lacquer and lacquer thinner, a hobby painting system that worked very well. The wheels were attached with a single wood screw after leveling the model to the proper attitude. Of course, I had to have whitewall tires, so I went one better and painted a pair of thin dual white stripes. Special features I included were air conditioning, rear view cameras, and a hydraulic system that opened the doors and canopies at the touch of a single button — all theoretical automotive features.

I quickly learned the importance of communication in order to successfully complete the various model construction tasks. I sent a blank of quarter-inch thick brass to a local

Tom Covert (age 17–18) holds the finished clay model for his 1961 Guild entry. The master clay model of the two-door sports coupe was transferred to kiln dried poplar, which was painted a DuPont cherry red with a pair of glossy black canopies. The aircraft industry of the time had a great influence on automobile design and this model incorporated aircraft type canopies, high door sills and a high front end to look like a jet fighter about to lift off. The design also employed concave coves behind the front wheel well openings and in front of the rear wheel well openings. The finished model entry won the 1961 Senior Division Third State Award for Indiana ($50) (courtesy Ronald J. Will).

machine shop from which they were to cut-out the brass bumpers for my 1961 model with the outlined shape of each bumper etched onto the brass. More than a week passed, so I asked my friend, "How is it going?" and he said he was still working on them. Finally, he had the project ready. I looked at the brass blank with a stunned look. Instead of the two individual brass bumper parts requested, I had a brass sheet with two bumper shaped holes. I thanked him and asked if by any chance he had saved the parts he cut out. No, he hadn't, he said, adding that it was a difficult job because he had to cut and weld his band saw blade three times. He had done exactly what I asked — he "cut out" my bumpers.

We "three amigos" swept Indiana: Ron won first place, Pat second, and I placed third in the 1961 Senior Division competition. Ron went on to win the 1961 First National Senior Scholarship Award of $5,000, and Pat won a 1961 National Styling Scholarship Award of

$1,000. These guys knew the secrets to winning a national scholarship and would be my consultants for the following year of 1962. That year, Ron and Pat were no longer eligible to compete because they had already won university trust funds. The coast was clear for me to try again to win a national scholarship.

1962 Entry (Ages 18–19)

My second car would be easier, I thought, because "knowing what you were doing" was half the battle. I considered doing an asymmetrical design, but I didn't want to copy Ron's 1961 national award-winning model car design, as would many Guildsmen. Knowing that the model car entries would be reviewed from above by the judges, an unusual plan view, or outline, was one of my first design considerations. Convertibles were often scored higher because of the detailed craftsmanship involved with the interior. So I set out to make a symmetrical, two-passenger convertible with a wedge-shaped plan view. This two-seater sports car was carved from kiln-dried poplar and coated with a polyester wood sealer. I painted it with light metallic blue lacquer called Wedgewood Mist and then trimmed it with 25 small rhodium-plated brass parts for a brilliant shiny appearance.

The interior instrument cluster was designed full-size then photo reduced to the proper scale of 1:12. The stainless steel trim surrounding the instruments was made at the custom kitchen plant where I worked full-time. To complete the back of the open windshield and canopy, a soft top would theoretically be stored behind the seats and would attach to the roll bar. The entire interior assembly was completed on one platform then loaded vertically through a rectangular opening in the bottom of the model.

I discovered that the head from a Schick electric shaver provided a perfect scale grille. This was the only found object that I used on either the 1961 or 1962 model. The taillight system consisted of a red neon bar made of polished red Plexiglas with theoretical sequential turn signals suspended from four brass rhodium-plated vertical supports. The wheel covers were made from a one and a quarter-inch diameter aluminum rod on my borrowed watchmaker's lathe. The wheels were attached with two wood screws each.

There were a number of special features I had thought about, and I conveyed these ideas to the Guild judges on a card attached to the bottom of the model. The occupant seats were heated and included built-in lumbar vibrator pads. There was a movable instrument pod and pedals as well as one lever for the driver to control steering, speed and braking. A center console housed the radio, air conditioning and seat temperature controls, while the headlamp system consisted of two high-intensity lights separated by a neon light bar for city driving. A white translucent shade would cover half the driving light and blend with the neon bar, although the driving lights would be fully exposed for highway use.

I was attending Purdue University's Extension Center in Michigan City and working full-time at Dwyer Kitchens as a blueprint operator and apprentice draftsman. Knowing the number of hours to build a winning car, I dropped out of college for the winter semester because something had to give. I would come home to eat dinner and then go straight to the basement to work until around midnight. My friends thought I had gone away to college. I set up a tight schedule in order to complete the model on time. The Guild executives must have been a little sadistic, as the model car competition deadlines always seemed to be in June — just when there were final exams, commencement exercises and graduation parties — but then I realized the judging would continue until August.

My 1961 and 1962 models were made from laminated, kiln dried poplar. My father's friend at church worked in a lumber yard, and I asked him where I could get some poplar wood. He said to come on over to the lumber yard. When I did, the salesman went in his office, up into an attic space, and came back with a look on his face like he found gold. In his hands was a four foot piece of clear, kiln dried poplar wood, a wonderful carving wood that sanded down to a beautiful finish. I transferred measurements and templates from the clay model to the poplar as I carved, and soon the 1962 body was complete.

Painting without a compressor was now my next challenge, as I found out doing my first car. Another friend at the local body shop showed me a touch-up kit that used a compressed air can with a paint bottle attachment. It allowed me to mix my own lacquers. At the local auto supply store, the salesman wondered if I needed two or three cans of compressed air — I told him I needed about "two cases" of air cans.

The cans did work great except they would freeze up when used constantly. The basement became my Rube Goldberg paint shop. A hot water bucket was used to thaw the frozen cans, damp sheets were used to keep the dust down, and a dehumidifier was used to keep the humidity down while a small heater was used to keep the temperature up. I did not use a respirator while working in the small, closed room — maybe the reason I think and act the way I do today! I applied a polyester resin wood sealer to boost the paint's gloss, put on twenty or so coats of color lacquer over five coats of primer, and then wet sanded between each color coat.

I had a distinct advantage making the small parts because I had worked at Blackmond's Jewelry Store. The watchmaker, Marshall Hanson, loaned me some equipment — a lathe and polishing tools — and gave me the know-how for using different kinds of rouge and buffing compounds for polishing. Before being chrome-plated, brass trim pieces had to be buffed and polished to a fine finish.

The small model parts such as wheel covers, knobs, interior parts, and headlight rings were turned on a lathe. The bumpers were cut on a metal cutting band saw belonging to a neighbor. The roll bar created one of the most impossible tasks: the Plexiglas windshield had to insert into and follow the contour of the roll bar, which my father said couldn't be done. Of course, this set me up to show him that I could do it. A poplar male mold of the windshield gave me a form with which to cut and fit the canopy trim parts made from brass. Using a ¼ inch wide strip of brass and a custom-made file, I cut a receiving groove, bent it to a rough shape and filed away everything that was not roll bar. Some fifteen to twenty hours later my masterpiece was complete and ready for final buffing.

Tom Covert (age 19) at his drafting table working on the final assembly of his 1962 model entry, which won the 1962 Senior Division First State Award for Indiana, Regional Award and Second National Scholarship Award worth $4,000. He would come home from work, eat dinner and go straight to his basement workshop until midnight. His friends thought he'd gone away to college, he was so consumed in building this model (courtesy Thomas L. Covert).

Beaming with pride, I showed my amazed father, who said he knew I could do it all along. Then I began with the high speed buff with a support board underneath. I am not sure what happened but the tip of the roll bar caught in the nap of the cloth buffer — ga-zing! — and in a split second my roll bar was all but destroyed. It had turned into a corkscrew. My knees got weak, and I had to sit down because I couldn't believe what I had done. I wanted to cry, but it was not in my schedule. I immediately started to bend the roll bar carefully back into shape. All was repaired except one small tear, so I took it to Marshall and he silver soldered the small flaw.

Now that the twenty-five or so small brass parts were completed, I was ready for chrome plating. There were no chrome platers in Michigan City so I went to a place recommended in Gary, Indiana, carrying the parts in a small custom jewelry box. I had this sinking feeling as I pulled up in front of this large factory that was obviously a heavy-duty plater. I went into the office and out of the back came a man who looked like the cook in the comic strip "Beetle Bailey." He said, "What chew want?" I showed him my brass car parts, and I thought he would never stop laughing. I said "What's so funny?" He said, "Foller me." He opened the door to this deafening roar of large machinery. They were doing some type of heavy duty bumper plating. He said, "See that there conveyor belt? It runs twenty four hours, seven days a week, and there is two to two and a half feet of sludge in the bottom of the tank [which was knee deep] which is three hundred feet long [length of a football field]." He asked me, "Where do you want me to put your parts?" Pointing to my little jewelry box, I kindly thanked him and went home.

With time running out, I again called on Mr. Hanson. He said he knew someone who did rhodium plated brass parts, a watchmaker named Joe. Again, I hoped to get back on schedule. I met Joe on a Saturday, and we began the plating process. All went well until we got to the bumpers. They were too large for the current flow in the tank. Spots were being left behind on the bumper where the inadequate current flow occurred. Multiple buffing and plating were now required to get an acceptable finish.

Plating completed, I headed home in triumph to fit the parts to the body of the car. To my horror, the plated bumper no longer fit to the body of the car. Repeated buffing had reduced the size and left about a three millimeter gap between the body and bumper. It was time for crisis management. I could not start over. I studied the situation and decided to make a gasket between the body and the bumper. I used some kind of black caulking to fill the gap and I was back in business. It didn't look too bad. Later, one of the Guild judges told me they liked the protective gasket for the body and gave me extra points. I said "Yes" as if I had purposely planned it that way.

The windshield, more appropriately called the canopy because it swept up and over the two-passenger occupant compartment without enclosing the rear, taught me the difference between clear Plexiglas and clear vinyl plastic. My grandfather, Don Laudeman, Sr., lived in Elkhart, Indiana, close to South Bend. He said he could help me by picking up the Plexiglas in South Bend.

I had made a male mold of the canopy from poplar covered with cotton cloth, a method also used to make the brass canopy trim. The Plexiglas was to be pulled and stretched by hand over this male mold — a four hand process with my dad supplying the other two. With only a male mold, the Plexiglas was heated in the oven until it was as soft as leather. Then, it was immediately placed on the mold and pressed down in every direction. It worked fine. I took our first experimental canopy part to try and buff out the scratches and scuff marks and to see how it finished. It buffed like chewing gum! The man in South Bend had sold Grandpa vinyl instead of Plexiglas, or "plex."

The next morning I called the South Bend store very upset. The salesman thought that he had sold us Plexiglas but "no problem," he would order some "plex" from Chicago that would arrive in one week. I explained I was down to forty-eight hours! It was crisis management time again. Dad and I decided that the car was going to have a vinyl canopy. The next problem was that dust or foreign objects would become imbedded or permanently affixed to the vinyl. The only dust free place we knew of was the freezer. The sheet stock would go from the freezer to the oven for forming. Some thirty to thirty-five pieces later, we decided to call it quits. We started grading for the fewest flaws and dust. We had windshield canopies everywhere. Finally, after grading from "unacceptable" to "AA," we had a winner, a "perfect" part. Actually, they would have made dandy sugar scoops.

To my amazement, the roll bar fit perfectly, but gluing it cleanly to the body was a real challenge. The other parts went one by one into place. The last two nights were all-nighters for me, with the last night's "push" a family team effort. My mother sewed a cotton cover, my father lined the shipping box with foam, and my big brother typed the legal paperwork needed for shipping. All was going well until about 4:30 in the morning when the final crisis erupted.

The epoxy holding the windshield was not setting up. This was the dark ages of epoxy technology, and it looked like it might harden sometime in July. We had become pros at dealing with Guild crisis after crisis. The determination was made that this model car was not going to be shipped by mail. It would be hand-delivered. Dad called work and took the day off. My mother, father, and brother loaded into our 1961 VW Beetle, and we headed for Detroit with my brother holding the 1962 model waiting for the epoxy to dry. The "stuff finally went off" with the catalyst and resin finally solidifying just as we arrived in Warren, Michigan, with an hour or so to spare.

Now the waiting began. A Guild telegram finally arrived, and I learned that I had tied with Allan Flowers of Ft. Wayne, Indiana, for the Senior Division First State Award for Indiana of $150. I was more than upset as I figured there was no way a "tied" state winner could compete successfully for a national award. A few weeks later, however, another Fisher Body telegram arrived informing me that Allan and I had both won the regional awards. I was going to Detroit!

Two General Motors designers met me at the train station. I was impressed! The convention was a marvelous, first-class treatment with lots of activities. We visited GM Styling with its Divisional Design Studios, test track and Styling Dome. We also saw a Detroit Tigers baseball game, attended a Shakespearean Festival in Canada, toured a GM assembly plant, and cruised around the Milford Proving Grounds' high speed test track. Probably the highlight of the week was seeing the proving grounds. We were in buses on the test track just as a test was underway. Someone yelled behind us. A 1963 split window "Vette" prototype passed us at lightning speed. The Guild tour planners had it right because the Corvette passed us on the upside in the high speed turn. Had it been on the downside, the bus probably would have rolled over as everyone shifted for a better view.

The culmination of the week was the awards banquet with an auditorium full of GM executives, designers, and associates. The entertainment was provided by Jerry Murad and his Harmonicats, who recorded with Columbia Records. We all marched into the Fisher Body Headquarters Auditorium proudly carrying our models and placed them on a long, elevated dining table with seating for 40 regional winners. The models would be the centerpieces of the table setting. We were served a fabulous meal that I still don't know how I kept down.

Then the awards began. The winners were announced by a picture of their car on a full-sized movie screen. Then the names were announced. Nervousness was a total understatement. It was all or nothing. I was 20 years old, and this 1962 competition was my last year of eligibility. Junior Division and styling awards were announced. Then they announced the Senior Division awards starting with the fourth place $2,000 prize. My Ft. Wayne, Indiana, nemesis, Allan L. Flowers, won the Fourth National Senior Scholarship Award. Harold H. Kram from Clayton, Missouri, whose father was a GM employee, also tied for fourth place and won a scholarship worth $2,000.

I thought I was sunk. Kaizo Oto from Fresno, California, took the Third National Senior Scholarship Award worth $3,000, and I was just settling into accepting my defeat when I looked up and saw my car was on the screen. I was numb. The guy next to me poked me; I was baffled until he told me, "Go get your money." I gathered my wits and went to receive my plaque for the 1962 Second National Senior Scholarship Award worth $4,000. Roy R. Dickey, from Kansas City, Missouri, and a student from the Art Center College of Design, took the First National Senior Scholarship Award worth $5,000.

Tom Covert with his 1962 Second National Senior Scholarship ($4,000) award winning model car. Guild judges were curious as to whom he knew at GM Styling, as the rear end design of his model looked very similar to the 1962 Buick Electra being released that fall. It was a coincidence of parallel design thinking, as no contacts or exchange of information had occurred (courtesy Thomas L. Covert).

My life had just taken a hard right turn. With funding for college in place, I completed academic studies at Purdue University and graduated from Art Center College of Design in February 1968 with a bachelor's degree in industrial design. After receiving offers of employment from each of the Big Three — General Motors, Chrysler, and Ford — I accepted a position with GM Styling, later known as General Motors Design until the carmaker's bankruptcy in 2009.

In April, I was getting settled as an intern designer when I received my army draft notice. Fortunately, I was able join a design group called the U.S. Army Exhibit Unit in Alexandria, Virginia. My job was to design traveling exhibits. My exhibit was "The Army's Basic Training" and I toured the country with it. After my tour of duty, I returned to GM Styling to resume my career. It was like one big family of Guildsmen; Chuck Jordan was an assistant to Bill Mitchell, who was then the head of GM Styling. There were Ron Hill, Bill Scott, Ron Will, Allan Flowers, Tom Semple, Terry Henline, and Elia "Russ" Russinoff— all direct descendants of the Fisher Body Craftsman's Guild. We came to work with a passion for cars that went back to our childhood.

Of my 38 years at GM, 13 exciting years were spent in the Pontiac production studio, most of which were under Terry Henline as studio chief. Later, as the Hummer, or H2, was becoming a hot design product, Terry asked me if I would team with him as lead designer on the production version. We both took on the sport utility vehicle (SUV); Terry later retired, and I completed the SUT, or sport utility truck, pick-up version.

In May 2005, I retired from GM Design in Warren, Michigan, where it all began. Ron Will had left GM to become the product design and planning manager at Subaru, while Allan Flowers had moved on to help start Nissan Design International in San Diego along with GM designer and Guildsman Tom Semple and GM designer and Guild judge Jerry Hirshberg. Kaizo Oto became a Raymond Loewy disciple working on safety cars at one point in his career, and Pat Saturday opened a printing business called the Whistle Stop.

I met my wife, Margaret, when she worked a summer job in the GM Styling cafeteria. She is now retired after 32 years of teaching. We have two wonderful and talented children. Our son, Adam, works in construction management while our daughter, Andrea, is an elementary teacher and mother — in both of whom I have successfully instilled the "curse of the perfectionist." Our son-in-law, Tim Stafford, is an accountant and talented woodworker. They have a beautiful daughter, Elizabeth and son, William. The Guild was truly an opportunity of a lifetime for me.

20

Like Father, Like Son: Two Creative, Inventive Spirits[1]

Paul F. Bonfilio (1958–1962)

My father had a passion for the more technical aspects of the automobile and followed his dream by patenting a revolutionary new automobile engine.[2] Years later, I had my own ideas for an innovative plastic automobile chassis which I patented.[3] At one point (1984), Chrysler's chief executive officer, Lee Iacocca, invited me to Detroit to meet with his engineers to discuss these ideas.[4] Regrettably, it never progressed beyond that point. Despite this, my father's and my automotive dreams have never died.

When I was 4 years old, my father built a plaster scale model automobile that he hoped would utilize his revolutionary engine. The quality of that model and its design was certainly worthy of submission to any Fisher Body Craftsman's Guild model car contest except that he was 38 years old and its scale was a little larger than 1:12. I couldn't help but be inspired by my inventor-father when I finally got involved in the Guild program.

Although I eventually became an architect and New York City urban planner, I always seemed to have a close affinity for automobiles during my early formative years and adolescence. In addition to naming the various makes and models I passed on the road, I drew cars in my school notebooks, and later on sketch pads, and I even named my car designs — such as Python, Panther, Cougar, Falcon or Puma.

I heard about General Motors' Fisher Body Craftsman's Guild at age 13 and sent for the competition instruction package: the choices of body styles, awards and prizes, the drawing specification sheets with overall dimensions, the how-to booklet, and a copy of the *Guildsman*, a quarterly newsletter devoted to the Guild program. The *Guildsman* helped me generate design and styling ideas as it often had pictures of the models made by previous Senior and Junior National Scholarship winners along with their success stories.

For the next five years, from 1958 to 1962, I produced a total of six plaster cast models, progressing up the ladder, from a "no award" junior division rookie contestant that first year to a First State and Regional Award winner five years later, and my first and only trip to Detroit and the Guild's annual awards banquet. To do this, I had two stalwart mentors, my father and my Uncle Frank.

The "workshop" was a corner of the basement with a table, an overhead light, and a slab of marble to work on. My tools were whatever was available from my father's own work area, and a coffee can full of clay sculpting tools given to me by Uncle "Frank" Fortunato, my mother's oldest brother. Uncle Frank was very skilled with his hands and had been a clay modeler for Chrysler in the 1930s; he later worked for various well-known industrial designers and model-makers such as Henry Dreyfuss, Norman Bel Geddes, George Nelson, and Raymond Loewy. Eventually, he worked on some of the terra cotta decorations of many New York buildings.

As skilled as he was, Uncle Frank shared his techniques and secrets with us very reluctantly. However, he did teach my father and me about five-piece plaster molds, and a release agent he used comprised of ground up soapstone and kerosene. Perhaps this concoction was one of his proprietary secrets. This mixture was made into a syrupy liquid and brushed into a plaster mold. When plaster was poured into the mold and hardened, the soapstone mixture allowed the hardened plaster to separate from itself. It took some practice to get the right proportions and there was always the fear that when I would use this release agent that I would end up with a solid chunk of plaster, mold and model, or broken mold sections. But miraculously, it worked every time. What actually occurred was that the kerosene was the medium to distribute a thin layer of soapstone on the mold, which would be left there as the kerosene evaporated.

I chose to design all my models in clay because it was the most forgiving. I could not carve wood that easily, and I feared either carving too much or too little. Clay allowed addition or subtraction and lent itself obviously to compound curves. My uncle taught me to work on a flat piece of marble, which lent itself to making clay models and castings in plaster. The plaster does not stick to the marble, and it allows molds with flat surfaces. I followed the Guild handbook and my uncle's guidance for making clay models utilizing a wood buck onto which you add clay and shape your design.

Two 1958 Entries (Ages 13–14), Junior Division

I naively decided to build two cars for the contest: a sports car and a four-door sedan. For many young people, a sports car was the preferred design to make — after all, what 13 or 14 year old kid didn't lust for the new Corvette! But I also liked four-door sedans and wanted to design one of those, too. My sports car was to be an expression of ideas related to streamlining. Although many innovative car designs were done in the 1930s and 1940s utilizing streamlining, I was also intrigued with the "teardrop" approach to form.

1958: Model No. 1

My first two-door sports coupe was a "cab forward" design with a rear-mounted engine and intake scoops just behind the rear window. Constructed from plaster, it had no grille in the front, just a rounded form. The rear intake scoops were separated by a split rear window which traveled over the rear deck. The split transformed into a mild centered vertical fin. There was also a pair of rear canted and slanted fender fins. An oval-shaped rear grille visually linked the three fins. Above each door, in the roof, was a fixed glass "courtesy window" which would theoretically bring in additional natural light into the cabin. The model was painted red enamel from a spray can with glossy black paint for the indicated windshield and window areas. A simple side molding made of wire traveled up from the bottom of the door towards the rear. The front and rear bumpers were built from the clay master, molten lead was poured into the plaster molds taken off the clay, and the lead castings were painted silver.

1958: Model No. 2

My second plaster model that year was a fairly conservative four-door sedan with very rounded front and rear ends. The front of the model had an oval grille, and the rear, once again, had canted fins which were shaped into a concavity above the rear fender. It also had upper greenhouse "courtesy windows." This model was painted copper with cream colored canted fins in rear fender "coves." The indicated window areas were painted black. Similar to 1958 Model No. 1, the front and rear bumpers were cast from lead and painted silver.

Fortunately, our next door neighbor, Mr. Frank Lane, was an executive with the Mayflower Moving Company. For each of the five years I competed in the Guild, he professionally packed my fragile plaster models in straw and folded paper. The models arrived and returned from Detroit via Railway Express without damage.

Front view of the 1958 Guild entries made by Paul Bonfilio. Shortly after his 1958 submissions were completed, this picture was taken of Paul at age 14, with his sister, Clelia, and his father, Roberto. Roberto was very proud of his Guild entries and was always supportive of Paul's efforts. They worked nearly side by side in the cellar in the evenings and on weekends; Roberto worked on his automobile engine, and Paul worked on his Guild models. Roberto learned about plaster mold making and modeling as Paul did, from Uncle Frank (courtesy Paul Bonfilio).

Despite my nine months of sustained effort in 1958 and designs I thought were innovative, neither model won any award, not even an Honorable Mention. The reality of the Fisher Body Craftsman's Guild competition soon took hold: the difficulty, the challenge, and, indeed, the important lessons to be learned. I vowed to myself to do better next year and to strengthen my weaknesses.

1959 Entry (Ages 14–15), Junior Division

This was going to be the year to make an all-out effort to win an award. I decided to concentrate on one model and make that the best entry possible. I realized that most of the winners were convertible sports cars with detailed interiors and clear plastic windshields. Therefore, this next entry had to be a two-door sports convertible constructed from plaster. I was still enamored of the "cab forward" idea with a rear or mid-engine layout, and essentially I designed an open convertible version of my first 1958 entry. The front of the model was similarly rounded in form like the 1958 models and the rear, without fins, had a large flattened rear deck which split at the back into two horizontal taillights. It was reminiscent of the rear of the new Edsel, but better proportioned.

I chose a black body color with a white interior for this model. The paint finish on the prior entries was not the best. Not having an air compressor for spray paint, I discovered that my mother's Kirby upright vacuum cleaner had a reverse air flow feature, enabling attachment of a spray gun. The paint finish was much better on the 1959 model, since I used lacquer with my new found spray equipment, and this was subsequently wet sanded between coats and hand-rubbed, or compounded, to a brilliant lustrous finish.

The bumpers were, again, cast lead elements, as were the taillight frames, except all these components were painted white. I believe that my choice of a white color for automotive trim cost me points, as silver-painted parts or chrome-plated brass pieces were the norm. My windshield was a clear plastic wraparound which I formed over a male plaster mold after having heated the plastic in my mother's oven.

I received a Junior Division Honorable Mention Award for New York of $25, but was beyond disappointed. At that point, with continuing high school adjustments and maintaining good grades, I wondered if I could continue in the contest. But my father's own persistence and stubbornness in pursuing his dream taught me to never give up. I re-applied for next year's contest.

1960 Entry (Ages 15–16), Junior Division

This was my sophomore year of high school, and I had some extracurricular activities that threatened to distract me from model building. Although I did not feel too inspired, I managed to build a newly designed convertible sports car model from plaster. My design looked like a Chrysler product, probably not the best kind of submission to GM. I loved the Chrysler Dart concept car, and my design was influenced by that automobile, with a high belt line becoming large canted fins towards the rear, ending in triangular tail lights. I seemed to have lost any originality and struggled through the design process. I was running out of time and apparently did not learn any secrets to making a winning design.

Some of the challenges in making a plaster convertible model included casting a place

A rear three-quarter view of the 1960 entry, a plaster cast convertible sports car, made by Paul Bonfilio (ages 15–16). He admired the Chrysler Dart prototype of the period and this had some influence on his design. The high beltline became large canted fins toward the rear of the model along with triangular rear taillights. The cast tin bumpers and the turned brass wheel covers were nickel-plated. The original model was painted a lavender body color, popular at the time and had a white interior. The model earned a 1960 Junior Division Third State Award for New York ($50) (courtesy Paul Bonfilio).

The 1960 entry made by Paul Bonfilio (ages 15–16) which won the Junior Division Third State Award for New York ($50). The windshield was made by forming a clay version on the finished plaster casting of the model. A plaster female mold impression was taken and pre-heated plexiglass was shaped and pressed into the cavity. Windshield trim around the perimeter was fashioned from tin and nickel plated. The finished plexiglass windshield fit into a pre-cast slot in the windshield cowl. Note the use of the single sealed beam headlights at the required height above ground level specified by the Guild rules. Originally painted a lavender body color with a white interior, the model has now been restored and repainted a silver metallic color with a tan dashboard and seats (courtesy Paul Bonfilio).

for the interior insert as well as forming wheel well openings to attach the official Guild wheels. For the first, the solution was to cast a rectangular hole in that portion of the body to accept the passenger compartment insert. In essence, you had a front engine compartment and the rear trunk compartment linked together by the doors. The doors were the weak link in the casting and needed to be reinforced with coat-hanger wire embedded in the plaster at the time of the pour. It was sort of like a reinforced concrete beam utilizing rebar. This allowed you to mount the seats, dashboard and transmission console on to a floor pan module and insert the assembly from the bottom of the model. The floor pan was a flat piece of ¼ inch thick plaster plate set into a recess around the opening and glued into the underside.

For the second challenge, the wheel wells were formed during the casting process by placing separate male, box-shaped forms into the mold with about ⅛ inch thickness for the outer fender wall. After the model was cast, these forms were removed, leaving four wheel well cavities into which to screw the GM supplied rubber wheels.

I made complex bumper castings for the 1960 model, but this time they were nickel plated and shiny. Regrettably, the castings were pitted. The hubcaps were simply turned brass pieces and nickel plated. I also managed to make a windshield trim piece. My father assisted me in making the steering wheel, as he did in 1959, and the seats were simply designed bucket castings.

The unique aspect of this model was that for some unexplainable reason I chose a lavender exterior color with, once again, a white seat interior. Although the color was called lavender, it was more of a pink shade popular on automobiles at the time. To this day, I do not know what possessed me to choose that color. The quality of the paint finish was very

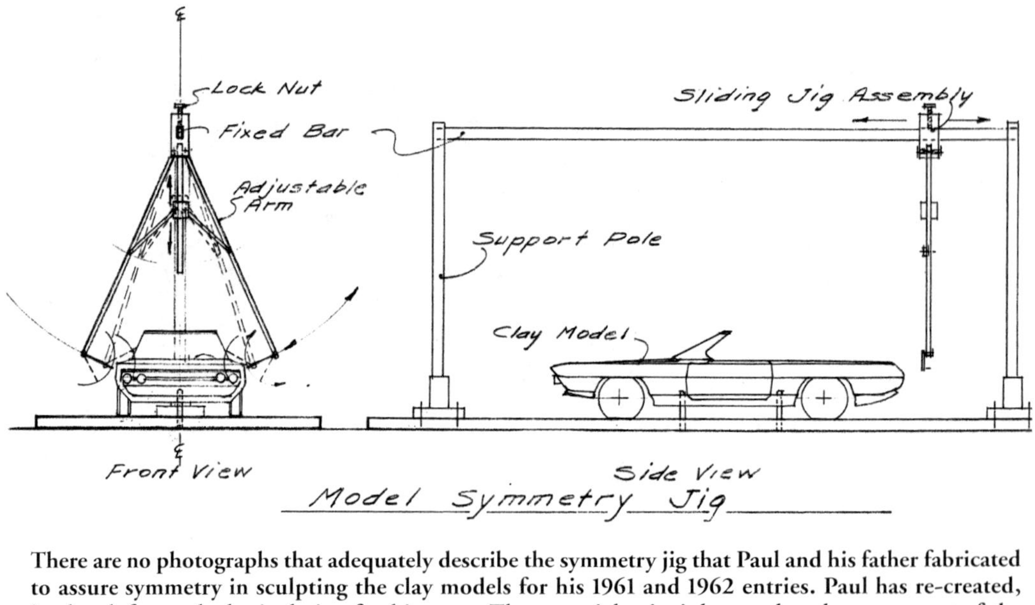

There are no photographs that adequately describe the symmetry jig that Paul and his father fabricated to assure symmetry in sculpting the clay models for his 1961 and 1962 entries. Paul has re-created, in sketch form, the basic design for his essay. The essential principle was that the movement of the arm on one side duplicated the movement of the arm on the other side. Used for "fine tuning" a clay model, the vertical and lateral location of a point on the left side of the body could be accurately located on the right side anywhere along the longitudinal center line of the model car. An additional rotating pin adjusted to the area to be measured, and a corresponding pin on the other side indicated the area of clay to be removed or added during the sculpting process (courtesy Paul Bonfilio).

good, but the overall presentation, I believe, suffered from the choice of colors and probably cost me valuable scoring points. I was too conservative, too practical and too boring. But somehow, I did win a Junior Division Third State Award for New York and was encouraged by my $50 prize.

1961 Entry (Ages 16–17), Senior Division

At the heart of every plaster cast model is a symmetric, sculpted master clay model or clay buck. When you build a clay model, it is very important to be able to make one side the mirror image of the other side. Working from a centerline, the Guild instructions recommended using cardboard templates from the finished side of the clay buck and then transferring the shape of each template to the corresponding longitudinal points on the opposite side. Up until this model, I never used that technique and did everything by eyeballing it. The models looked correct and symmetrical, but I am sure that careful measurements would reveal slight discrepancies.

To address this, my father and I devised a jig to help make one side of the clay model identical to the other side. My father helped me build this machine, as it was fairly complex, and I used it to fine tune my 1961 and 1962 clay models. We built a plywood base, with a ½-inch diameter vertical tube column support at both ends of the centerline. A rectangular horizontal bar placed 16 to 18 inches above linked the two supports. On this bar two lateral extended arms pivoted on a lockable support which would slide along the horizontal bar. The two arms had adjustable metal points or "fingers" at the ends.

The clay model was placed between the supports and under the bar aligned along the

Pictured is a three-quarter, top view of Paul Bonfilio's (ages 16–17) 1961 plaster cast Guild model entry. The 2-door convertible sports car, painted turquoise with a baby blue interior, won the 1961 Senior Division Second State Award for New York ($100). Paul is very fond of this design with its large rounded "pontoon" fenders and sweeping voluptuous lines. The dashboard is an asymmetrical wing floating within the compartment supported by a central console pedestal. The windshield is a large sloping design curving around onto the doors. The bumpers are chrome-plated brass and the taillights were embedded precariously in the rear bumper. Paul describes the design as "original, sensual, exciting muscular yet lean" (courtesy Paul Bonfilio).

From this angle, Paul's 1961 entry displays an aggressive oval shaped bumper grille with elliptical headlights. These were not sealed beam headlights. At the time, Paul believed this design would provide a wide swath of light onto the roadway and be more efficient because they were lower to the ground. This may have cost him points in the judging since they were not the conventional round sealed beam headlights. The design was characterized by a concavity or crease beginning at the front wheel well opening, at the front fender line, and continuing along the side in a continuous line defining the rear fender winglets and ending at the rear bumper (courtesy Paul Bonfilio).

centerline. The model had two pins on the bottom of the wood armature that were placed into two corresponding holes on the centerline. This assured that the removal and replacement of the model would always be in the same position.

When you moved one arm, a 4-bar linkage attached to the lockable support allowed the other arm to move the same amount laterally. It looked like a vertical pair of spring-loaded ice tongs. The linkage mechanism reminded me of an old rotating speed governor (a 4-bar linkage with a vertical sliding collar) from a steam engine. The adjustable "fingers" at the ends of each arm helped locate the exact point on the clay to replicate, both laterally and vertically. I could now measure one side of the model at a particular point (laterally, vertically and longitudinally) and transfer that exact location to the other side, and then, could either add or subtract clay. As I slid the arms longitudinally along the bar, I could check other points of the model, such as the top of the hood or trunk, points along the beltline, corners of the windshield, and wheel well opening lips. The model was still shaped by "eye," but fine-tuned with this device.

Paul Bonfilio as a 17 year old, in a photograph taken after he submitted the entry that would become his 1961 Senior Division Second State New York winner (courtesy Paul Bonfilio).

For this 1961 entry, I designed a very voluptuous two-door convertible sports car constructed from plaster which featured swooping lines, rounded pontoon fenders, and concave front wheel well coves. It had a large oval air intake for the front grille with inset elliptically shaped headlights and a tapering V-shaped rear end. The windshield was a large sloping design curving around onto the doors.

The dashboard was a floating asymmetrical wing on a console pedestal, and the seats were thin buckets consistent with the overall design. The bumpers were made of brass, simply shaped and chrome-plated. The paint was a metallic turquoise color with a baby blue interior. The hubcaps were chrome rings with brass cone-shaped inserts.

It was beautiful, original, sensual, and exciting. It was muscular yet lean. I won a Senior Division Second State Award for New York of $100, and I had a taste of victory.

1962 Entry (Ages 17–18), Senior Division

This was to be my last year in the Guild contest. I was a senior in high school, and college was looming. Although I was still interested in automobile design, I was also interested in industrial design. I had to make a decision about my future. I agonized over which direction to follow. After much consultation with my father and friends of friends in different professions, I decided that architecture as a career choice would provide the discipline and a basis for designing almost anything. Many famous architects have not only designed buildings, but furniture, lamps, appliances, flatware, and even cars. I focused on applying to various schools that had an architecture program.

This last model, a two-door convertible sports car, would incorporate all the lessons I learned over the last four years using plaster construction techniques. Although I felt that my 1961 model deserved a better award, I am reminded that despite the various categories used for judging, the competition could also be subjective. There are different fashions and trends in the car design field, and the judges were susceptible to these shifting winds. The 1962 entry would not be as flamboyant as the previous one. It would be a high quality, pleasing design with flat surfaces, crisp fender lines, and good proportions. It would have what I characterized as a "prismatic" design. The surface development of the body would be like a prism or a crystal in shape.

The bumpers were simple straight pieces with a slight curve at the ends. The dual round headlights were mounted on a horizontal bar suspended below the front bumper. The taillights were also at the ends of a horizontal bar suspended below the rear bumper. The hubcaps were chrome rings with a brass bevel gear inset. The rocker panel was a concave section on which I had planned to mount a chrome-plated exhaust tube. I later rejected the idea, and the resulting concave rocker panel gave the car a lighter, more nimble appearance. The color scheme was metallic silver blue with a darker silver blue interior.

The design was elegant and simple, but it was probably too conservative to be a scholarship winner. As I look back at some of the national winners that year and prior years, they appeared oddly impractical and flashy with no room for an engine, barely a trunk and wheels that couldn't turn. Perhaps to some degree that was the purpose of the contest, to push the envelope of design. I was designing cars that could really be built and work, with very few changes required for production.

On July 2, 1962, I received a Western Union telegram informing me that I had won a Senior Division First State Award for New York of $150 and a Regional Award for the area — Region II — that encompassed the competitive State of New York. The Fisher Body Craftsman's Guild Convention would be held July 31 to August 3. I was later told by one of the Guild judges who remembered the model that my design was considered for a National Styling Scholarship, and that my model entry was practical enough to be mass produced. At last, my five year journey bore fruit, and I was pleased to have finally received proper recognition for all of my hard work.

Paul Bonfilio's (ages 17–18) 2-door convertible sports car painted metallic silver-blue with dark silver-blue interior. This model won the 1962 Senior Division First State Award for New York ($150) and Regional Award. It features very crisp and clean styling with what he called "prismatic" design for the surface development, which consisted of flat surfaces with sharp edges. The design is pleasingly proportioned and Paul was told by one of the design judges during his Detroit convention trip that it was considered for a National Styling Scholarship. The design also featured a floating asymmetrical dashboard shaped like a wing. Both the headlights and taillights were supported on a floating horizontal bar suspended under the integrated front and rear bumpers. The hubcaps were made of brass bevel gears set into a concave ring which reflected the gear teeth in a circular pattern (courtesy Paul Bonfilio).

Epilogue

Persistence and determination are qualities that have sustained me throughout my life, and I owe it to my Guild experiences. I regret to this day disposing of the first three models I built — the 1958 and 1959 entries — because of the space constraints of apartment life. I should have never done that in hindsight, because eventually I acquired a home with lots of space. I did keep the last three models and have them in my possession to this day. Interestingly, the lavender 1960 entry was recently restored to repair a broken fin, and I repainted it in silver with a tan interior. It appears much better in this color scheme and might have scored better had I chosen these colors in the first place.

After the Guild, I was accepted at the Cooper Union, considered one of the best schools of architecture. After five years of participation in the Guild contest, I was approaching the end of my eligibility and knew I was entering a new phase to my life. I did well in high school, graduating cum laude, but a five-year college architectural program at Cooper Union would demand a lot of my attention.

After college and graduate school with bachelor's and master's degrees in architecture as well as a master's of city planning, I pursued a career in public service as a city planner, an urban designer, and later as an architect. Concurrent with this period, I pursued a 20-year career as a professional architectural model maker and built models for the Museum of Modern Art in New York City and other museums. This was in my spare time and weekends, not unlike the Guild's car contest — only for buildings. These were models of the classic works by the great architects of the twentieth century. I built a model of Fallingwater by Frank Lloyd Wright as well as other famous buildings of Le Corbusier, Mies van der Rohe, Richard Neutra, and Philip Johnson. The techniques of my Guild experience easily carried over into this field, and the discipline of meeting a deadline also served me well. I wrote a book, *Fallingwater: The Model,* in which I share many of the techniques used in building that model.[5]

Paul Bonfilio in 2009 with his 1962 Senior Division First State Award for New York and Regional Award winner. Trained in architecture and urban planning, Bonfilio pursued a public service career and retired in 2001 as commissioner of a major New York City agency. As a hobby, he enjoyed a 20 year avocation as a professional architectural model-maker for the Museum of Modern Art (MOMA) and other museums. His scale models accurately replicated the great classic works of the 20th Century such as those designed and engineered by Frank Lloyd Wright, Le Corbusier, Mies van der Rohe, Richard Neutra and Philip Johnson (courtesy Paul Bonfilio).

When I retired from city service in 2001 as a commissioner and vice-chairman for an important agency, I reverted back to architecture and have been in private practice since. My Guild experience proved invaluable. To this day, I am still enamored of automobiles and automotive design. I have owned a few classic cars and appreciate all the years of automotive production. I often wonder what could have been, had I chosen to be an automobile designer.

Notes

1. Paul Bonfilio says this of his chapter title: "My Guild experience was really about my father and myself, working together, he in an adjoining room in the cellar with his lathe, drill press and work bench, and I, in the main area near the boiler working on a table, with an overhead lamp, near a window. As much as I loved my Uncle Frank (Fortunato Duci), my father's influence as a designer and tool/die maker with his ability to solve problems, make jigs, go from a drawing to the three-dimensional form, taught me a great deal. We literally worked side by side on weeknights after schools and on the weekends. His work ethic was extraordinary. He was so obsessed with his revolutionary automotive engine and invention, that my mother painted the rooms, shampooed the rugs, did the gardening, while we were creating and inventing in the basement. She appeared to do all the chores we were supposed to be doing. In October 2010, she will be 100 years old; and this says much about an active lifestyle. Certainly Uncle Frank taught me the basics about plaster and mold-making, a flat marble pouring surface, and his 'secret' parting compound formula (soapstone and kerosene). But, there was more to making a successful Guild model that I learned either on my own or with my father's help (e.g., metal finishing; heating, shaping and polishing plastics, gluing plastic materials together; reliable fastening methods, etc.) Therefore the title of my essay accurately reflects my life with my father and honors his memory."

2. Rotating Cylinder Internal Combustion Engine, Inventor Roberto Bonfilio, U.S. Patent No. 4,612,882, issued September 23, 1986.

3. Modular Chassis and Body for Motor Vehicles, Inventors Paul F. Bonfilio and Richard Stobe, U.S. Patent 4,422,685, issued December 27, 1983.

4. Bonfilio elaborates, "At one point, in 1984, Chrysler and its chief executive officer, Mr. Lee Iacocca, met with me and one of my associates for a presentation about my father's new, revolutionary, automobile engine and my innovative, modular, plastic chassis. My father had moved back to Italy from New York by this time. Chrysler had been working on a 'Concept 1990' proposal that was very similar to my patented chassis, except (to paraphrase Mr. Iacocca's words) 'You are way ahead of us, you can show my boys something.' The meeting was a held at the Waldorf Towers in New York City and was precipitated after years of fruitless communications with Chrysler Corporation. Vic Damone (the singer), a friend to Mr. Iacocca, arranged the meeting for me. After my initial meeting with Mr. Iacocca, various Chrysler executives called and invited us to Detroit to present my father's patented, revolutionary engine idea and my patented, innovative, chassis idea. Although there was initial interest, the 'Concept 1990' Program was dropped and both the engine and chassis ideas never went any further. Despite this, my father's dream and mine have never died."

5. *Fallingwater: The Model*, Rizzoli International Publications, 300 Park Avenue South, New York, NY 10010.

21

From Automobiles to Air Force Aeronautics

George G. Herzog (1955–1959, 1963)

In the sixth grade, I wrote a report on the topic of "automotive engineering." It coincided with my first year of competition in the Fisher Body Craftsman's Guild model car contest in 1954–1955 at ages 12–13. Obviously, I had a strong interest in automotive design, but as a perversity of the way young people can think, I thought that a career had to be serious work and that designing cars would be too much like fun. If only I could have chosen a career that so neatly combined fun and work! In high school I was known for winning Guild contests. When I graduated, our class president wrote in my yearbook, "George — Maybe one day in the near future I'll be driving a car designed by you."

As a youth, I succeeded in winning three Regional Awards in the FBCG design competition, and some classmates thought I was destined to end up at General Motors crafting cars. Instead, I went on to study physics in college, was commissioned as an officer in the United States Air Force and completed a civilian career at Wright-Patterson Air Force Base, Ohio, supporting Pentagon programs.

I built six model cars to compete for college scholarships in the Guild as an adolescent competitor from Virginia during the years 1955–59 and 1963. The contest was originated to recognize boys who displayed the best craftsmanship skills in building a miniature model Napoleonic Coach, the trademark for the Fisher Body Division of General Motors. In the late 1940s, the famous contest evolved into building a 1:12 scale model of a futuristic car of one's own design, and scholarships were awarded based on both design and craftsmanship skills and abilities.

1955 Entry, Ages 12–13

I carved my first model entry in 1955, at ages 12 and 13, from a block of balsa wood sent to me by my grandfather, who worked in a boat yard in East Moriches, Long Island, in New York. The car was a four-door sedan with a hand-brushed, two tone paint job of bronze and ivory. The grille had large conical bumper sections, reminiscent of the era's "bosomy" Cadillacs, at both sides of an oblong grille opening. The grille sections were fashioned from balsa and painted silver. Headlight openings above these protuberances were canted rearward 45 degrees with inset headlights fashioned from a clear plastic bobby-pin case and etched with horizontal and vertical lines. The fenders had a slight arch from the top of the headlight hoods to the rear, designed to break the optical illusion of a sometimes "sagging" perfectly straight line. My automotive design career was off to a promising start, as I won the Junior Division Second State Award for Virginia ($100).

1956 Entry, Ages 13–14

My second attempt was in 1956 with another sedan, carved from a laminated block of wood, made up of ¾-inch thick boards glued together, and painted a royal blue and a deep bronze color scheme. The model was carved directly from outlines drawn on the wood for both the side and top views. Symmetry was maintained by using the glue joints as a guide. The full forward edge of the hood projected as an air scoop above the grille. The 1956 model's grille shape served simultaneously as the front bumper, and the "mouth" exposed a metallic surface with several staggered rows of holes. The "chrome" grille/bumper and rear bumper were made of wood and painted silver.

The headlights were wooden dowels "turned" to bullet-shaped appearances in a drill chuck, painted white and set to protrude from forward-canted hoods. Square aluminum stock was used for the greenhouse support pillars. The lower side panel and a "dart" section above the rear wheels were painted a royal blue color, and the remainder of the body was a deep bronze. Buttons with concentric circular patterns were used for hubcaps. This entry also won the Junior Division Second State Award for Virginia ($100).

The 1957 Junior Division First State Award for Virginia ($150) and Regional Award winning model made by George Herzog (ages 14–15) of Falls Church. This 4-door hardtop design was carved from laminated poplar and was painted black with light gray indicated window areas. A full width "light bar" across the front contained the headlights, license plate, turn signals and parking lights. The model had a card attached to the bottom which described its conceptual features such as overhead ventilation system and gear selection push buttons in the center of the steering wheel. George entered a total of six times (1955–1959 and 1963) and this was the first of his three Regional Award winners (courtesy George Herzog).

1957 Entry, Ages 14–15

For my third try, in the 1957 competition at ages 14–15, I won the Junior Division First State Award for Virginia ($150) and the Regional Award (Region 6 — Virginia, West Virginia and Washington, D.C.). The model car was a 4-door hardtop sedan, carved out of laminated poplar wood and painted black with light gray indicated window areas. It had a "light bar" the full width of the front of the vehicle to house the headlights, license plate, parking lights and turn signals. The side panels had a concave "V" contour expanding from the front light bar to the full fender height at the rear of the car. The grille/bumper was a rounded trapezoid shape with a darkened grille opening, using black emery paper, and a central array of narrow bars represented by straight pins.

The rear treatment was a concave "V" from the side view, as in the two preceding entries, with a "W" contour from the top view, a shape which constituted quite a carving challenge. The red and white painted sections of the tail/backup lights were split by a bent straight pin to simulate a chrome divider. As allowed by Guild rules, a card was attached to the bottom which described the vehicle's overhead ventilation system and gear selection pushbuttons located in the center of the steering wheel.

1957 Guild Convention Trip

As a 1957 Regional Award winner, Fisher Body Division made arrangements for my train trip to Detroit. I lived in Falls Church, Virginia, and departed from Union Station in Washington, D.C., and was accompanied by James R. Pierce from Washington, D.C. who was the Senior Division Regional winner. We had individual sleeper rooms, enjoyed a lavish meal in the dining car, left our shoes at our doors in the evening to be polished by the porter, were awakened the next morning as we neared our destination, and had breakfast before being met by Fisher Body Division officials in Detroit. I was 15 years old and this was my first solo, out-of-state trip.

Upon arriving in Detroit, we were all fitted in blue blazers with the Guild emblem and then photographed. The four-day convention included meeting the other regional winners and being greeted by automobile designers and GM executives. We stayed at the Sheraton Cadillac Hotel, and among our many activities, we toured the General Motors Technical Center, where we saw full-scale prototype cars of the future, cruised on a private yacht to an exclusive island club house, enjoyed a buffet at Bloomfield Hills Country Club, toured the Oldsmobile plant, and were given the "keys to the city" at a civic luncheon.

The highlight, however, was the Guild Awards Banquet. Guests included Charles F. Kettering, GM's director of research and development, and the event was broadcast nationwide with Walter Cronkite, renowned radio and TV news commentator as well as the master of ceremonies. This VIP treatment rivaled any honor I have received since.

1957 Score Sheet for a Regional Winner

The 1957 score sheet, which I received after the Guild convention, helped me identify my weaknesses and prepare for the 1958 competition. Scoring was in two major categories, each having three sub-categories. Under the craftsmanship category were scale fidelity, workmanship and painting/finish. Under the design category were originality, artistic merit and practicality. The cover letter accompanying my score sheet gave the national averages for that year. There were 250 points allocated to design and 200 to craftsmanship.

Table 3.
Score Sheet for George Herzog, Model #1945, 1957 Junior Division, First State Award for Virginia ($150) and Regional Award (Region 6 — VA, WV and DC)

	Available Points	*Herzog Points*	*Herzog % of Avail. Pts.*	*National Avg. Points*	*National % of Avail. Pts.*
Craftsmanship					
Scale Fidelity	50	48.0	96.0	43	86.0
Workmanship	80	37.5	46.9	40	50.0
Painting/Finish	70	44.5	63.6	35	50.0
Sub-total	200	130.0	65.0	118	59.0
Design					
Originality	100	72.0	72.0	45	45.0
Artistic Merit	80	65.5	81.9	34	42.5
Practicality	70	60.5	86.4	49	70.0
Sub-total	250	198.0	79.2	128	51.2
Total	450	328.0	72.9	246	54.7

Reference: Original score sheet for George Herzog. *How Your Model Scored in the FBCG Competition*, Year 1957, Model #1945, Junior Division, Region 6 (Virginia, West Virginia and Washington, D.C.) compared with national contestant results distributed by the Guild technical staff director.

The Guild score sheet showed that as a Regional Award winner, my design scores exceeded the national percentage in all three sub-categories and my craftsmanship scores exceeded the national percentages in two out of three categories. But clearly, the data showed that my lower "workmanship" score was sinking my dream of a national scholarship.

1958 Entry, Ages 15–16

The 1958 Guild entry I designed was probably the best from the point of view of originality and risk-taking. Most entries were sports cars because they permitted more flashiness in design and a grander display of craftsmanship with a finished open cockpit and clear plastic windshield. Station wagons were the least often attempted body-style by Guild contestants. Asymmetric designs, with dissimilar design lines on each side, were a bold departure from the ordinary, and could end in a disastrous design outcome.

My 1958 Guild entry was an asymmetric station wagon design. I used the same sculptured sheet metal shapes on each side of the model, but in reverse directions. These were integrated with asymmetric front and rear shapes that were also similar, but oriented differently left to right. The model was constructed of plastic wood molded onto a balsa armature. A plastic metal compound that came in a tube was directly applied to the wooden model where metal trim was to be located. The plastic wood and plastic metal had roughly the same density and were finished simultaneously in place. The color scheme was a two-tone coral and ivory, which my mother persuaded me to use as appealing to female customers.

Additionally, the car was designed to be a convertible. This was not a functional feature on the model, but the contest allowed "Special Features" to be explained on a card attached

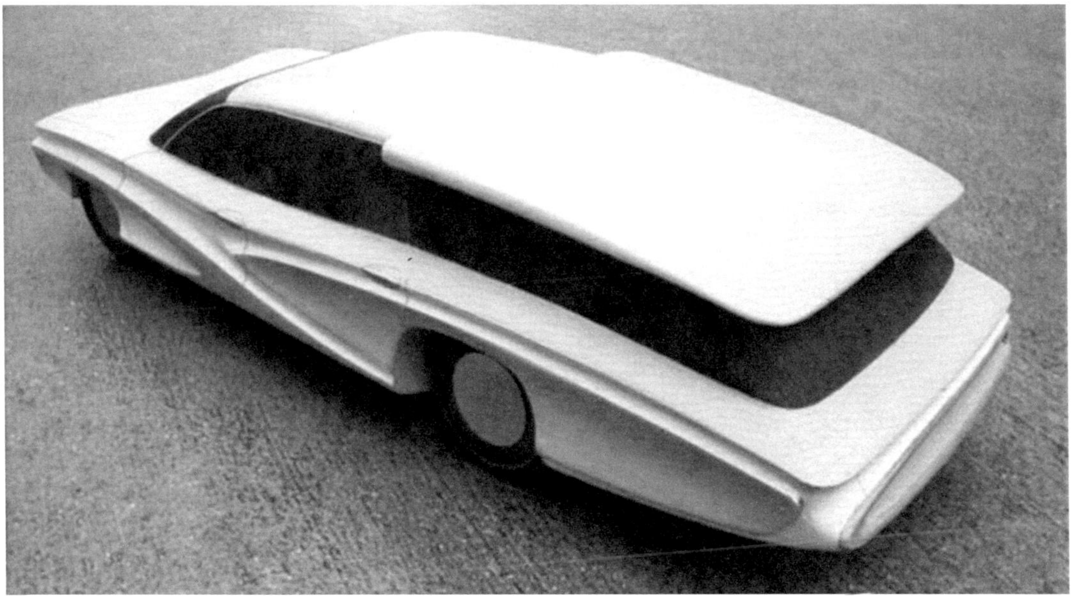

George Herzog (age 15–16) of Falls Church, Virginia, won the 1958 Junior Division First State Award for Virginia and Regional Award with this asymmetric station wagon design painted coral and ivory, a color scheme his mother recommended to appeal to the female consumer. George used the same sculpted lines on both sides except in reverse directions. It was constructed of plastic wood and plastic metal hand-molded on a balsa wood armature. The model also had a convertible top feature which was non-functional, but was explained on a card attached to the bottom of the model, as allowed by Guild rules (courtesy George Herzog).

to the bottom of the model. My description explained how the forward roof section retracted into the aft section, and the side windows slid rearward to clear the rear wheel wells, so that the whole assembly could be lowered with the retracted roof becoming the trunk lid. I won a Junior Division First State Award for Virginia ($150) and Regional Award with this station wagon design. One of the Guild judges told me, at the Convention that I had come close to receiving one of the National Styling Scholarships ($1,000). This 1958 station wagon was my last entry in the Junior Division.

1959 Entry, Ages 16–17

My 1959 entry was a sports car and won a Senior Division First State Award for Virginia ($150) and Regional Award and a third trip to the Guild Convention in Detroit. One of the design tips in the *Guildsman* newsletter suggested how a vehicle design could be made to appear dynamic and in motion by borrowing from organic shapes such as the mako shark or the manta ray. The mako shark was the inspiration for my 1959 model entry. Its gaping grille wrapped from the clenched center of its mouth, back around and over the front wheels. The widest point of the body at the front wheels tapered rearward in muscular curves to a horizontal fin-like integrated rear bumper and lights. Separate teardrop-shaped roof sections and a subtle valley between them suggested individual bubble canopies for the driver and passenger. The solid-top canopy similarly tapered all the way to the rear bumper. Because of the tapering rear shape, the rear wheel tread width was narrower than the front tread width.

George Herzog at age 16–17 with his 1959 sports car model entry. This picture and the inset photograph are the only remaining images of his 1959 Senior Division First State Award for Virginia ($150) and Regional Award winning sports car he made for the Guild. Carved from laminated wood and painted candy apple red with a black canopy, this fastback sports car design was inspired by the mako shark. The whole body from the side view and top view tapered toward the rear end where it came together in a narrow bumper. The teardrop-shaped canopy flowed in the same direction. As a result the rear axle was narrower than the front axle. The front and rear bumpers were shaped from coat hanger wire. Etched grooves were used to indicate door, hood, trunk and retractable headlight openings. Regrettably, George discarded this model when an attempt to remodel it for the 1960 competition was abandoned. Also shown are one of his Regional Award trophies (top shelf), 1956 Guild model entry also on the top shelf, and his 1957 Regional Award winning model on the bottom shelf. Note the front light bar on the 1957 model entry. Neither the model nor a photograph of his 1955 Second State Award winner exist, except in this picture obscured by his head (courtesy George Herzog).

The model was carved from laminated wood and painted candy apple red. The integrated rear bumper was carved as part of the overall shape and painted silver. The taillights and backup lights were outlined by using carefully etched grooves and painted red and white, respectively.

The grille was chromeless with a series of free-flowing vent lines cut into its black concave opening. Fabrication involved very few exterior pieces to be attached to the model. The only external metallic brightwork pieces were the two front bumper guards fashioned from shaped and polished coat-hanger wire, and thin wire used as door handles installed above concave hand recesses. All door, hood, trunk and hidden headlight openings were defined by grooves carefully etched into the wood. My disinclination to commit to a lot of detail was reflected in the simple exposed slotted head of the wheel mounting screws as the sole representation of hubcaps.

1959 Guild Convention Trip

We were able to view and analyze all the entries submitted nationwide in a massive display that filled a very large auditorium at the Fisher Body headquarters. Regional winners got to examine the interior of a partially finished GM vehicle as we toured a production line. We had the opportunity to see many mock-ups of futuristic automobiles, crash test dummy set-ups, full-sized and reduced scale clay mock-ups as well as GM's Proving Grounds. The entire Guild convention was rich with experiences, but a scholarship was also not to be for this 1959 entry.

1960: Missed Opportunity, Ages 17–18

I intended to enter again. It was the 1959–1960 competition year and would have been my last year in high school. I was busy with school work and college applications, and procrastinated. I made a last minute attempt to "remodel" my previous year's entry, like the "big boys" in Detroit often did with model year facelifts. I was reminded of this decision and endeavor by Jacobus's first book about the guild in an essay of another Guild alum who made the gutsy decision to cut off the front and back of his winning 1957 Regional Award entry so that he could re-use the same cockpit for his 1958 entry. I stripped the paint from my 1959 Regional Award winning mako shark–shaped sports car with indefinite plans of what form the remodeling would take. I eventually despaired and the redesigned 1959 model car for the 1960 competition ended up in the trash.

1963 Entry, Age 20–21

My last entry was in 1963 with a 2-door fastback sports coupe cast in plaster, painted a deep, robust Air Force blue with black indicated window areas. It had crisp, sharp design lines inspired by European sports cars. Guild instructions emphasized creating a master model in clay that could be used to change and refine the design, and then having the option of making a plaster mold from which the final permanent model could be cast. This was the technique I used.

The side view had a sharp-edged fender line that swept in a singular continuous arch from the hood to the rear and featured a highly raked windshield, an efficient and compact "greenhouse," and a roof line that blended into the rear deck. A similar more sharply arched line defined a deeply sculptured lower side panel that ran continuously from the front wheel well to the rear wheel well. The lower edge of this concave panel flared out at the door sill level like a dorsal fin, and had a spear-shaped metallic cap.

Taillights and tubular bumpers were mounted in an oblong cut into the body below the rear edge of the roofline. The overall appearance of the model from the rear was that of a hydroplane racing boat. The hood was a smooth flat surface contoured only with a gentle arch from the windshield to the front edge. The grille opening was a single arch spanning the full width of the model. The grille treatment was a flat piece of aluminum with three large holes progressively diminishing in size outward from each side of the central license plate location, and protected by tubular bumper bars. Concealed headlights were indicated by grooves etched in the hood.

Whereas the "brightwork" or trim on previous cars was typically silver-painted wood pieces, most of the trim for the 1963 entry was shaped from metal stock and highly polished. I used coat hanger wire to represent tubular front bumpers, with a thin slice of black plastic wire sheathing to simulate the rubber boot where the bumper joined the body. The rear tubular bumpers were also coat hanger wires. The lower side dorsal fin trim was aluminum stock laboriously filed and polished to the final compound curved shape. I used a red-handled toothbrush for fabricating the rear taillights. For hubcaps, I hand cut and filed aluminum sheet stock into circles, drilled perimeter arrays of six ventilation holes and, for the hubcap centers, used octagonal miniature buttons taken from my mother's collection and painted to match the car. The entry sustained major structural damage while in transit as well as some scattered cracking in the finish due to paint incompatibilities. Despite these anomalies, I won a Senior Division First State Award for Virginia ($150).

Newspaper Article Triggers Memories

Several years after my third Regional Award, I was asked for information on my Guild experience for a newspaper article for the *Alexandria Gazette*, which served the Virginia community where my home was located. Recently that *Gazette* article helped refresh the

The Guild model built by George Herzog (ages 20–21) which won the 1963 Senior Division First State Award for Virginia ($150). Cast in plaster from a clay master model, this two-door fastback sports coupe was painted a deep Air Force blue color (influenced by his AF-ROTC training) with black indicated window areas and emulated crisp European sports car inspired design lines. It featured a sharp-edged fender line, or shoulder, that swept in a continuous arc from the front to the rear of the model, a highly raked windshield and a roof line that blended into the trunk area. For trim, the model had highly polished steel wire bumpers, a low rocker panel dorsal fin enhanced with polished aluminum, and wheel covers fabricated from hand-filed aluminum sheet stock and accented by Air Force blue color buttons in the center. This was his last Guild entry and was completed during breaks from studying physics at Duke University. The plaster model was badly damaged in transit to or from the Guild competition, but was recently restored (courtesy George Herzog).

George Herzog was a physics major at Duke University and was commissioned as a U.S. Air Force officer upon graduation. He enjoyed a civilian career at Wright-Patterson AFB in Ohio supporting Pentagon programs. He took early retirement in 1993 after having served as deputy director for a Department of Defense program to acquire aircraft to modernize the USAF's training fleet. This picture shows Herzog recently with his 1958 asymmetric station wagon and Regional Award winner (courtesy George Herzog).

faded memories I had of my 1963 Guild entry. The *Alexandria Gazette* article was published June 1, 1964, right before I graduated from Duke University with my bachelor's degree in physics and a commission in the Air Force. About my Guild entry, the article stated, "After modeling it in clay, Herzog molded the car in plaster." When asked if any of my models had won national awards, I was quoted as saying, "No, but I kept hoping they would until this last year when I knew my time had run out." The article went on to say, "Not only was the model judged to be the best among all those submitted in the senior group from Virginia, but in its final form as a sleek hardtop painted in 'Air Force Blue' it appears as if it would give the Memorial Day racers at Indianapolis a rough time if rendered full size and fully equipped."

Retirement

I completed a federal career at Wright-Patterson Air Force Base, Ohio, as deputy director of a $2 billion program to modernize and revamp the Air Force's training fleet of some 200 aircraft. My wife, Lee, and I now enjoy biking from the timber frame home we built in Beavercreek, Ohio, adjoining a bike path and nature preserve. With time to visit the past, and motivated by the preparations necessary to attend the 2008 Guild reunion in Boston, I retrieved the models I still had in my possession, located my "small parts" box containing their missing pieces, and restored the 1957, 1958 and 1959 models to their former glory.

Based on my reading of the *Fisher Body Craftsman's Guild: An Illustrated History*, published by McFarland & Company in 2005, and comparing my Guild experiences with those of the National winners, I came away with a greater appreciation for the level of effort required and the high quality necessary to succeed in the competition. Arthur Russell's stunning sports car that won the First National Senior Scholarship (worth $5,000) in 1957, and his 1956 sedan which won a National Styling Scholarship (worth $1,000), are still vivid in my memory. Cars of this caliber are reported to have involved well over 1,000 hours of labor. One of the newspaper articles about my 1957 Regional award winner quotes me as saying I had put about 75 hours of effort into my entry. This comparison clearly shows how deserving the National winners were of their scholarship awards and why my craftsmanship scores were lower than my design scores.

To most of my peers, I am sure that the FBCG was fundamentally an outlet for automotive enthusiasm and a possible means of facilitating a college education. But, culturally and from GM's business perspective, it was much more. GM very capably promoted the contest through the media and public school systems as a means of developing in young American males individual creativity, pride in workmanship and the "stick-to-itive-ness" to see their projects through to completion, working to a deadline. For GM it was a means for ferreting out budding automotive design talent, creating brand loyalty before these buyers even entered the market, and likely providing many instances of fresh design concepts for their products. Similarities between some of the entries and later production cars are evident. Even today many of the FBCG designs remain appealing, and many of the winners did go on to work for GM, Ford and Chrysler.

22

Guildsman Designed Rational, Urban Mini-Car

Samuel T. Kjellman (1959–1962, 1964)

As a youth in northern New Hampshire, I built Erector set machines with gears and pulleys, assembled model kits and carved model boats out of scraps of pine. At an early age, my mom tried to foretell the future by saying I would be an architect. It turned out that I would be just as good — a designer of automobiles and new consumer products.

I became interested in automobile design when my dad drove his brand new 1955 Chevy into our driveway. Bored at times in school, I took to sketching new car design ideas. During class breaks, I would stand outside school and watch for a new production model to drive by. I collected new car brochures and learned what every make and model looked like. When the announcement for General Motors' Fisher Body Craftsman's Guild competition appeared on the school bulletin board, I could not get my application in fast enough.

When the Guild's wheels and plans arrived in the mail I was excited to give it a try. I read the *Guildsman* newsletters, which explained how to laminate and glue ¾-inch thick pine boards to make the right-sized block to be carved and sculpted. We did not have power tools, but the neighbor's relative, Uncle Oscar, ran a machine shop; he turned out to be a great help. He allowed me to use his patented, custom designed "Versatile Saw." It was unique because you could either move the wood through the blade or secure the wood and move the blade through the wood. In subsequent years, Oscar permitted me to use one of his metal turning lathes to make trim parts.

1959 Entry (Age 14)

By Guild standards, my first model in 1959 was a pretty crude attempt. The model was a two-door convertible, carved from white pine, hand painted with a white enamel exterior and a red interior. The windshield was shaped using clear acetate plastic and trimmed with a plastic electric wire covering. I applied aluminum filled epoxy over wood and polished it to make the wheel covers and bumpers. The epoxy came in a can with filler that made it look like aluminum when dry. It could be sculpted by hand, filed and shaped and then burnished to a high luster with silver polish.

My 1959 design was all about fins that ran from the front to the rear. The fin theme was a direction that Harley J. Earl, GM's styling vice president from 1950 to 1958, promoted with the Firebird I, II and III experimental cars. I thought that if a car was to have fins, then the overall design should reflect the development of the fin form rather than just a development on the rear quarter panels. My fins developed from the front fenders and overhung in the rear and spread horizontally across the rear deck lid.

The design had obvious practicality issues, and despite the shortcomings in the craftsmanship, its bold expression was enough to earn me a 1959 Junior Division Third State Award for New Hampshire. When the announcement and the check showed up in the mailbox, I was shocked. GM had put me in the news and recognized my efforts. My world was forever altered. The possibilities for greater things in life now seemed a reality that required only focus and effort. The energy and enthusiasm from the Guild experience translated into new energy for life. My school teachers that next year even noted a difference in my class work. I read the Guild newsletters, studied the winning designs and postulated greater ambitions.

1960 Entry (Age 15)

My second model in 1960, a two-door light blue sports car with full interior and clear canopy, was my first real attempt at entering the Fisher Body design competition with a desire to win. My first Guild entry was primarily for the love of building a model. On this serious attempt, I multiplied the scope and breadth of my efforts and the *Guildsman* newsletters became my mentor. This model was carved from a laminated white pine and spray painted with lacquer over primer and wood sealer. It featured a complex, fabricated Plexiglas canopy as well as lathe-turned parts and leather seat covers.[1]

I was excited to learn from the *Guildsman* how to form clear Plexiglas, or acrylic plastic, into compound surfaces. This information led me to design a model that incorporated a clear canopy roof with intricate aluminum trim. The canopy was made from five clear acrylic plastic parts that were formed over a wood pattern. I had to wear leather gloves to protect my hands from the heat while stretching the plastic to form the required compound surfaces.

The *Guildsman* newsletter further told where to order soft aluminum and how to form and polish it. This was soft 1/16-inch thick aluminum stock that could be honed using fine silicone carbide sand paper and then polished to a high luster with a buffing wheel powered by an electric drill. With Plexiglas having been formed in five sections, the aluminum pieces basically covered the joints and simulated a structure for the canopy assembly using plastic adhesives.

Oscar set me up with a lathe for turning wheel covers, headlights, and dash panel instrumentation. I ordered fine white glove leather to wrap the carved pine seats. The model was painted and finished with automotive lacquer. The effort paid off big. It earned a 1960 Junior Division First State Award for New Hampshire ($150) and a Regional Award representing Maine, New Hampshire, and Vermont. The country boy was off to see the Motor City.

The Guild convention was more than I could have ever dreamed. The exposure to the inner workings of GM as well as other national winners from across the country was an education. On the tours we were divided into groups. I had the good fortune to travel with Ace Fogarty, who went on to earn the 1960 Junior Division Second National Scholarship of $4,000. The prior year he had won a national scholarship in the Soap Box Derby for the

This Guild entry designed and constructed by Samuel T. Kjellman (age 15) won the 1960 Junior Division First State Award for New Hampshire ($150) and Regional Award. Inspired by his first Guild award, Sam Kjellman set his goal much higher with this 1960 light blue sports car to include a full interior and a fabricated clear Plexiglas canopy. This model was carved from a block of laminated white pine and spray painted with lacquer over primer and wood sealer. The canopy was formed from five pieces of clear acrylic which were heated and stretched over a wood armature to create the compound surfaces. Leather gloves were needed for hand protection during this process. Soft aluminum stock was cut, formed and polished to fabricate the metal trim for the multi-piece clear canopy structure. Cylindrical parts were lathe turned. The seats were carved from pine and covered with supple white glove leather (courtesy Samuel T. Kjellman).

same amount. Touring with Ace, I was learning from one of the best. He was totally focused and disciplined in his approach to competition. Tales of his experience gave me many pointers and motivated me to take my efforts to the next level.

1961 Entry (Age 16)

My third model in 1961 entered in the next level, the Senior Division. This was a challenge greater than I had expected because I would later discover I was going up against two other excellent model builders in my own state: Harry E. Schoepf of Manchester and Richard J. Johnson of Portsmouth. My 1961 model was a bright red convertible sports car, carved from laminated poplar, featuring dual Plexiglas windshields, an acrylic lacquer paint job, stitched white leather seat covers, and nickel-plated copper trim.

When I began to work with the poplar, I was immediately taken by the improved qualities of the wood. Although it was somewhat harder, it was still easy to carve, and the close grain finished to a much better surface than the pine. The laminated poplar wood block had a permanently visible centerline which aided symmetrical carving.

The "double bubble," or dual windshields, for this open convertible design were formed from acrylic plastic over a wood pattern. I used supple white glove leather to form seat covers over foam for seat cushions. Stitching was sewn into the white leather door panels and seats. The fiberglass floor panel for the occupant compartment was covered in black velvet. The model also featured nickel plated copper trim, as I had consulted a local electro-plater about brighter metallic finishes that would be superior to polished aluminum. I was also able to give the model a Fogarty-inspired paint job with the assistance of the local Chevy dealer's body shop. The fellows in the paint department taught me how to lay down a fantastic finish. The bright red paint job, the Plexiglas windshields and the black and white interior made it a stunning model and, as some would point out to me later, a very sexy looking design.

Frankly, I expected the model to be worthy of a Senior Division Regional Award, and I was amazed to find I had only won a Senior Division Third State Award for New Hampshire ($50). What happened? I discovered the competition was in my own back yard: my archrivals Johnson and Schoepf. The competition had set a high bar for me. Based upon the news articles about other local Guild winners, I tracked down Harry E. Schoepf and paid him a visit. Harry had earned a 1961 National Styling Scholarship for his very uniquely designed open convertible. Its fluid lines inspired me to think of more sculptural form possibilities.

1962 Entry (Age 17)

What to do next? My 1962 design would be a highly sculpted design, two-door sports coupe with cream color body and black bubble canopy or top. The fenders would be influenced by early vintage automobiles, body contoured grilles would beckon to the designs of the late 1940s, and external running lights would be independently suspended or hung due to the lack of a flat or vertical fender mounting surfaces.

I made a big decision for the 1962 competition to stop carving wood and begin sculpting clay for casting the model out of plaster. This would lead to a fiberglass/polyester model. I had some experience with fiberglass on my 1962 convertible model where I laid up a flat floor panel. My dad worked for a concrete block manufacturing company that was experimenting with polyester face coatings. He introduced me to a contact of his at a company manufacturing sprayed up fiberglass caskets. I was inspired to think more about the potential of working in fiberglass. Finally, I decided that this was the way to go.

The Plasticine clay allowed me to design and sculpt an automobile body that was much more organic and made for easier changes. Rather than being guided by a boxy block of wood, I was able to feel the form develop in my hands. I loved it. This was a more natural process than carving wood. I made plaster molds from the sculpted Plasticine clay model. A release agent or parting compound was applied to the inner mold surface. A black polyester gel coat was applied prior to the commercial application of glass fibers and sprayed polyester. The hollow black shell was stripped from the mold and this became the body of my 1962 model. The bottom was enclosed with a flat fiberglass panel.

I sacrificed the full open interiors, previously embraced by the Guild, for the experience of a new medium, materials and processes. The trim parts were formed, fitted and polished aluminum. Auto body filler was used to fill gaps for precisely fitting the trim parts. Cylindrical parts were turned on a lathe.

Score sheets from prior year models showed I was high in craftsmanship, but I was losing points on originality and artistic merit. So for this 1962 model entry, a two-door sports car with a cream white body and black dual canopies, I tried to improve styling originality and aesthetic merit in a direction inspired by Harry Schoepf's 1961 curvaceous scholarship winner. In my 1962 design I incorporated classic automotive themes, boldly sculpted and fluid fender lines, dual body contoured grilles, and external running lamps. This seemed artistic to me while taking the classic theme to the next level of design expression. The labor saved for not building the full interior went into the complex casting process and finishing the delicate trim pieces. The result was a stunning looking model of high craftsmanship. I was very pleased with the result, but a regional award was not to be. I took Senior Division Second State award for New Hampshire ($100). I began to wonder what I was missing. It was like the famous line in Butch Cassidy movie. "Who are those guys?"

The Guild entry designed and built by Samuel T. Kjellman (age 17) which won the Senior Division Second State Award for New Hampshire ($100). Looking to design originality and artistic merit to boost his Guild scores, Sam created this highly sculpted 2-door sports car with cream color body and black canopy. This truly transcended the mental constraints of the rectangular carved block approach. His highly fluid design bears influences of fender lines of early vintage cars. The model was sculpted in Plasticine clay, cast in plaster and sprayed up in fiberglass with the assistance of personnel at a company that manufactured fiberglass casket products. A black gel coat was applied prior to the fiberglass spray up. The hollow shell was stripped from the mold and finished for painting. A flat panel of fiberglass was used to close in the bottom. Sam sacrificed the full interior type model for the experience of the new medium, materials and processes. The trim parts were formed and fitted from polished soft aluminum and some were chrome-plated. Auto body filler was used to fill gaps for precise fitting of the trim to the body. The cylindrical parts were turned by lathe (courtesy Samuel T. Kjellman).

That year it was Richard J. Johnson of Portsmouth, a second year student at the University of Cincinnati, who had taken the Senior Division First State Award for New Hampshire, the Regional Award and the 1962 National Styling Scholarship worth $1,000.[2] I would cross paths with Richard years later at Ford Motor Company and again in Massachusetts, where we both lived after leaving Ford.

The following competition year, 1962–1963, I entered college at Syracuse University to pursue a major in industrial design. The first year of the five year program was basic art courses and academics with the primary curriculum starting in the second year. I had begun a clay model influenced by the trend toward asymmetry, but I lacked access to the shop facilities necessary to build a competitive model. Also, I was not happy with the asymmetrical direction. It did not feel right to me, and I was not energized enough to complete the model for 1963 competition. This was a lost Guild year.

Samuel T. Kjellman (age 19) created this Open Category, 2-passenger, micro city car, painted chestnut brown with cream color indicated window areas. It won the 1964 Senior Division First State Award for New Hampshire ($150) and Regional Award. This model was designed and constructed as a student project during Sam's first year majoring in industrial design at Syracuse University. The work he did was recognized and contributed significantly to his earning a Ford scholarship. As a student he had come under the influence of "rational design" professors where "form followed function" rather than emotions. The micro city car form was restrained considerably from his previous highly expressive and emotive Guild models. The form was sculpted in Chavant clay and cast in plaster using "Hydrocal," a professional dental casting plaster. To make the body of the model, catalyst-activated Bondo (polyester body filler) was poured into the Hydrocal mold over a wooden core. The trim parts were formed and fitted of soft aluminum and polished. The cylindrical parts were lathe cut. The concept defined a detachable power train module for maintenance. The controls were envisioned to be mounted on joysticks, leaving the front door wide open for entry and egress. There was no traditional dashboard (courtesy Samuel T. Kjellman).

1964 Entry (Age 19)

The second year of college, and the first year of my major, was my last chance to compete. This time I had access to the shops and interest from my design professors who were eager to counsel me. Designing and building this model became like a special elective course. That was the good news. The bad news was that despite one professor being a former designer for GM and the appliance maker Frigidaire, the industrial design ethos at Syracuse University was decidedly anti–Detroit. This was an ideologically driven program based on the Walter Gropius/Bauhaus School of Design approach. I was there to understand and learn this philosophy, but I never became a Bauhaus disciple.

Although the Guild experience was in my blood, I pleased my professors with a Bauhaus-influenced rational design model for the 1964 Guild competition. It was an Open Category, two-passenger "micro" city car painted chestnut brown with cream white window areas. Occupant entry was from the front, not the sides. The steering controls were envisioned to be joysticks to make front entry and egress possible.

It had smooth and harmonious surfaces cast from Chavant clay using Hydrocal cast in plaster and poured with catalyst-activated Bondo, a type of polyester resin, over a wood core.[3] The Bondo casting method permitted a technique wherein parts were coated with mold release and cast into the body block to achieve mating surfaces. The parts were then removed for final finishing and later assembled permanently. The trim parts were formed, fitted and polished soft aluminum. The front and rear bumpers were beautifully sculpted and integrated pieces of aluminum with cut-outs for the license plates.

I took the Senior Division Regional Award with near perfect craftsmanship. Scores for design indicated I did not especially please all the judges. I believe it was Chuck Jordan in a group critiquing session who described my design as "an overgrown Isetta-Volkswagen." Another designer was much more complimentary noting the harmony of form development — a compliment that I credit to my Syracuse professor Peterson, for his counsel. In a subsequent discussion, Peterson remarked, "To win this, you have to design in the GM image." I could not help but think that my professor had totally missed the idea that the object of the competition was to win.

At school, I was recognized as an outstanding sophomore in industrial design and awarded a full scholarship from Ford Motor Company for $1,500 per year for the final three years of my program. Ford at that time was granting scholarships to industrial design students at a number of colleges, and these awards came with the enthusiastic support of Professor Peterson. Certainly, the Guild experience was a big part of earning that award. After graduation, Ford came recruiting, and I was hired that summer to fulfill my childhood dream. I was to become an automobile designer, an experience that truly rounded me out professionally.

I don't recall Ford designers discussing the Guild, and I was surprised to learn years later that I had worked next to two Guildsmen, Ken Dowd and Darwin Hawthorne. I had also crossed paths with Virgil M. Exner, Jr., and Bud Magaldi at the Ford Motor Company as well. I spent five years working in the Ford Design Center in Dearborn, Michigan, before moving into the field of major appliance design that allowed me much more involvement in the total product.

Looking back on two Guild national conventions I attended and the experience of building five Guild models, I developed great interest in manufacturing and fabrication. I was especially interested in plastics, stampings and casting as methods that permitted design-

Samuel T. Kjellman with his 1962 fiberglass Guild model which won the Senior Division Second State Award for New Hampshire. As a young man he was named the Outstanding Sophomore in the Industrial Design program and was awarded a Ford Motor Company scholarship sufficient to cover his final three years at Syracuse University. After completing his bachelor of science degree in industrial design he worked for the Ford Motor Company as an auto designer (1967–1972), appliance product designer (1972–1974), Parker Brothers toy and game designer (1974–1989) and independent designer (1990 to present). He heads "Design Resolutions" of Weare, New Hampshire, a CAD/CAM design consulting service and Pro/Engineering software licensee. Distillation Technologies is another company Sam Kjellman formed for licensing his patented water distillation process (courtesy Samuel T. Kjellman).

ers to capture sculptural forms in manufacturable processes. Touring the various facilities on convention tours, I was intrigued by the GM research labs where materials were studied and by watching molten metal being poured and later destined to be stamped into body parts. I was amazed at the welding machines in body assembly plants wherein boxes filled with stamped parts were whipped into assembled bodies.

Production machinery made it seem that anything a designer imagined could be tooled up and manufactured. Industrial design was the right field for me because it combined my love of art with my fascination for manufacturing. I could explore sculptural forms while developing machines, processes and manufacturing concepts. The Guild experience opened a door for me that would otherwise not have been likely and offered an education without parallel that led to a career of my heart's desire.

The 2008 FBCG Reunion, Exhibition and Retrospective at the Museum of Fine Arts in Boston was like a homecoming. It was a once-in-a-lifetime event that brought together Guildsmen from across the country with their award winning model entries for a timeless Fisher Guild retrospective. Bud Magaldi filled me in on some of the people I crossed paths with at Ford. Coordinating the speakers program put me in touch with Virgil M. Exner, Jr., Meeting Ace Fogarty at the exhibition once again closed the loop. Ace and his wife, Bonnie, had become toy inventors and regularly called on the management that I worked for at Parker Brothers. I even got reacquainted with my old nemesis Harry Schoepf.

My most prized model, the radically sculpted '62 cream color model, was selected for the back cover of the Guild-MFA exhibition catalogue. Comments from my fellow Guildsmen confirmed an important lesson in retrospect. The Guild was about passion for form and expression in automotive design. This design is very special to me, as it was driven purely by artistic intent and the love of the form. Now, as a professional designer, I always design for others rather than to please myself. In my mind, this is what makes the Guild models an art form. One national winner was amazed to learn that this model scored 52 out of 100 for originality. Was it too radical and sculptural for the time? Was it not in the GM image, as Professor Peterson has commented? Seen in a 2008 perspective, I was able to see it as an art object that was unique for the time period. The retrospective renewed my trust in my artistic intuition.

I thank all of the Guild members for their comradeship and efforts to keep the Fisher Body Craftsman's Guild legacy alive with their exhibits and stories. I still have and preserve all the models I made for the Guild competition and enjoyed photographing them for this essay.

Notes

1. Plexiglas is a brand name of acrylic plastic that can be heated as a thin sheet and formed by hand. Plexiglass is the common, generic name for acrylic plastic.

2. In addition to Richard J. Johnson (age 19), of Portsmouth, New Hampshire, the Guild captured the imagination of many African-American youths. Other award winners were Roosevelt Harris, 1949 Honorable Mention, Kansas City, Kansas; James M. Toller (age 14), 1954 Regional Award winner, Boston, Massachusetts; Alton E. Johnson, 1955 and 1956 Regional Senior Award winner and 1957 National Styling Scholarship Award winner, Melrose, Louisiana; and Edwin L. Wilson, Jr., 1963 Regional Junior Award winner from Nashville, Tennessee.

3. Chavant clay is a reddish-pink professional brand of automotive styling clay consisting of beeswax and sulphur; Hydrocal is a brand of casting plaster used in casting dental prosthetics, and Bondo is a polyester-based filler used in automobile body paint and repair shops which can be finished to a smooth glass-

23

From 1:12 Scale to Full Scale: Ambition to Design Detroit Iron Fulfilled

Frederick J. "Bud" Magaldi (1958–1960, 1962, 1965)

The year 1966 was an exciting time for me. I graduated from the newly named Art Center College of Design, married my high school sweetheart, Mary Ellen, and moved to Detroit to launch a more than 32-year career at Ford Motor Company as an automobile designer. It turned out to be a bountiful journey. I got to work on the styling of some of the automobile market's exciting production cars, including the 1986 Ford Taurus and the 1994 Ford Mustang, and enjoyed various special assignments in California, Brazil and Europe. Next to my wife, children and grandchildren, designing cars and trucks at Ford was the highlight of my life. And it was the Fisher Body Craftsman's Guild that made it all possible.

In 1957, when I was 13 years old, all of the boys at the Clarence R. Edwards Junior High School in Charlestown, Massachusetts, were required to go to the auditorium for a presentation by a scientist from General Motors Corporation. The most exciting part of the presentation was the discussion about how Detroit designed and built cars and how we could build futuristic scale model cars for the Fisher Body Craftsman's Guild Model Car competition, possibly winning a college scholarship.

Afterward, the Guild representative handed out a packet that included a how-to instruction booklet on designing and building an original 1:12 scale automobile. I was hooked on the idea of this creative undertaking, and that little 5½-inch by 8½-inch brochure became my model car design bible. Soon, I developed various designs of my own to build.

I was a B-student with an interest in cars and a natural ability to draw. I lived fairly close to a ship terminal in Boston, where most of the exotic European cars of the time were unloaded from huge freighters, mostly Jaguars and MGs. They offered inspiration and stoked my interest in cars at an early age with their low, sleek looks and great color schemes. With these observations and my booklet, it was not long before I imagined various designs of my own to build.

After drawing up my first Guild model design plans at age 13, I visited the various lumber yards in Charlestown trying to find appropriate carving wood to sculpt into my future model. According to the Guild's booklet, this was supposed to be close-grained soft wood like bass, poplar or pine. I was guided to a furniture-maker's shop and explained to the owner what I needed. From his finest stock of "clear white pine," he cut enough material for me to build my first Guild model; subsequently, he became my supplier for the next three model entries. Each time, I would bring him the finished model so he could see what I had created from scratch.

The process used for building and crafting the first four models was basically the same. My workshop was in the bedroom I shared with my younger brother, Bobby, and my workbench was our shared wooden bedroom chair. My seat was a small footstool. My tools consisted of a handsaw, a hacksaw, a wood chisel, half round gouge, a hand drill, various metal and wood files, plenty of sandpaper and a jack knife. The paint came from spray cans, bought at Sears Roebuck and Company, with my earnings from delivering newspapers and shining shoes at the local pubs.

Because my tools were limited, it was necessary to use my creativity and ingenuity to find or make components like bumpers, wheel covers, headlight bezels, and steering wheels. For example, the wheel covers used on the third and fourth models were made from round plastic checker game pieces that were painted with silver spray paint. These were just the right size for the depression that was molded into the rubber tires supplied by the Guild. On the third model, each checker game piece/wheel cover was enhanced with a nickel-plated thumb tack in the center. For the fourth model's wheel covers, I drilled evenly spaced holes around a chrome button center. Also, the local sewing shop provided plenty of other small accessories that were modified and used to adorn the models such as red beads for taillights.

One of the most exciting moments in building a model was being far enough along in the construction process to send for the set of four free rubber tires supplied by the Guild. For this, I had to sign a form, along with my Dad, and state that my model was ready for tires. Of course, when these were added to the model, the design took on a completely different character, a whole new dimension, because now it had a set scale in which to judge the proportion. In retrospect, properly designing and building a Guild model required a set of old official Guild wheels in-hand to serve as visual references throughout the entire process.

The FBCG also sent out newsletters on a quarterly basis called the *Guildsman*, which periodically contained feature articles on current or previous Guild national winners and how they built their award-winning models. The *Guildsman* often included tips for making or fabricating the variety of components that might be needed. My favorite part of the newsletter was the concept sketches from the GM Design Studios, all of which I used to develop my own designs.

1958 Entry (Ages 13–14)

My first Guild entry was a bench seat, two-tone, red and white convertible sports car with a white interior. Carved from clear white pine, this model had protruding front fenders with vertical dual-mounted headlights, a sloping hood leading to a simple textured grille and a straight, horizontal polished aluminum bumper located at the bottom of the grille. The exterior side surfaces were simple, highlighted by concave rear quarter panels. The rear of the model had a simple horizontal polished aluminum bumper. The wheel well openings were raked rearward from top to bottom, inspired by Plymouth's "Forward Look" design philosophy.

The interior was fitted with a carved contoured bench seat (like two bucket seats coupled together) covered with white leather from one of my mom's old handbags. The gear shift was made from a white hat pin and was mounted in the center section on the forward edge of the seat. The steering wheel shaft was made from a section of an old pen, and the steering wheel itself made from the circular part of an old earring fitted with spokes fashioned from aluminum. The floors on all of my convertible models, including this one, were made from sheets of aluminum or tin screwed to the bottom of the model from the outside, so the interior was accessible from the top and bottom during construction, assembly and finishing. The interior floor of this model was covered with a red velveteen fabric that had a very fine, carpet-like texture.

The hardest part to make was the windshield frame, cut from one piece of 1⁄16-inch aluminum sheet. The design process started with a paper template, after plenty of iterations, to get the right configuration for the wraparound windshield. The design was transferred to the sheet of aluminum using carbon paper, and then the inner and outer design lines were etched into the aluminum with my jack knife. After the blank was drilled all the way around the inner perimeter with holes, a hacksaw blade was inserted and carefully used to cut between the holes. The hardest part of the process was drilling and cutting. Then, a series of files of different sizes and weights were used to complete the inside shape of the windshield frame.

Next, the outside shape of the windshield frame was cut using a hacksaw blade, allowing for tabs that would later be used to attach the unit to the model. This was further shaped with different files. The next step was polishing prior to shaping, as I felt it would be easier to polish it while it was flat. The last step was shaping the windshield frame piece, a process that required a wooden pattern I made from scrap wood. The acetate sheet was heated in our oven and shaped over the wood pattern, then trimmed and attached to the aluminum trim frame carefully with contact glue. The inside was covered with a piece of black 1⁄8 inch wide architectural tape. The acetate windshield and aluminum frame were fitted to the model prior to painting.

After finishing the model, the next step was to build a crate for shipping via Railway Express to Fisher Body in Detroit. The information for building the crate was also found in the *Guildsman*. It was made from 1⁄2-inch plywood and fitted with two small dowel rods placed on the bottom that corresponded with two holes of the same size in the bottom of the model. Soft sponges were also added to the inside of the top, sides, and ends of the box to keep the model from moving. This all took place in May ahead of the early June deadline. I used this particular box for all my other models as well.

Once the model was sent to GM to be judged on design and craftsmanship, the wait for my scores seemed like months, when in reality it took just weeks. My first clear pine model did not do well at all. My design scores were okay, but my craftsmanship left a lot to be desired. I was not prepared for this news. Along with the scores, however, came a note informing me I had won a Junior Division Honorable Mention Award for Massachusetts ($25). Mom thought it was great!

1959 Entry (Ages 14–15)

Afterward, I had a chance to review a traveling display of about 20 Craftsman's Guild models from the 1957–1958 competition at a downtown Boston department store. I went every day after school. It was a great opportunity—I then knew what I was competing against with my first try and how beautiful the models could be. They were the first real

This metallic Evinrude blue roadster, with blue and white interior, designed and built by Frederick J. "Bud" Magaldi (ages 14–15) of Charlestown won the 1959 Junior Division Third State Award for Massachusetts ($50). The construction materials included white pine, aluminum, marine spray paint, leather, plastic and acetate. A bucket seat convertible with scalloped coves behind the front wheels and in front of the rear wheels, the design featured massive front hood and rear deck areas, dual headlights spaced evenly across the front end, dual taillights and high mounted dual exhausts. The shaped aluminum bucket seats were covered down the middle with white leather. The model had a clear acetate plastic windshield and aluminum bumpers. The original model was lost or destroyed and these photographs show the replicated or reproduction model Magaldi completed after he retired in 1999 from the Ford Design Center. Using old photographs and his memory, he replicated the original model using carved closed cell foam. The model is true to its original design (courtesy Frederick J. "Bud" Magaldi).

Guild models I had seen close up, and they provided plenty of inspiration and motivation for me to improve on my 1958–1959 entry.

The 1959 model, carved from clear white pine, was a very simple, bucket seat convertible design with scalloped coves near the front and rear wheel well openings, both trailing off to bladed fenders. It featured a massive front hood and rear deck area, dual headlights spaced evenly across the front end, dual textured grilles between the headlights, and a place for the license plate in the middle. There were dual taillights, high mounted dual exhaust pipes, and again, a high center mounting area for the rear license plate.

The recommended paint for FBCG models was lacquer. This model was painted with Evinrude Blue, a metallic blue lacquer from a local marina. It was the only place in the Boston area at the time where I was able to locate such a color in a spray can. The interior was a two-tone white and blue, finished with bucket seats shaped and polished from aluminum sheet stock covered down the center with the white leather, again from one of mom's old handbags. It was finished with a white, full-length, transmission console with blue accent stripe and white floor gear shift.

The design process for this model was about the same as the first, except more care was taken in developing the details. For instance, the wheel covers on this model were sheet metal blanks, hand hammered over a wooden form, detailed with a hole punch, and hand polished to a high gloss. The taillights were turned from red plastic toothbrush handles using my dad's electric hand drill and files. The headlights were formed from acetate heated in the oven and shaped over a slightly rounded end of a ½-inch dowel. The clear plastic windshield and molding were made using the same technique as the first model. The bumpers were fashioned from ⅛-inch thick aluminum and polished to a chrome-like finish.

This model scored a little higher in the design and craftsmanship sub-categories than my first model and won the Junior Division Third State Award for Massachusetts ($50).[1]

1960 Entry (Ages 15–16)

I knew the next year's competition in 1959–1960 was going to be a bit tougher, and I committed a lot more time to the project. The design I came up with was much more exciting and dynamic than my other models, with open wheel wells and mild fins that stopped short of the rear end of the model and sported round, elongated taillights.

Carved from clear, white pine, this design was inspired by a beautiful and graceful fish — a tiger shark. The grille was under-slung, and the nose of the model car came to a point. The headlights were defined by cut lines on the hood, and the exterior sides of the body had highly sculpted wheel well openings.

While the polished aluminum front bumpers resembled teeth, the rear end was simple with its high mounted taillights and its polished aluminum bumper stretching across the back with a slight indentation to highlight the license plate area. The wheel covers were made from injection molded checker game pieces with thumb tack center pieces and painted silver. The exterior was painted a metallic green, and the interior was predominately white.

The seats of this model were fashioned from aluminum and leather as in the second model, with the addition of four nickel-plated common pin heads, cut off and glued to the leather, resembling buttons on the seating surface. The steering wheel shaft was made from a ¼-inch dowel, fitted with handle bars for steering control, shaped, filed and polished from

The metallic green sports roadster, with a white interior, designed and built by Frederick J. "Bud" Magaldi (ages 15–16) which won the 1960 Junior Division First State Award for Massachusetts ($150). The construction materials included white pine, aluminum, injection molded plastic and acetate. The front end design of this model was inspired by graceful tiger shark and featured fully exposed wheels/tires with highly sculpted wheel well openings and mild fins with integral high mounted tail-lights that stopped short of the rear bumper. Other features included a handle bar instead of a steering wheel, aluminum seats covered with leather, polished aluminum front and rear bumpers and wheel covers made from round injection molded "Checkers" game pieces painted silver (courtesy Frederick J. "Bud" Magaldi).

aluminum wire and fitted with a center ornament that was just the right scale. The front part of the center console was made from a scrap piece of hard foam painted white with a gauge in the center. A shaft exited from the back of the console and extended to the rear of the cockpit. This was made from a serrated dowel and covered with aluminum foil, then fitted with a white hat pin for a gear shift.

The shaping and sculpting of this metallic green sports roadster required 400 hours to complete, and for my efforts on this model, I received a Junior Division First State Award and $150.

To my surprise, a nice large trophy was presented by a GM representative to my school and a smaller trophy to me at an assembly of students and teachers in the high school's auditorium. My picture receiving the Guild trophy, along with my parents, Fisher Body representative Anthony Gilbert and school officials Martin Keane and Louis Welsh appeared in the next issue of the *Charlestown Patriot,* our local newspaper.

I did not compete in the following Guild competition of 1960–1961. It was my last year of high school, and studying was crucial for me to attend college. I also had a job to earn college money.

1962 Entry (Ages 17–18)

For the next Guild competition year, 1961–1962, I found time to compete again, this time with a very simple solid-top design carved from clear white pine. This model featured flowing body lines and exposed wheels and I also envisioned a removable glass roof panel like a modern day T-top or sunroof.

The sides of the body, although highly sculpted with fully exposed tires, were clean and well organized with defined highlights using natural reflected light. It was a dark metallic red sports coupe featuring polished aluminum bumpers and a very thin and simple grille

Guild trophy awards ceremony for Frederick J. "Bud" Magaldi, age 16, at his Charlestown High School around September or October 1960. Left to right are Louis Welsh, head master, Anthony Gilbert, Fisher Body Representative, and Guildsman Frederick J. "Bud" Magaldi. On the table is the metallic green model Bud Magaldi designed which won the 1960 Junior Division First State Award for Massachusetts ($150). This is a cropped, enhanced newspaper image from the *Charlestown Patriot* newspaper (now called the *Charlestown Patriot-Bridge*.) (courtesy Mr. Stephen Quigley Independent News Group, *Charlestown Patriot-Bridge*).

The dark metallic red sports coupe, with white indicated greenhouse glass, designed by Frederick J. "Bud" Magaldi (ages 17–18) which won the 1962 Senior Division Second State Award for Massachusetts ($100). Construction materials included white pine, aluminum and plastic. Made with solid-top construction and designed with flowing body lines, this model featured exposed wheels/tires, polished aluminum bumpers and a simulated removable glass roof panel like a sun roof or T-top. The recessed headlights and removable glass roof panel were indicated by scored grooved lines. The polished aluminum grille had chrome bullet-shaped teeth equally spaced. The 1962 model wheel covers were injection molded plastic checkers game pieces with holes drilled evenly spaced around a chrome button hub (courtesy Frederick J. "Bud" Magaldi).

above the horizontal bumper with equally spaced, chrome, bullet-shaped grille teeth across the front. It also had recessed headlights indicated with simple grooves in the body. The rear of the model was a repeat of the front except the taillights were acetate cut to fit into thin openings and painted red with fingernail polish. The only brightwork on the model was the aluminum grille, bumpers and wheel covers. The greenhouse glass was indicated with white paint and featured a removable roof panel indicated by scored lines.

This model proved to be too plain, and the details were too few to compete successfully. Although I liked the design, the score sheet showed the Guild judges did not. Still, it won a Senior Division Second State Award for Massachusetts ($100).

1965 Entry (Age 20–21)

My next Guild model opportunity came my junior year at the Art Center School in Los Angeles. It was my last year of eligibility. This final model was another sports coupe, painted a 1965 Cadillac Fire Frost Metallic Maroon color with a glossy black greenhouse accented with chrome-plated trim. This model was the first one I made that was preceded by a clay model, which the Guild had always recommended.

This model, a two-passenger sports coupe, was inspired by the Jaguar XKE with a long hood, or a long axle to cowl ratio. It boasted an efficient, compact greenhouse area and a short rear overhang. It had a mild V-shaped front-end highlighted by a contoured and polished aluminum bumper, dual under-slung vertical grilles and a sweeping wraparound V-shaped windshield. There were no 'A' pillars, as the integral roof and windshield canopy structure was supposed to "swing up" as one piece for passengers to enter or exit. The rear was tapered both in side view and plan view, with a contoured full width taillight system. It was finished with two separate, tapered, polished aluminum bumpers.

Since I was well on my way to becoming a designer, making clay models at school was part of our training. We had professional advice on the proper way to develop a design, sculpt a clay model, prepare the molds from the clay, and turn it into a fiberglass model. We had been working on larger scale automotive clay models, and, although the techniques were the same, a 1:12 scale clay model proved to be a challenge. The tools used for clay are specialized, and we made our own at the Art Center from various widths of piano wire, hack saw blades, and wood.

When I was happy with the clay model design for my final entry, I cast the fiberglass molds in Hydrocal plaster and used hemp cloth — akin to dried grass — for strength and reinforcement. This required multiple mold parts and pieces due to the complexity of the clay model. The individual mold pieces were coated with shellac and waxed with common car wax. A thin fiberglass resin and talcum powder gel coat mixture was applied to each mold piece independently. While still tacky, a layer of fiberglass fibers and then cloth were applied in various width strips that conformed to the concave shapes of the mold parts.

After all this was complete, another layer of resin was applied and left to harden. The excess glass was then trimmed from the perimeter of each mold piece while still in the mold to eliminate rough edges. All the fiberglass sections were then welded together with more resin to form the completed body of the fiberglass model. This was a lot like assembling a Revell or AMT plastic model kit. The process of final sanding and finishing began when the glass was cured and hard.

The metallic Firefrost Maroon sports coupe with glossy black greenhouse glass, crafted by Frederick J. "Bud" Magaldi (ages 20–21) from Charlestown, won the 1965 Senior Division First State Award for Massachusetts ($150), Regional Award, and National Styling Scholarship Award ($1,000). The construction techniques and materials included a master clay model, a multi-piece plaster mold, fiberglass and polyester resin lay-up for the finished body of the model, aluminum, highly polished stainless steel wire and lacquer paint. The paint was a 1965 production Cadillac color. The design featured a V-shaped front end with polished aluminum bumper, a sweeping wraparound V-shape windshield, a "swing-up" canopy, and dual under-slung vertical grilles. This model was inspired by the Jaguar XKE, in that it has a long hood, a very personal greenhouse look, and a short rear overhang (courtesy Frederick J. "Bud" Magaldi).

The finished model turned out superbly and secured a Senior Division First State Award and Regional Award of $150 cash that also included a trip to the Guild's convention in Detroit. It became a trip that left me with some of the most exciting memories of my life. While boarding the plane from Boston to Detroit, I was introduced to two other Guildsmen-winners from the New England area, James Miga and Stephen Paulson. A GM photographer also took pictures of us at the Boston airport and made us feel special.

Once in Detroit, we were escorted to our hotel and fitted for a new sport coat, dress slacks, a shirt and a tie. From there it was a trip to the GM Tech Center and introductions to some of GM's design executives and designers, then a chance to see, touch and ride in some of GM's new advanced concept cars. The rest of the week included a trip to the Milford Proving Grounds and test track, a visit to the Henry Ford Museum and Greenfield Village, and a luncheon at the Oakland Country Club, including a group photograph. Of course, the biggest thrill was the awards banquet, sitting and talking with other Guildsmen and distinguished guests.

Since this was my last shot at a national award, I was more than a little surprised when my name was read off for a National Styling Scholarship worth $1,000 and trophies to be awarded to me and the Art Center School. The highlight of this journey was that my model was put on display at the 1964–1965 World's Fair in New York City. Of course, my mom,

Guildsmen from Boston area disembarking at the Willow Run Airport in Detroit and are greeted by Guild representatives before attending the 1965 Guild Convention. At the top of the stairway is Bud Magaldi (1965 National Styling Scholarship winner) from Charlestown Massachusetts, middle of the stairway is Stephen Paulson (1965 National Styling scholarship winner) from Upton, Massachusetts and at the bottom of the stairway is James W. Miga (1965 Junior Regional winner) from Winchester, Massachusetts. Names of all persons on tarmac are unknown. (courtesy Frederick J. "Bud" Magaldi).

dad, and fiancée traveled to New York in the summer of 1965 to see the FBCG display and a glimpse of what the future looked like in the eyes of aspiring auto designers like me.

I joined Ford Motor Company as a designer in 1966 after graduating from the Art Center School.[2] There I contributed to both production and concept car, truck and tractor designs and traveled for other special design assignments. I contributed to many production car designs such as Jack Telnack's 1986 Ford Taurus design; Ford's first "Aero" car; the Ford Explorer Sport Trac Concept SUT called the "Adrenalin"; the Lincoln Town Car and the

A recent photograph of Bud Magaldi holding his 1965 Guild model. Magaldi built five Guild models over a period of years spanning 1958 to 1965. After graduating from the Art Center School, he joined Ford Motor Company as a designer in 1966 and over his career contributed to production and concept cars, trucks and tractors. Most notably, he managed the 1994 Mustang design team and contributed to other production Ford products (Jack Telnack's 1986 Ford Taurus, Ford's first "Aero" Car, Ford Explorer Sport Trac Concept SUT called the "Adrenalin," Lincoln Town Car and Mustang Mach III). He retired in 1999 as design manager, Mustang Studios, Ford Design Center. This photograph was taken at his home in Canton, Michigan, in October 2009 (courtesy Frederick J. "Bud" Magaldi).

Mustang Mach III. I managed the 1994 Ford Mustang design team and retired from Ford in 1999. I continued to design concept cars on a freelance basis until 2004. I still love automobiles and am a member of the League of Retired Auto Designers. I also attend many classic car events and cruise nights with my 2002 T-Bird, enjoying retirement with my wife, Mary Ellen. I guess you could say I have gas in my veins.

Notes

1. The original 1959 Evinrude Blue model was lost or destroyed, and the image in this book is of a replica or reproduction model made by Bud Magaldi from old photographs using rigid polyurethane foam. The replica Guild wheels were resin cast by a model shop buddy from Ford Motor Company as a gift.

2. The Art Center School was renamed the Art Center College of Design during the 1966–1967 period.

24

Lucky Number 7

Geza A. Loczi (1959–1965)

How would you like to wake up every morning and feel as though you were going to your garage to work on a favorite project? Well, that is how it has been for me working as a car designer. The love affair I had with the Fisher Body Craftsman's Guild competition lasted seven years, with seven very different model car designs. But the love affair I've had with automobile design, as a professional auto designer, has lasted my entire adult life.

In 1959, cars were big and flashy, with huge tail fins and massive front grilles. Everyone knew when the new cars were coming out, and we waited with awed wonder to see what they would look like. Mesmerized by the red glow of automobile taillights, I was age eight when the car bug got me. At that point, I began drawing cars, then carving them out of wood a few years later. My parents had watched me draw cars for years, so telling them that I wanted to enter a model car in the Guild competition wasn't a surprise. I couldn't wait to find out how to enter the Fisher Body Design Competition.

1959 Entry (Ages 14–15)

My father knew a pattern maker and brought home a block of sugar pine wood. It was wonderful to carve and sand because it was soft, but solid, with an almost omni-directional grain where the cross grain presented little resistance. As I got older, I discovered something elemental in the simple act of carving and sanding, and I found that I would carve and sand that sugar pine until my hands ached. I began to realize how really satisfying it was to see the final shape take place.

Designing, carving, and sanding didn't present a problem. The challenge for me was priming and painting my model. Not having experience, I wrote to other Guildsmen and asked them about appropriate techniques. Their feedback was most valuable, and after many tries I completed my model with a decent finish.

I submitted my first Guild model car design during my freshman year of high school in South Bend, Indiana, after seeing a school presentation where the possibility of designing and building your own Fisher Body model car was introduced. The Guild representatives brought pictures and past winning models to show as examples. I said to myself, "I have got to build me one of these." One of the cars they showed us had a beautifully sculpted shape and a fantastic paint job. I thought it was made out of red plastic, not wood. To this day, I love that non-metallic red that looks like plastic and even produced one for myself with my 1961 Guild entry.

Geza A. Loczi (ages 15–16) of South Bend, Indiana, won the 1960 Junior Division First State Award for Indiana ($150) and the Regional Award with this 2-door roadster carved from sugar pine. It was painted a light metallic green with a neutral interior color. The exterior paint job was selected to represent a "Hammer" finish, suggesting the dimpled surface of a golf ball for better aerodynamics. The model also featured dual windshields that swept up and over the occupants, highly sculpted and organic side sheet metal, a steering lever instead of a steering wheel and body colored bumpers. Stereo buttons were cut down and painted body color for wheel covers. Mr. Loczi was happy with the sportiness of his second model (courtesy Geza A. Loczi).

My first Fisher Body model was a 2-door sports coupe, carved from sugar pine, painted dark blue with a silver teardrop-shaped canopy over both the driver and passenger. It was a balanced design, with a black painted air intake vent, made out of metal, running up and over the hood and through the center between the large rear vertical stabilizers or tail fins. The black painted vent on the rear deck was an exhaust vent, also metal. The canted, V-shaped stabilizers were higher than the passenger's canopy and red stop lights adorned the tips of the stabilizers. The model featured projection beams for headlights.

The late 1950s were all about large and flashy tail fins, and, of course, I was influenced by that emotional statement of design. My biggest challenge in creating that look on my 1959 model was that I carved my fins from the same block of wood as the car body and top. That was a huge challenge, but the model was awarded a Junior Division Second State Award for Indiana of $100, and I was inspired to do even better on the next one.

Were I to face the same design challenge again, I would have built the fins out of separate sheets of material, and fit and glued them to the main body of the car. But, as a boy of 14, I was determined to carve them out of the only single, solid block of wood that I had been provided! Like I said, it was all trial and error. That is how I learned, and, looking back, it was a lesson that I never forgot.

1960 Entry (Ages 15–16)

My second model was influenced by the 1960s, which had toned down, sculpted body forms compared to the previous flamboyant decade. I began to appreciate the interior design in cars and realized that I needed to design and build my own expression of what an interior would look like. I loved drawing and designing. It took about three months until I was pleased with the design.

The second model was a 2-door convertible or roadster carved from sugar pine, with a light metallic green paint job to represent a "hammer" finish and a light neutral interior color scheme.[1] Dual windshields were shaped from clear Plexiglas and joined seamlessly together to sweep over the heads of the driver and passenger, and then open in the rear. Highly sculptural and organic side sheet metal shapes as well as V-shape forms on the tonneau surfaces were used to emphasize the driver and passenger positions. Inside, the seats were sculptural and body-fitting while the steering wheel was replaced with a steering lever on a protruding center console. Information for the driver was displayed on a green screen in front of the driver. The wheel covers were stereo buttons cut-down and painted. The front and rear bumpers were body color and integrated into the body form like the soft fascia of today.

The 1960 model entry had a full, carved interior, so the top needed to be formed from transparent plastic. I had made some spoons in school shop in South Bend, and gained some experience making wooden molds and heating sheets of plastic to be formed in between the molded shapes. After carving an inside shape and an outside shape for the top, I spent hours melting plastic. After sustaining a number of burn marks on my arms and hands, I was able to form a top canopy to fit with the lower body design which allowed me to show off the interior design.

My hard work and diligence brought with it a Junior Division First State Award for Indiana of $150 and a Regional Award. Each new model car entry brought new challenges but also greater rewards. I was so excited to start my next model for the coming year.

Geza A. Loczi (ages 16–17) built this 2-door sports coupe from sugar pine and won a 1961 Senior Division Honorable Mention Award for Arizona ($25). During the 1960–1961 Guild competition year his family moved to Scottsdale, Arizona. His 1961 model borrowed aerodynamic Formula race car performance themes such as a front air dam, a mid-engine air intake behind the passenger's heads (with a rounded waterfall-shaped canopy) and sharp rear trailing edges to reduce drag. It allowed him to look at proportion differently, utilizing a center driver position, cab forward design (a term not invented at that time) and the engine behind the driver and passenger. The non-metallic, red enamel body was hand-rubbed with polishing compound to look like plastic. The canopy was painted black (courtesy Geza A. Loczi).

1961 Entry (Ages 16–17)

Realizing the incredible effort that designing and displaying the interior took on my last model, I decided to concentrate on a solid-top model construction and a new theme based around something I read relating to how air passed over shapes, or aerodynamics. This was about the same time that I realized the influence that race cars had on me, and so I set out to design some pretty cool shapes to help direct the air in and around the car as well as into the cooling area.

The third model car was a 2-door sports coupe, carved from sugar pine with solid-top construction and finished in a red and black color scheme. This was a mid-engine design that borrowed aerodynamic race car themes such as a front air dam and integral rear air intake behind the driver's compartment. The air flowed over the engine air intake and down a waterfall-shaped rear end. The rear wheel openings were covered with sheet metal, and there was a sharp rear end trailing edge to encourage air separation. Wanting that "non-metallic red that looked like plastic" paint job, I used enamel paint and hand rubbed the paint with a military-like "spit-polish" technique. The solid-top canopy carved from sugar pine was painted black and fitted as a module rather than integral to the body.

During this time, my family moved to Scottsdale, Arizona, in time for my senior year of high school. This was my first year competing with the "big boys" in the Senior Division. I received a Senior Division Honorable Mention State Award for Arizona with this model car, which was all the encouragement I needed to begin to think about next year's entry.

1962 Entry (Ages 17–18)

Being that I was a little older, I wanted to make a mature design statement, and my fourth model was to be sleek with complicated shapes. This was to be a 2-door convertible sports car or roadster, carved from sugar pine, painted a metallic champagne light gold with a single conventional wraparound windshield. It had a strong arching shoulder, or design line, running from front to back with concave wheel accent forms designed around each wheel. Plan view surface shapes emphasized the dual interior compartments. Even though the design of the car was symmetrical, the driver's cockpit was ahead of the passenger's for better visibility. The steering wheel was replaced with a steering lever. It was my "gold car"—both the exterior and interior had a golden hue.

I carved out dual headlight shapes and added a lot of detail on the sides and rear of the model—an example of what I felt represented a balanced architecture in a beautiful, simple combination of silver and gold metallic. Looking back, the color was probably the simplest aspect of the model. I was rewarded with the 1962 Senior Division Second State Award for Arizona ($100).

As I carved this 1962 Guild model, I began to think seriously about where I could receive the best education that would allow me to continue to design cars for a living. I wrote to the Guild asking about the best automotive design schools, and the Art Center School in Los Angeles was the school recommended most highly. That became my goal.

1963 Entry (Ages 18–19)

The western lifestyle was more laid-back than the Midwest, a lifestyle that I took to like a duck to water. My next model reflected my new attitude and had a much simpler

Geza A. Loczi (ages 17–18) won the 1962 Senior Division Second State Award for Arizona ($100) with this 2-door convertible sports car or roadster. Carved from sugar pine, the model was painted a metallic champagne color (light silver-gold) with gold tinted interior accents. The styling featured a strong arching shoulder, or design line, running from front to the rear with wheel well accent coves designed around each wheel/tire and a raked wraparound windshield. There were dual passenger compartments with the driver's side slightly forward for better visibility and a steering lever rather than a steering wheel. For this fourth Guild entry, Geza Loczi wanted to do something dynamic and to emphasize the details. Even though the proportions of this model were well balanced and good, he was determined to simplify the design and details on his next Guild model entry (courtesy Geza A. Loczi).

design, although it embraced asymmetric themes. For my fifth model entry, I designed a silver metallic 2-door sports car with dual black canopies and carved from sugar pine. The black painted solid-top bubbles or canopies were designed to cover the occupants' heads and shoulders, but most of the body or torso was below the beltline. There were asymmetric front and rear treatments, and some top view asymmetry with the body section changing from front to rear. The side design emphasized flowing concave forms around each wheel below a strong shoulder, changing in section from front to rear. Similar to Ronald J. Will's 1961 and Roy R. Dickey's 1962 designs, I finished my 1963 model with one canopy on the driver's side—not as a shortcut but to further emphasize the asymmetrical aspect of the design.

The Guild styling judges were not as pleased as I was with this model, and I won the Senior Division Third State Award for Arizona of $50. This was to be the last wooden car, because I was soon off to Art Center School in southern California, complete with an entire array of new adventures and a growing seriousness about designing and building real automobiles.

Days of Sugar Pine Are Over

California was a unique place, and a good influence. Going to school in California in the 1960s, I realized that the outdoor ambient lighting was different and that car colors glowed in the reflection of those amazing western sky sunsets. People took particularly good care of their cars and great pride in their appearance. It was, and continues to be, a place where trends start and where new ideas are welcome. It was just the right atmosphere for a young man with a constant stream of automotive visions swimming around his head to believe he could imagine the future. With my open, optimistic, and endlessly positive attitude, when combined with the concept of "California Dreamin'," I was ready to roll!

At the Art Center, we learned to design and shape models in clay, cast them into molds, lay up the molds with fiberglass, and create a light, sturdy shell which was the body of the car's design. All of those years carving models out of wood really helped me understand three-dimensional design. Although adjusting to new techniques was quite simple for me, learning new techniques and having new materials does not by itself achieve a great end product.

Without a good, solid design first, I can't imagine any model evolving into a successful product. Good design always wins out. I think I spent close to half the time, if not more, designing and sketching, and the remaining time building my cars. The tough part was finding the time to work on the Fisher Body projects, with Art Center's homework schedule lurking in the background and taking first priority. Knowing that I had to work on my Fisher car was a good incentive to finish my school assignments early.

1964 Entry (Ages 19–20)

For my sixth car, my horizons were opened, and I was embracing the idea of creating something entirely new. Everything was, "What if I did this?" or "What if I did that?" This became my "What If?" car.

I pulled out the stops and dove into envisioning future modes of transportation. For

Geza A. Loczi (ages 18–19) won the 1963 Senior Division Third State Award for Arizona ($50) with this asymmetric design. The 2-door sports car was carved from sugar pine and painted a metallic silver color. A single black canopy was used to emphasize the asymmetry. The solid-top canopy was designed to cover the driver's head and shoulders, with most of the body below the beltline. The asymmetric design lines and styling included an ever changing body section on the top surface of the model car. With a single seat and canopy for the driver and a tonneau cover for the occasional passenger, this design combined simplicity with sculptural forms (courtesy Geza A. Loczi).

the first time, I began working in the open category. A car design has two great design opportunities: one is the proportion and shape in side view, and the other is the shape of the plan view. So, what the car looks like from side and what it looks like from the top can be combined in making a very dynamic statement.

It occurred to me that, in the future, for long trips, it could work to have a car that ran only on two wheels, much like a motorcycle. This would be more efficient and save on fuel with the two side wheels folded up into the body of the car for better aerodynamics. When needed, they would return to their original position. I settled on an integrated diamond tread pattern design for the wheels, which was defined by having two wheels front and back, and two on each side. This pattern also allowed for a center driving position, which is unique with great visibility as well as a great "fun factor." There was room easily for two passengers behind the driver. The top was designed as a large canopy that lifted up for entry and exit. The doors opened in a more traditional manner, either swinging out or sliding open.

This "diamond" design was built first as a clay model in two halves, a top form and a bottom form, with the bottom having as much surface shape and unique form as the top. Each clay part was cast into a plaster cast mold, and then fiberglass was laid up into each mold. When hardened, each part was cut, filed and sandwiched between a brass and plastic sheet, which was shaped to match the perimeter shape of the body. This was the separation between the upper and the lower body and included the front lights and the rear tail lamps.

Wheel designs are very important to the overall feel of the car. Most of my wheels were repurposed from other goods: buttons, metal gears, bearings, and washers, among others. I would take weeks or months to find just the right fit. Otherwise, I would turn the wheel shape on a lathe to a specific design, as was the case with the diamond tread design. Both front and rear wheels needed to have a suspension system to make it feasible. So I built what was needed. The 1964 wheels were a blast to turn on one of the biggest lathes I'd ever seen, and since the front and rear wheels were like those of a motorcycle, the suspension was built accordingly. Brass rods and tubes brought it into reality. I kept the color of the sand blasted brass wheels for the submitted model.

The Final Touches

I chose a silver color for the body because it defined the design surfaces and has what we call "flop," which is light and dark separation that shows off the form beautifully. The top was painted in the blackest of black to represent a glass surface. By this time a very fine sandpaper had been invented to use either wet or dry.

After sanding the fiberglass surfaces with a 400 grit sandpaper and filling in any pin holes in the surface, I sprayed on a couple coats of primer paint which were then sanded with 600 grit wet. The model was smooth as silk and was ready for color coats. After spraying several lacquer coats, which looked pretty good, the orange peel from spraying needed to be sanded out of the final coat of color. I used the highest grit sand paper available, which was close to 1000 grit wet. A polishing compound, water, and hand rubbing the painted surfaces brought out the highest luster. After that, applying a little wax protects and preserves the paint. I wanted the wheels to remain brass looking so I sprayed them with a clear coat. The brass and black bumper details gave the design a finished appearance.

By this time, I was finishing up my final projects for Art Center, so less time was being

Geza A. Loczi (ages 19–20) won a 1964 National Styling Scholarship ($1,000) with this Open Category, 2-door sports car with 1 plus 2 seating, while attending the Art Center School in Los Angeles, California. The model was built in the tradition of the Art Center School's famous model-making program. Designed with a diamond-shaped tire tread, the front and rear wheels/tires had front and rear motorcycle-type suspension systems for "at speed" operation, while the two outrigger wheels/tires were retracted. When not needed, the side wheels would fold up, and the car performed similarly to a motorcycle. Carefully conceived and designed in clay, this was a fiberglass lay-up model with an upper and lower half. The body was painted a silver color and the wheels were sand-blasted brass with a clear coat. There were brass and black bumper details. With a diamond tread pattern and a center driving position, the architecture of the car led to an exciting and dramatic design for Loczi's sixth Guild model entry. With architecture and proportion being about 80 percent of the design, the rest was just frosting on the cake (courtesy Geza A. Loczi).

devoted to my Fisher car. Luckily, semester break was near, and I shifted my finishing model work to my family's home in Arizona. As the deadline approached, I had to tackle the painstaking task of rubbing out the painted surfaces and polishing the metal parts. After so many models and so many deadlines, I must have been operating on sheer adrenaline and instinct, because I actually made a date with my girlfriend for the night of the deadline for submitting my car. What was I thinking, when, in fact, I spent the last three days and nights leading up to that night without sleep, in a determined effort to complete my model?

Deadline night found me riding in the back seat of my car as my date drove feverishly to the mail station at the Phoenix airport, which was open until midnight. I was gluing the top, the brass and black moldings and the bottom together with quick-set glue, and counting the minutes as I packed my car into the box that I had customized and built for the model's transport. What a ride! We managed to arrive with less than 10 minutes to spare. I was honored to receive a 1964 Senior Division Second State Award for Arizona ($100) and a National Styling Scholarship Award of $1,000 for my sixth Fisher Body model. This was a truly futuristic car design.

1965 Entry (Ages 20–21)

"Lucky Number 7" was my last car, and it proved to be the biggest challenge of all. I had been going straight through Art Center School on a tri-semester schedule with no breaks, and it came down to my last semester. No one would have argued had I decided to forgo competing during this, my last year, of eligibility. But my friend and classmate, Tom Semple, had built a First National Senior Scholarship winner just the year prior. It was a very impressive and beautiful model—complex and perfectly crafted with a full, finished interior. Inspired by Tom's effort, I had to give it a shot. I realized that it wouldn't be possible to do an interior, considering my school workload during these final months. Instead, I concentrated on the design. My plan was to build something so advanced, futuristic, and beautifully sculpted that it was bound to get attention.

Two-thirds of my effort was spent drawing and fine tuning the design of this car. Once I had the design, all that was needed was to execute the build process. This car qualified in the open category, with two wheels wide in the front, and two wheels closer together in the rear. Working with the aerodynamics that I understood, I designed a sleek, graceful plan-view shape and profile. I knew that I wanted to paint it yellow, with a black top. It needed to be something very graphic, an attention-getter, and it needed to look very dynamic.

This was my favorite model because its design was simple yet very dynamic. The top view was wide in the front and narrow in the rear. Its side view was pointed in the front with its mass in the rear, giving the car the appearance of being in motion. It was also good for aerodynamics having the mass in the rear and having a small wake at the rear trailing edge. It is important when having warping or twisting shapes that you control the highlights on the surfaces. I wanted the front wheels to be partially covered for aerodynamics and the rear wheels, although close together, to give the impression of power. The bottom surfaces are as important in design as the top surfaces and I wanted to make them very sculptural. Conceptually, the accent chrome bumper bars acted as radar systems, keeping the car from hitting objects and avoiding accidents.

As before, I cast the model in clay and laid up each molded piece, the top and the bottom, in fiberglass. I pulled the molds and bonded the two halves together. I made sure that any gel coat surface pin-hole imperfections were filled in with epoxy resin, to assure that there would be no shrinking (sink holes). I laid out all the door cuts and molding lines around the windows by filing in the cut lines with a small jeweler's file. The window moldings were shaped and formed prior to final painting. The window moldings were filed flat on one side of the round section for better fit when gluing.

I then painted the primer, followed by two coats of the soft yellow body color, and finished up with two coats of black for the glass. I chose a pale yellow color because of its reference to being premium as well as being sporty, while the black represented the window surfaces. I finely sanded the fiberglass with 400-grit wet to prepare the model for primer. After priming and sanding, I sprayed two coats of color. To remove any "orange peel," I then sanded with the finest of grit sand paper using plenty of water.

Once the painting was completed, I began hand rubbing all surfaces, a task that is very time consuming, but critical, because there is always the danger of rubbing through any of the paint surfaces. Polishing compound, water and hand rubbing the painted surfaces brought the colors to the highest shine and luster. There was simply no time for re-painting. Once this was completed, each cut line for a molding was masked with tape to protect the final paint surface. I applied wax to protect and preserve the painted surfaces. It took over eight hours to glue in all of the pieces of bright, shiny half-round window molding. If I learned nothing else, I learned to be patient with my process and to realize just how time-consuming each step of the process was.

I am gratified that I chose to compete in this last year of my eligibility, and I was determined to make it my best effort. Even to this day, I love the design of this car. This time, I mailed my model early enough so as not to panic at the last minute. What I had control over was finished, and it was in the hands of the expert judges as well as General Motors Vice President of Design Bill Mitchell.

I won the Senior Division First State Award for Arizona ($150) and the Regional Award, and was invited to attend the Fisher Body Guild convention in Detroit. The awards banquet was the grand finale where they announced the final awards. It was very special indeed. As I sat waiting expectantly to hear the results, I remembered my first experience attending the Fisher Body Awards ceremony in 1960 when I competed in the Junior Division. This 1965 event trumped all others.

Geza A. Loczi (age 20–21) won the 1965 First State Award for Arizona ($150), Regional Award and First National Senior Scholarship ($5,000) award with this Guild entry. Advanced, futuristic and beautifully sculpted, this 2-door hardtop, Open Category model, painted pale yellow and black, was made from fiberglass in the best tradition of the Art Center School. The graceful top view had a wide front end and a tapered rear end. The result was a wide track width front axle and a narrower track width rear axle. From the side view, the design flowed from a forward point, up and over the passengers' heads, and down to a graceful tapered rear end. This sculpture required warped and twisted surfaces which had to be carefully controlled with "highlight" studies. Loczi considered the bottom surfaces just as important as the top surfaces. The chrome-plated bumper bars housed a theoretical crash avoidance radar system. The original half-round shiny metallic window trim or bright work was lost during a Fisher Body exhibit which toured the country with the 1965 national scholarship winning models promoting the Fisher Body Craftsman's Guild. Loczi stated, "Twisting and warping surfaces, with a unique tread pattern and architecture and yet simple in design, led my seventh model to achieve a timelessness in its design. Believe me, during that time frame, unique surfacing, like this car had, was not the trend. Lucky number seven was the culmination of all seven models that I had built, and all that I had learned while designing and building them" (courtesy Geza A. Loczi).

They announced the winners in the various divisions, the ten Styling Winners, and continued with the Fourth National Senior Scholarship Award (Jeffrey B. Wehking [21] Ypsilanti, Michigan, $2,000 award). By now, I was beginning to wonder just what was unfolding. My name had not been called. The winner for the Third National Senior Scholarship (Tommy L. Strine, Mansfield, Ohio, $3,000 award) was announced, and then the Second National Senior Scholarship Award (Paul H. Peterson [21] Pittsburgh, Pennsylvania, $4,000) was also announced. Well, I knew I had a good car model, but I was still thinking and wondering—was it good enough? When I saw my model flashed on the big screen in the banquet auditorium, I had my answer. I had won the First National Senior Scholarship of $5,000. I walked up to receive my award, proud to have been a part of a journey that had taken seven years and seven model cars to achieve.

Career and Conclusions

After graduation from the Art Center College of Design I joined General Motors Design Staff in 1965. The next 15 years were spent designing some of the most emotional cars of that period, such as the Pontiac Firebird, Trans AM, Grand Am and GTO, as well as other muscle cars like the Olds 442 and the Judge. I contributed to designs mainly in Pontiac, Oldsmobile, Chevrolet and Advanced Studios and left GM as an assistant chief designer.

In 1980 I joined Volkswagen of America as design manager and worked both in Michigan and in Germany until 1983. My last contribution was to the design of the Corrado.

I then joined Volvo Car Corporation in Sweden, where I relocated my family. After one year abroad we moved to California in late 1984, where I joined Charles W. Pelly in his Designworks/USA Studio as a direct consultant to Jan Willsgaard, head of Volvo Design in Sweden.

When the Volvo Monitoring and Concept Center (VMCC) was created in 1986, I was asked to join as chief designer. Since its inception, VMCC has contributed to a number of concept cars and production models, including the Environmental Concept Car, the Volvo S-80, S-60, V-70 and XC-70, the Safety Concept Car, tandem designs, plug-in hybrids, the Volvo C-70, XC-90 and crossovers such as the XC-60.

Having been named director of design at VMCC in 1997, I was asked to join Volvo Design in Sweden as chairman of design in 2004, where I served until I chose to return to California in late 2005. I am sketching and designing on the computer to this day. I officially retired on December 31, 2010, after 45-plus years and plan to continue with Volvo as senior advisor and pursue other projects as the principal of Geza Loczi Design.

Working internationally also has been a wonderful experience, not only for me, but also for my family, my children—J. Paige, Magda, and Zachery—and my wife, Barbara. My passion has taken me to England, Spain, Paris, France, Japan, Australia, Italy, Czechoslovakia, Norway, Germany, Budapest, Hungary, and Sweden. Besides Volvo car designs, my automotive passions are continually stirred by my favorite rides: a Ducati Monster, a Lotus and my antique MGB.

Looking back, now I feel it all seems to be a bit fussy. In the 1960s, my slacks had to be pressed with a sharp crease, and my shoes were spit-shined. My entire life was about being fussy. What made me think that wouldn't influence how I designed a car?

So what did I learn through the Guild?

Portraits of Geza A. Loczi in 1959 at age 15 (left) with his first Guild entry (a 2-door sports coupe) which won Junior Division Second State Award for Indiana and more recently with his 1965 First National Senior Scholarship winner. He became the director of the Volvo Monitoring and Concept Center in Camarillo, California (courtesy Geza A. Loczi).

First of all, I learned a term called "Stick-to-itiveness," meaning to not give up, and to stick with something that you've started. I came to understand how to find answers to problems I needed to solve by doing research, and talking to other Guildsmen around the country. I grew unafraid to step outside my own comfort zone to find these answers.

I also learned how to judge good proportion and how to reach into the emotion bank and pull out a design that was exciting to work with, and to bring that design through to fruition through the building process. I began to tune my senses with hand-eye coordination through carving and modeling. I refused to accept a mediocre result and instead push for perfection at every stage of the designing and building process.

I learned about planning a year's worth of work. I discovered one could not look at the whole project, but rather break it down into smaller bits and sections. I used this same technique when running marathons, that is, you break your workload down into the smallest of tasks each day. When running a marathon, it was about breaking my stride into the smallest denominator and concentrating on each footfall. Soon, six miles had passed, then ten, and so on. Looking at the entire package and knowing what needs to be done can be overwhelming. The same was true with building a Fisher Body model car for competition each year. By defining and completing a small amount of concentrated work each day, I was soon able to reach each milestone. I learned that working this way makes life a happier place and the journey a great experience.

Most of all, I learned how important it was to enjoy the process and have fun creating an emotional statement that until then, didn't exist. Designing and creating the future was a blast!

I owe a lot to the Guild. I was honored and proud to have been a part of it. I feel fortunate to still be working in the automotive industry and also to have had the opportunity to teach others automotive design at the Art Center College of Design, which I did from 1986 to 1996. The best lesson I learned was to be generous with what I have come to understand so that others can carry on the legacy of design, knowing that each step to excellence builds upon the steps that were taken before. I learned that my Lucky Number 7 was just the beginning of my adventure — and oh, what a ride it has been!

Notes

1. A "hammer" finish appearance was desired for the light metallic green 1960 model. Imagine sheet metal having a similar surface to a golf ball with indentations but on a smaller scale. Loczi wanted the body surfaces to appear to have a systematic pattern of small, spherical indentations as an aerodynamic performance statement. Such indentations reduce aerodynamic drag. Upon closer examination of the 1960 model with an enlarged photograph, swirls of paint color can be observed below the surface. America's Cup yachts are known to have textured slippery surfaces under the water line designed to reduce friction and increase speed. Similar surface texturing in some areas of the automobile may reduce drag coefficients and increase mpg.

25

Guildsman Served Up Plexiglas Sandwiches and Candied Apples

Jeffrey A. Jones (1962–1966)

I was a country boy and budding wood carver living in the middle of nowhere known as Fountain Springs, California, when the Fisher Body Craftsman's Guild came into my life. By the age of 10, I had won several "best of show" awards at the local county fair for my carvings. I won a grand champion award from the fair one year with a gorilla carving made from the design in the February 1956 issue of *Science and Mechanics*. By age 12, I began carving my first Guild entry. Although I never won a Guild award during the five years I participated from 1962 to 1966, the program helped hone the master craftsman skills and abilities that I enjoy today.

When I found the Guild I was in awe. I think I first became aware of it from ads in *Boys' Life* magazine since I was heavily into Boy Scouting.[1] The clincher was an article in *Science and Mechanics* magazine that chronicled the amazing Guild careers of Thomas F., Joseph S. and Daniel C. Greene, the three brothers from Medina, Washington, who won the 1955 First Junior National ($4,000), 1956 Fourth Senior National ($2,000) and 1957 First Junior National ($5,000) Scholarship awards, respectively, for a total of $11,000.[2] The article showed some "in-progress" or "how-to" shots of the brothers' cars, and the story made me believe that I could do that too.

I carved all of my Guild entries in wood, as that was my expertise. The first, second and last cars were made from sugar pine; the third came from black alder, a semi-hard wood from the Pacific Northwest. My fourth model was carved from mahogany. I had never heard of basswood and had no source for it. In my long wood carving career, since those Guild days, I still prefer sugar pine.

Although I ravenously consumed the Guild's how-to booklet information on model construction methods, I could see no benefit from creating a clay model before starting construction. I actually tried building a clay model, but didn't have the talent or patience for that. I finally got frustrated and just started carving the wood body from a few sketches and my mental vision. That is the process I used for every car I built, and I never had any problem with the car not evolving exactly as I envisioned it.

I achieved symmetry without the benefit of templates from a master clay model. I did all the preliminary carving work by eye, using a ruler to make comparative measurements, and paper line sketches traced on the body with carbon paper. These techniques got me really close to a respectable form of symmetry. For final correlation and fine tuning, I used pieces of lead solder bent to conform to one side of the body to form a template then used to check the other side.

My plastic bubble canopies or tops were very good. I made them using modifications of the methods recommended by the Guild, which involved carving a wooden male pattern and stretching hot Plexiglas down over that mold. In the years since the Guild, as a model-maker, I have made similar clear canopies or greenhouses by cutting a passenger compartment-shaped hole in a piece of plywood, clamping the plastic to that hole and 'blowing' the top using compressed air. That method is much superior since it yielded an optically perfect bubble, compared to the stretch method, which required a lot of sanding and polishing of the plastic to complete the part.

Brightwork, or trim, was one of my biggest problems, as I lacked the metal working skills and experience. I generally made the brightwork from polished aluminum, except for the last car, which I made from polished and chrome plated brass. For one model car I used just polished brass since the aesthetic theme was gold trim instead of chrome. Two years I stopped and used chrome-looking vinyl tape.

I used my imagination to adapt common, ordinary household items for each model entry. For half a decade, I was on a quest for unique materials. I was constantly measuring round objects that might become hubcaps. Anything that was exactly 1¼-inch diameter instantly became a candidate. I ended up making hubcaps from two styles of drawer pulls, and the ones on the first and last cars were old, size "D" battery ends! Hair curlers, golf tees, glass necklace beads, watch gears, electrical wire, bracelet chain links, straight pins, and "AA" size battery ends for headlights were just a few found items used on my cars. Forty years later, I still find myself sizing up widgets I run across for possible use as a part of a Guild car.

My designs had some strong influences from the pop car culture of the day. Many of the oddities exhibited in my models were the result of my reading the newsstand hot rodding magazines of the time. Custom car builders like Gene Winfield, Bill Cushenberry, George Barris, and especially Ed Roth were producing really wild concept cars for the show circuit. These builders spoke to my spirit and were a huge influence for much of my Craftsman's Guild career. I laid down great Guild paint jobs using pearl and candy custom color "rattle can" paints from the hobby shop.

However, when the Guild introduced the "Open Category" in 1963, I was from then on consumed by the futuristic concepts and designs. To call my Guild designs far-out, radical, unorthodox, rebellious, or blue sky would be complimentary to me. Common features on all my cars were bubble tops and body colored bumpers, which were ahead of their time. I gravitated toward custom colors such as candy apple red, candy burgundy, candy lime green, Pagan Gold Pearlescent and Orange Candy Apple. Three of the five model cars, if built for the road, would have required technology that didn't even exist at the time.

1962 Entry (Ages 12–13)

My 1962 entry, a two-door sports coupe, made during the eighth grade and carved from sugar pine, was the most practical and well proportioned of my five Guild entries. With a conventional wheel base, it was in retrospect "boring" compared to my other 1963 to 1966 "blue sky" thinking. My inspiration for 1962 was Ford's recently released Falcon

Right side view of Jeff Jones' first model entry in the 1962 junior competition. The Fountain Springs, California, 8th grader was ages 12–13 at the time. Carved from sugar pine and painted a Sky Blue and Dove Gray color scheme, the design and styling were inspired by the compact Ford Falcon. Size AA "Eveready" battery (negative ends) were used for headlights and Size D "Eveready" battery (negative ends) were used for wheel covers. He purchased a kit of stock aluminum from a Guild vendor to finish this and his other Guild model entries. Red glass necklace beads were used for taillights. This was the only model he made for the regular competition with a conventional wheelbase. Jeff was very ambitious and was planning a full interior and plastic top for this design. At the last minute he realized it was not in the cards so he carved and glued on the top shown. If the top were pried off today, a fairly complete interior would appear (courtesy Jeffrey A. Jones).

compact car; I loved the styling of that car. The two-door compact car template that the Guild offered was a perfect platform as my design was a sportier two-seat interpretation of the Falcon.

I used layers of ¾-inch sugar pine glued together for the body. I had intended to build this car with a full interior and clear plastic top, but I way underestimated the time and way overestimated my skill level with this first model. I finally decided to carve a wood "solid top" and glue it on the body. The model car had a partial interior under that block of wood!

I bought the smallest kit of aluminum stock that was offered in the *Guildsman*. It was all I could afford at the time, and it lasted for the duration of my Guild career. I cut and filed some aluminum bars for the trim on the hood and used a piece of heavy gauge aluminum electrical wire for the rear bumper. The hubcaps were Eveready size D battery negative ends that popped right into the Guild tires. The headlights were Eveready size AA battery negative ends that were exactly ½-inch diameter per Guild specs and had a shape that simulated the bulb filament. They were filled with Elmer's white glue, which dried translucent and is a fair simulation of a diffusion lens. A perforated and flattened sheet metal aluminum hair curler made polished grilles behind the headlights, and red glass beads from a necklace made the round taillights.

I envisioned the color scheme very early with this 1962 car. After sanding and patching defects with wall spackling paste, the body was brushed with a couple coats of hardware store sanding sealer. The sealer I used was lacquer-based and sanded very nicely. I bought hardware store rattle cans of sky blue and dove gray lacquer paint and sprayed them in my bedroom on my craft card table. Final finish was sanded with 600 grit wet sandpaper and polished with DuPont rubbing compound then toothpaste. I couldn't find pumice that the Guild recommended, and I didn't know a source for fancy auto polishing compounds. It was waxed with DuPont carnauba wax.

1963 Entry (Ages 13–14)

This second model car carved from sugar pine celebrated the possibilities of the Open Competition category that the Guild introduced in 1963. That, combined with the influence of custom car builder Ed Roth, encouraged me to design something really different. In my heart I felt that Guild entries had to be so unique that they had no features in common with any other design. My designs not only couldn't look like anyone else's, and they couldn't even look similar to my previous designs. In retrospect, a better approach might have been to stick with one basic design envelope and refine it each year, as some successful entrants appeared to have done.

This design was inspired by jet aircraft with wing-shaped front and rear fenders and open wheels. It was a single-seat, experimental car with hidden, flip-out headlights and a single stick control that handled steering, braking, and acceleration. It was designed with body-colored bumpers and the rear axle was cantilevered from just behind the driver's seat on a separate T-shaped strut. That strut was painted gold to carry on the gold trim theme. I wanted a unique finish to go along with my original forward thinking design, so it was painted in burgundy candy apple and had gold tone brightwork instead of chrome.

This model entry was designed to have a clear plastic canopy. I carved a truncated cone-shaped pattern out of pine, and per the Guild's "how-to" instruction guide, I covered it in felt. I made two ½-inch thick plywood frames with identical holes in these boards about 1 inch larger than the canopy's pine pattern. A 1⁄16-inch thick Plexiglas plastic sheet was sandwiched between these two boards and screws were installed at 1 inch spacing all around the perimeter. This sandwich was heated in the kitchen oven to 325 degrees Fahrenheit. The idea was to heat and stretch the plastic sheet over the pattern. The wood frame prevented burned fingers and gave a lot of structural stability to the process.

Once the plastic, suspended in the plywood frame, sagged from the heat I quickly removed it from the oven and stretched it over the conical pattern form. This method yielded a horrible, unacceptable finish with the felt grain imprinted on the plastic. I tried several different cloth coverings for the pattern to no avail. I tried painting the mold, which gave a better finish, but the paint stuck to the plastic in places, and that wasn't good. Frustrated, I finally formed one windshield over the bare pine pattern and that worked. All I needed to do was to sand it with 600 grit, then 1000 grit sandpaper. I then polished it with some DuPont rubbing compound followed by toothpaste, and I got a spectacular-looking clear top.

Interiors were another problem area for me. The interior was upholstered with pearlescent white vinyl upholstery from an upholstery shop. To simulate button tufting, I laid out a grid on the vinyl with a pencil and pushed in common straight pins at grid intersections for the buttons. I assumed I needed cloth upholstery in my models with interiors, and my efforts with miniature upholstery were less than glorious. It never dawned on me to build painted interiors.

The brass trim pieces came from several sources. The hood ornament was a link from a lady's bracelet glued in a cavity carved in the hood, the conical brass hubcaps were cabinet drawer pulls with the shafts cut off, and the trim around the back of the bubble top was a shaped piece of brass HO gauge model train track with two flanges filed off to form a C-channel and polished.

The taillights were from an old Revell plastic model kit. The intermittent thin white-walls for the tires were cut from white cloth medical adhesive tape using an X-Acto knife blade taped to my drafting compass, then adhered to the tires.

My unique, transparent candy-colored paint job was done using rattle cans of custom enamel paint obtained from a hobby shop. I sealed the body with sanding sealer, then sprayed on a coat of gold metallic base coat and several coats of the transparent burgundy color coat. I finished the painting with a couple coats of clear lacquer. It was sanded and polished as my first car. The finish was striking back then, and candy-colored hues still give me a thrill when I see them today.

1964 Entry (Ages 14–15)

Continuing my self-imposed constraint of totally unique design, I again took advantage of the Guild's Open Category and designed this car with an offset wheel base. The driver's side wheels would be about 10 inches at full scale forward of the passenger side. I didn't work out the steering geometry, and I still don't know if it could be done. The automotive advantages I perceived at the time, for this type of wheel base, escape me now. I don't remember any particular inspiration for the car other than the offset wheel base and the possibility of needing computers to steer the car. It was a two-passenger car with dual pods, or cockpits, and no rear facing windows. I envisioned a television screen and camera system for 100 percent rearview access.

This car was carved from a solid block of black alder wood, a common hardwood in the northwest United States that was very nice to work with. It looked a lot like a dark maple but was quite a bit softer. My aunt and uncle, who lived in northern California, supplied me with an alder block from their stash of firewood. I carved the exterior of the dual pod body in one piece with canopies in their design positions. I then cut off, or excised, the solid-top canopies with a hand saw to use them as patterns for another set of clear plastic canopies. Making the top patterns this way and using the same plastic-forming method as the 1963 car, the fairly complicated shaped tops fit the body perfectly. I was proud of how well I performed this task.

The interiors of the dual pod body were carved with seats, instrument panels and crash pads, and then covered in a yellow and white corduroy cloth to simulate pleats. The dramatic black body and yellow interior color contrast was what I wanted, but the corduroy texture was disappointing. I lacked the metal working skills necessary to form metal trim pieces — so, I stooped to using narrow chrome tape strips for the metal trim around the plastic tops.

The car was brushed with sanding sealer and painted with hardware store rattle can of black lacquer. There was a small, rear-facing panel on the back of each pod painted silver with an aluminum trim made from my Guild aluminum stock kit. A set of front grille trim pieces were also made from sheet aluminum. It had body-colored bumpers. The hubcaps were chrome cabinet drawer pulls, and the floating headlight cells were common golf tees painted silver and filled with white Elmer's glue for the lenses.

This is the 1964 sports car design with dual pods, or cockpits, made by Jeff Jones (ages 14–15) for the Junior Division Guild competition. It was an Open Competition model, with dual plexiglass canopies, carved from western alder, a soft carving wood. The driver's side pod is slightly forward of the passenger's pod. There were no rear windows and a theoretical TV camera system would have been required for rear vision. The carved interior with seats, instrument panel and crash padding was covered with yellow and white corduroy cloth to simulate pleating and the body was painted with black lacquer. Narrow strips of chrome tape were used around the clear canopies. Jeff carved the car in western alder then cut the tops off with a hand saw and used those as patterns for the canopies, thus ensuring perfect fit. He employed numerous clear plastic canopies in his Guild entries and had most success with stretching the hot plastic over smooth, bare, dry wood patterns (courtesy Jeffrey A. Jones).

1965 Entry (Ages 15–16)

I was a junior in high school, worked all summer in construction with my dad, made the varsity football team at school, and had a lot of distractions I never expected. The Guild competition year snuck up on me, and I was totally unprepared. As a result, the model design suffered that year.

I carved the body using laminated mahogany. The model design had a "short cowl to axle ratio" but featured a long nose protruding in front of the wheel wells, placing it in the Open Competition category. Conventional category designs had front and rear overhang limits. Due to my distracted life, I decided to forgo the clear plastic top and interior and opted, instead, for a solid-top design. Again, bumpers were cut from more of my Guild aluminum stock kit in a simple design.

One of the features I liked best on this model was the headlight treatment. The prism-like lenses were a departure from the Guild's directive of round headlights. I don't remember if I rebelled or if the Guild allowed custom shapes that year. I made them using a homemade vacuum-forming tool built from an old vacuum cleaner motor and some wood. I made a box of thin plywood, about a 6-inch cube. The top was made from ⅛-inch thick composite peg board for the suction pad. I carved the lens shape in wood, placed the pattern on my machine and vacuum-formed heated thin plastic over them. It worked great. I made the taillights the same way and painted them candy apple red enamel.

The wheel covers were made from white pearlescent vinyl upholstery and glued to the recesses in the officially issued Guild rubber tires. The chrome-like tripod hubcap was made from thin aluminum sheet I got from the local newspaper press. They sold used newspaper page masters for pennies, and the aluminum in them was great for model work. All they needed was a wash with lacquer thinner to clean off the ink, and the metal was perfect. I cut the shape with scissors and used a large carriage bolt as a forge. I placed the disk on a piece of soft pine and hammered on the threaded end of the bolt with the head on the disk, and constant light tapping formed the disk into a perfect concave dish shape.

The finishing method I chose for this car was less than stellar. Since mahogany has a very aggressive grain, I decided to coat the whole body with spackling paste to seal the pores in the wood and achieve a smooth surface. It worked very well, and lasted a couple of years. But, eventually, the spackle dried out and delaminated from the wood — and the entire paint job fell off! The model in the image submitted for this essay has been restored; otherwise, there wouldn't have been anything to show. Oh, if I had only known about Bondo back then!

I painted the body with a coat of rattle can white lacquer, then several coats of rattle can Pagan Gold Pearlescent and Orange Candy Apple lacquer that was sold by "Cal Custom," which was a big hot rod parts supplier in the 1960s. The glass area of the top was painted with white lacquer base and plastic model transparent orange candy enamel from a hobby shop. This was a special feature I envisioned: a low-E coating for ultraviolet ray protection. All I could think of for the coating was an elemental gold vapor deposit, thus the orange color of the glass.

Carved from laminated mahogany by Jeff Jones (ages 15–16) for the 1965 Junior Division Guild competition, this solid-top, Open Competition model had a short axle to cowl ratio, but had a long nose forward of the front axle. The 2-tone body was painted Pagan Gold Pearlescent lacquer and Orange Candy Apple lacquer. Custom-shaped and vacuum-formed prisms were used for headlights. The wheel cover design consisted of pearlescent vinyl upholstery as the background in the official Guild wheel/tire recess with chrome-like tripod hubcaps made of aluminum. He painted the model with custom candy (top) and pearlescent (body) spray lacquer rattle cans made by Cal Custom, a popular hot rodding accessory manufacturer in the '60s (courtesy Jeffrey A. Jones).

1966 Entry (Ages 16–17)

The 1966 model car was to be my last since I was a senior in high school, heading to college, living in a dorm and, most probably, unable to work on an entry. I saw the trend in the late 1960s was to go to what I call football-shaped bodies, so I used that as my starting point. I was into racing 1:24 scale slot car dragsters, and a design feature I developed was to set the motor on a vertical angle, placing more weight on the rear wheels for more traction. That was the genesis of the V-shape of the body on this car as viewed from the side. Carved from laminated sugar pine, this three-wheeled vehicle was suited only for the Open Competition category. This sports sedan was a three-seater, with two passengers in front and one slightly elevated in the rear, called a "2+1." The engine I envisioned was a front mounted pancake type, similar to the Corvair. Since it was a three-wheeled car, I presumed it would require a gyro for stability. It had no rear facing window area so would have required a video rear view vision system.

It was a very simple shape so carving went quickly. Trim was minimal; there was a gas filler on the front fender made from watch gears, there were four ¼-inch diagonal sliced brass tubing exhaust tips deployed vertically on the back driver's side edge of the body, and the rear had a row of five huge conical shaped taillights all across the back panel between the wheels. They would each be 6 inches in diameter at full scale. The lights were made from a mold made from a golf tee. I shortened the tee, made a plaster of Paris mold of it and cast the lights using polyester resin from the hardware store. These jewels were painted with candy apple red enamel paint.

This is the only car I made with chrome-plated trim. I read about this brightwork option in the *Guildsman*. All the metal pieces except for the wheels were chrome-plated. I felt guilty about not doing the chrome myself. I talked my parents into buying me a chrome plating training course from an ad in a do-it-yourself magazine. I think the course cost about $200, which back then was a lot of money. I never set up a system, but I learned a lot about the subject.

One pet peeve I had, after looking at the marvelous winning Guild entries in the *Guildsman*, was what I perceived as the lack of inspiration in designing wheel treatments. Some contestants did go above and beyond, but the majority seemed to be content with some version of drawer pull, a simple lathe turning or the ever-present beveled gear.

I decided to build my own 'mag' wheels, again drawing from the drag racing world. First, I had a local blacksmith — a real blacksmith who owned a lathe — turn an outer rim in aluminum. Then I carved five wooden spokes with a wood dowel hub and used epoxy to glue all that into the aluminum rim, making the pattern for the wheels. I then took plaster-of-Paris molds from that pattern. Using a torch and steel pan, I melted some aluminum scrap from some old casting and poured it into the molds, producing my wheels. I didn't have a lathe to clean them up and already felt a huge amount of guilt having used the blacksmith for the original turning, so I finished the wheels with files and sandpaper, and then I painted them flat aluminum for the magnesium look. They screwed to the body with ⅜-inch chrome plated brass tubing stub axles. I hollowed out the Guild tires and epoxied the wheels into them.

The front single wheel has Eveready D-sized positive battery ends for caps on both sides. There wasn't much wheel or hubcap showing, as only about one-third of the tire was used, but the bump on the battery ends simulated the axle for the wheel.

The finish for the model car was more of the good old "Cal Custom" rattle can transparent candy-colored lacquer. This time I painted the whole car with metallic silver undercoat. The top was masked, and the body was sprayed a transparent lime green. Finally, the masking paper was taken off the top, and the whole car given a couple coats of clear lacquer so the silver left on the canopy would stay bright. All was sanded with 1000 grit wet paper and polished with the DuPont rubbing compound, toothpaste, and carnauba wax.

This model car was one of my best efforts in terms of workmanship and finish, but when I opened the crate after it was returned from Detroit, a ¼-inch piece of the nose had broken off. I couldn't see how it could have happened in transit, as my crate was bullet proof. I assumed it was a casualty of Detroit combat.

Lone Guildsman but No Regrets

Needless to say, I had few playmates or close friends living in the countryside, and absolutely no one was as interested in crafting like I was. Now in my 60s, I still have yet to meet another Guildsman or to see another Guild car in person! In retrospect, that is one aspect of the Guild experience that was truly a hardship for me, given that the California region with its southern California car worship mentality represented the toughest model car building competition in the nation.

Building a car and submitting it to a prestigious competition was both a daunting task and a huge honor. Even though I never won any awards, I took much away from the experience. I was encouraged to pursue a college degree from California State University, Fresno. I was the first in my family to go to college, and I have worked as a professional petroleum

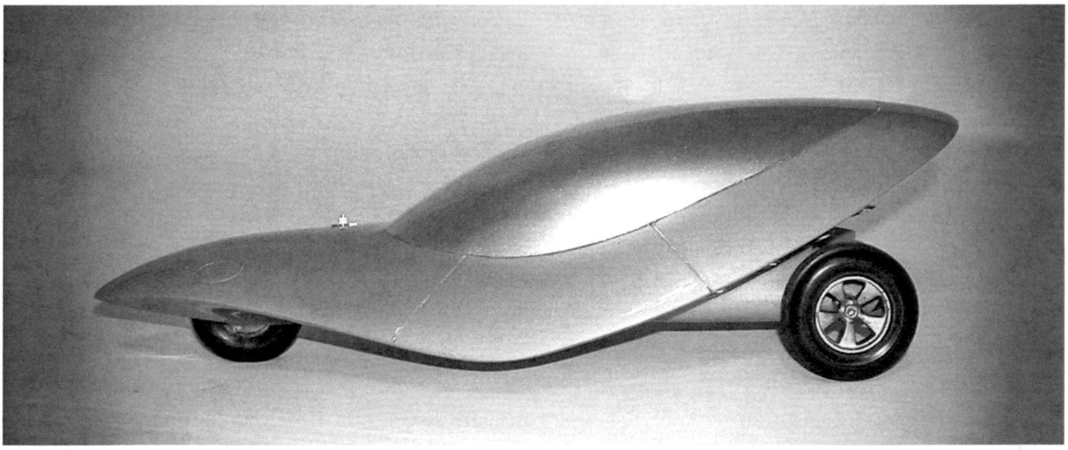

Side view of Jeff Jones' 1966 Senior Division model entry. Carved from sugar pine, this 1966 three-wheel design was made at ages 16–17 for the Senior Division. An Open Competition model with a 2 plus 1 seating arrangement, it was painted a transparent lime green color which required a metallic silver undercoat, the same color as the canopy. The whole model was painted with clear coats. Jones had noticed over the years that regardless of how great the Craftsman's Guild car styles were, most had uninspired wheels. He took a clue from the drag racing fraternity of the day and cast the rear wheels on this car in aluminum. He made his own "mag" or magnesium wheels popular in the aftermarket at this time. He built a pattern from a professionally turned aluminum rim and added five wooden spokes. He made plaster of Paris molds from this and cast his wheels from melted, liquid aluminum scrap. This was his final entry and he had succumbed to the football-shaped designs that had won awards in prior years of Guild competition (courtesy Jeffrey A. Jones).

Jeffrey A. Jones of Bakersfield, California with his 1963 model entry. He still treasures his Guild experience these many years later. He credits it with the inspiration to earn a mechanical engineering degree from California State University–Fresno, leading to a very enjoyable career as a petroleum engineer in the oil fields of central California. Jones has been quite successful authoring many technical papers. He secured four U.S. patents on oilfield equipment, authored the "Steam Enhanced Recovery" chapter in the *International Society of Petroleum Engineer's Handbook*, and was selected as the Production Engineer of the Year by the society in 2003 *Note*: Several of the photographs in this chapter show restored models (courtesy Jeffrey A. Jones).

engineer for 35 years. I give much credit for this to my exposure to the Guild, which taught me to persevere and endure whatever sacrifice and effort it takes to achieve my goals. I lament the fact that my son and grandsons have no Fisher Body Craftsman's Guild to hone their characters.

I have been an avid wood carver for most of my life and do not specialize in a single carving style.[3] My projects range from whittled caricatures to carousel animals and classic carvings. This has always been a hobby, with my true career being petroleum engineering. I am "hobby poor" with a 1,700 square foot workshop that includes a welding area, two car bays, a car hoist, a woodshop, and a fully functional paint booth. There I build hot rods, do woodworking, carving, metal work—you name it! My current project is a full restoration of a 1959 Chevy El Camino with my 15-year-old grandson.

Also, I have one bedroom of our home converted to a workshop and showroom where I do carving and model car building. The walls are lined with glass display cases for the cars, die-cast Willys model collection, Marx toy train collection, Fisher Body Craftsman's Guild memorabilia and other assorted treasures. I guess it's genetic, as my great-great-grandfather was a master wood carver in the Chicago area around the turn of the 20th century. My family still has several exquisite pieces of the furniture he made. Despite my expertise today as a wood carver, I have never approached his level of craftsmanship.

All these years later, I still treasure my Guild experience. I credit it with the inspiration I needed to earn a mechanical engineering degree that led to an enjoyable career as a petroleum engineer in the oil fields of central California. Unfortunately, I didn't keep the score sheets from the Guild competition so I don't remember what categories disqualified my cars. Fortunately, I did keep my Certificates of Design and Craftsmanship, all dated and signed by C. W. McClellan, the legendary administrator of the Guild during the 1960s—as they are treasures. Did the nose falling off my 1966 sugar pine model keep me from becoming the next John Z. DeLorean protégé? I will never know. But one thing I firmly believe is that if I had been able to have contact with other Guild members I could have gained the knowledge necessary to make the next step in model quality and merit an award or two. One can dream can't he?

Notes

1. The submission of a model to the competition satisfied the requirements for earning the model building merit badge, thus helping Jeff earn the Eagle Scout rank. He stated, "A recent acquisition I made from a www.ebay.com auction is a copy of our beloved Craftsman's Guild 'Designing and Building a Model Car' pamphlet, reprinted in a small hardcover book by the Boy Scouts of America as a model building 'how-to' guide."

2. The Greene Brothers from Medina, Washington, were a Guild aristocracy; Thomas F. Greene won the 1955 First Junior National Scholarship ($4,000), Joseph S. Greene won the 1956 Fourth National Senior Scholarship ($2,000) and Daniel C. Greene won the 1957 First National Junior Scholarship ($5,000) for a total of $11,000. If you include Norman Greene, a relative from Medina who won a 1958 National Styling Scholarship ($1,000), the total for the family was $12,000 in GM trust funds.

3. Jeff Jones, "Recreating a Master Piece: Turn of the Century Table Inspires a Life Long Love of Carving, Three Generations Later," *Wood Carving Illustrated*, Spring 2007, Issue 38, p. 42.

26

So, "Which end is the front?"

Robert W. Lawhn (1961–1964, 1966, 1967)

For as long as I can remember, I have been interested in automotive styling. As a youth, this fascination was fueled by the brilliant automotive designs that appeared in the market place annually and led to my incessant sketching of advanced conceptual car designs. In fact, I once sent Chrysler drawings suggesting improvements for their "Forward Look." I frequently attempted to redesign the cars in model kits, with varied results.

While thumbing through a *Boys' Life* magazine in November 1960, I discovered an advertisement showing model "dream cars" and offering a book titled *Designing and Building a Model Car*. Thrilled at the prospect of designing and building my own models, I promptly mailed the request form. I received the Fisher Body Craftsman's Guild package of information and quickly realized this was a competition. I had no idea such a program existed. Since entries were due the following June and it was already December, I seriously doubted I could complete a model in time to enter the contest.

1961 Entry (Age 14)

For my 1961 entry, I sketched a car that incorporated styling themes from two of my favorite cars — Ford's Thunderbird and General Motors' Firebird II concept car — for what was to be a 4-passenger, 2-door sports coupe. I purchased balsa wood and applied the skills of a 14-year-old plastic kit model builder with the help of a vise, a grinding wheel and a few hand tools.

After shaping the body with saws and files, I used a blackboard eraser wrapped in sandpaper to smooth the body surfaces. I cut a section from a hacksaw blade then ground the teeth to create a tool to score body cuts. The bumpers were crafted from an aluminum rod while chrome furniture leg casters were used for wheel covers on the Guild tires. Metal screen was used to fill four separate grille openings and red toothpaste caps were glued to a chrome-plated backing and mounted as pontoons behind the rear wheels to serve as taillights.

I painted the body with four coats of a GM production metallic rose red lacquer called "Sunset Glow" and completed the model in time to enter the 1961 competition. I won a Junior Division Honorable Mention Award for Texas ($25). Astonished that I could win anything on the first attempt with just six months to build the model, I promptly began designing an entry for the 1962 competition.

1962 Entry (Age 15)

I focused on originality for my second entry, a rear-engine sports car, using the "thought starter" sketches provided by the Guild. My design incorporated the shape of a door wedge and featured a flat, continuous surface that ran from the pointed front to the peak of the roof. The model was built using the same materials and construction methods as my first entry. Other than the personal stories published in the *Guildsman* newsletter, I had no sources of new model-building ideas, as I didn't know of any other Guildsmen in school or my community. My winning strategy was simply to improve on the use of familiar materials and methods.

The bright red paint finish was improved over my first model, since I took more time to seal the wood and apply six coats of lacquer. Like my first entry, the design featured retractable headlights, space for license plates and a screen grille, but no additional details. Thirty days before the submittal deadline, I had yet to make the bumpers and realized I had failed to adequately plan the project. I quickly crafted bumpers out of an aluminum sheet and submitted the model, uncertain if I would win anything. Someone must have had a higher opinion of the model than I because it received the Junior Division Second State Award for Texas ($100).

I had yet to meet another Guildsman or see another Guild model entry up close, so I had no way to gauge how my models compared with my competition. I spent hours studying pictures of scholarship winners, wondering how those kids built such beautiful models. Shortly after my model was returned to me in July 1962, a newspaper reporter from the *Houston Chronicle* called requesting an interview about my Guild award.

When I arrived at the reporter's office, I was greeted by three other Guildsmen from Houston. Two were college students (Robert L. Straight and John M. Steiner) whose models had impressive metal work, trim detail and paint finish. The third student (Michael S. Reese) had previously topped me in the Junior Division and won the Regional Award. Chatting with these Guildsmen, I learned the benefits of using core materials other than balsa wood, how to use Plexiglas for lights, how to adapt chromed automobile dashboard knobs for wheel covers and tips for better paint preparation of the model surfaces. As a result, I completely changed the way I built my next entry.

1963 Entry (Age 16)

I moved to the Senior Division in 1963 and entered a 4-door station wagon, thinking it would stand out among all the sports cars. For a while, I tinkered with the concept of a 9-passenger station wagon with an elevated third seat row to allow passengers to view passing scenery through an elevated "greenhouse." But my designs of this concept appeared awkward, so I shelved the idea. In 1964, GM introduced its Vista Cruiser wagons with a stylish, although not so functional, version of the concept.

I utilized "guided missile" design themes to create a station wagon with a pointed front and rounded body and glass surfaces. The design featured full-length side glass with no visible pillars and dual grille openings to accommodate a conceptualized Chrysler-style turbine engine. Part of the roof would open to allow tall objects to be placed in the back. I switched to pinewood as the core material and found some attractive chromed Chrysler dashboard knobs that readily fit the Guild tires to serve as wheel covers. Lights were crafted from pieces of Plexiglas I fished out of a trash bin at a plastics shop, while bumpers were fashioned from scrap pieces of aluminum.

From a jeweler, I acquired two plastic watch faces to use for headlights and a winding knob to simulate an exposed gas filler cap. I used aluminum wire to craft door handles placed at an angle on the body to mitigate twisting your wrist while opening the doors. I painted the body with seven coats of metallic red lacquer with black-indicated window

Robert W. Lawhn's (age 18–19) 1966 Senior Division, First State Award for Texas ($150) and Regional Award winning model. A sports car with a wraparound "greenhouse" was envisioned with no visible pillars and a cantilevered roof line. The fastback model was constructed from white pine, coated with a wood sealer and painted dark turquoise automotive lacquer with black indicated window areas. This was the original color scheme. The model had a full-width rear taillight; bumpers were chrome-plated brass, window moldings were stainless steel wire and the wheel covers were cut Ford chromed dashboard knobs. In response to a Guild judge's comment that the color scheme lacked contrast, Robert Lawhn later tried a silver and black color scheme and recently (February 2009) settled on a dark turquoise and white color scheme (courtesy Robert W. Lawhn).

areas, and then, touched off the rear end with four exhaust pipes. The finished model was in another league from my first two entries and it earned a 1963 Senior Division First State Award for Texas ($150).

The same year, Robert Straight and John Steiner finished Second and Third State, respectively, behind me in the Senior Division, while Michael Reese won another Junior Division Regional Award. There was just one other entry between me and the award I wanted most — the Regional Award — and that was Richard Bergez from New Orleans. Under the Senior Division, first place winners from Texas and Louisiana were Region 16 competitors, and he had bested me. I had no confidence that I could ever build a scholarship-winning entry.

1964 Entry (Age 17)

Feeling that my craftsmanship skills had sufficiently improved for the 1964 competition, I chose to enter the Open Category with a sports car that I hoped would improve the originality of my entry design. The Open Category allowed for unusual passenger placement configurations, unique methods of vehicle entrance and exit, accommodations for passenger comfort and convenience, and various placements of the tires, thus generating a unique array of body styles.

The shape of a helicopter, with its bubble-like passenger compartment and tapered body, was my inspiration. The layout featured a back-to-back, or a tandem seating configuration with the engine located between the passengers that made access to all interior spaces very challenging.

As with my previous entry, I built the model out of pine and used Plexiglas for lights and aluminum for bumpers. I incorporated provisions for an exhaust, door handles and a gas cap, but these features were simplified in the hope of appearing more futuristic. I departed from the usual red body color and chose light metallic tan with black-indicated window areas. The model placed a disappointing Senior Division Third State for Texas and garnered $50. My Houston nemesis, Michael S. Reese, won the 1964 Fourth National Senior Scholarship worth $2,000.

No 1965 Entry (Age 18)

Discouraged that I had not been able to win the Regional Award, I focused on my freshman year in college and declined to build an entry for the 1965 competition.

1966 Entry (Age 19)

My absence from competition was short-lived. During the summer of 1965, I decided to make another attempt at winning the Regional Award, but felt I needed to significantly improve my design and details. Instead of beginning with four tires on a "shoebox" sketch, I began with a sketch of the windshield. I envisioned a two passenger conventional sports car with a wraparound "greenhouse" with no visible pillars and a cantilevered top. I sketched a glass structure resembling a curved sun visor with tapered ends, drew a body of similar shape, then added the tires. A full-width taillight spanned the space between the tapered rear fenders. Suddenly, I was looking at a design unlike any I had ever created.

I retained pine as the core material but used plastic reflectors from an auto parts store for lights. Bumpers were made from chrome-plated brass instead of aluminum, and stainless steel wire was added for window molding. The wheel covers were made from Ford chromed dashboard knobs. I painted the body with ten coats of a GM dark "Twilight Turquoise" lacquer with black-indicated window areas, a decision I would later regret. The design and craftsmanship of this model was noticeably advanced over my previous entries. When a large envelope arrived from the Guild in late June, I knew I had won the 1966 Regional Award (Region 16,Texas and Louisiana).

1966 Guild Convention

For the first trip in 1966, GM provided me with tickets on an airline notorious for losing luggage, and I became another casualty. When I arrived at the hotel empty handed, I was ushered to a room where a tailor took my measurements. I then joined other Guildsmen on a trip to a movie. When I returned to the hotel, there was four days' worth of casual clothing and toiletries waiting for me. What a way to start the trip!

On the first day, we visited the GM Proving Grounds in Milford, Michigan, and saw the new 1967 Chevrolet Camaro. We ate lunch outdoors adjacent to a "bone yard" of wrecked vehicles. Intrigued by a strange profile under a car cover, one Guildsman reached through the wire fence and lifted the canvas, unveiling a wrecked 1957 Corvette SS race car. Boys rushed toward the fence but Guild officials quickly restored order. After lunch we traveled to the GM Technical Center where the Mako Shark, Corvair Monza GT and Corvair Monza roadster concept cars greeted us and the Stingray race car waited to take us on exhilarating rides. As I entered the Monza GT under its canopy opening, I decided right there my 1967 model would incorporate a canopy design. The awards banquet that evening was a marvelous event that celebrated what the Guild was truly about. It was also the first and only time we were able to view the other Regional Award models.

On the second day, we visited a manufacturing facility and spent the afternoon at a lakeside park, releasing steam by playing baseball and swimming. That was a prudent action by the Guild because that evening we went to the Design Center, where we became ecstatic admiring all the models, drawings and the Pontiac Banshee concept car.

The next day, we boarded a train for Kalamazoo to visit a body stamping plant. We returned late in the day so I had plenty of time to confer with other Guildsmen and the designers who had judged my model. The designers complimented my design but one suggested the minimal contrast between the dark turquoise body and black windows did not allow the main features of my design, the wraparound glass and cantilevered top, to be as prominent as they could be. On Thursday we headed home, heads filled with memories that would last a lifetime. A few days later, my lost luggage appeared.

I did not win a scholarship at the 1966 Convention in Detroit, but I acquired several new design and craftsmanship ideas to apply to what would be my next and last entry due to my age. I also gained a sobering appreciation that my competition would likely include students attending prestigious design schools. I had paid a price for the time I devoted to my Guild model — I failed a computer science course that haunted my GPA until graduation. But the Regional Award was well worth it.

1967 Entry, Age 20

For my last entry, in 1967, I returned to the Open Category because I felt it offered the best opportunity to create an original design. The wedge-shaped, aerodynamic design featured sweeping lines from the front over the roof which emphasized the aerodynamic flow. The body platform was a standard rear-engine sports car with the front tires moved inward toward the centerline. This configuration enabled the car to have a pointed "leading edge" from which every surface would emanate, flowing back, out and up. Air would enter the engine through a ribbed grille along the top of the car. A special feature of the design, from a conceptual point of view, was radar-enhanced rear vision.[1]

Access to the interior would be provided by a canopy opening that included the top and window surfaces. The entire front and rear sections would open to provide access to storage and the engine.

I allowed time in the schedule to seal the pine wood with fiberglass resin and applied ten coats of lacquer. The result was a drastically improved paint finish. I returned to metallic red as the body color with black windows for sufficient contrast. Chrome-plated brass was used for bumpers as well as the grille. The full-width taillight was made from red reflective plastic, while stainless steel wire served as window molding. I used Ford chromed dashboard knobs again, but painted the centers black to create a different look.

Again, I won the Senior Division First State Award for Texas ($150), Regional Award and was satisfied having twice achieved my objective. The trip to Detroit's national convention

With this design, Robert W. Lawhn (age 20) of Houston, Texas won the 1967 Senior Division First State Award for Texas ($150), Regional Award and National Styling Scholarship ($1,000) award. What he created was a rear engine, wedge-shaped, aerodynamic Open Category sports car design. The front end styling was tapered, resulting from a narrower front tread width compared to the rear. The model was carved from white pine, covered with fiberglass resin and painted in metallic red with black indicated window areas. Bumpers and grille were chrome-plated brass, window moldings were stainless steel wire and wheel covers were cut Ford chromed dashboard knobs. Occupant entry was by way of a canopy which included the top and window glass surface. The designer envisioned a radar enhanced rear vision system (courtesy Robert W. Lawhn).

Robert W. Lawhn with his 1967 National Styling Scholarship model in-hand at the Museum of Fine Arts, Boston, model "Dream Car" exhibition and retrospective held August 9–10, 2008. Lawhn participated in the Guild competition six times between 1961 and 1967. Four of his entries won State Awards for Texas, while his 1966 and 1967 models won Regional Awards and the 1967 entry also won a National Styling Scholarship ($1,000). Lawhn was attending the University of St. Thomas in Houston, studying physics, when he constructed his two regional award winners. Professionally, he is celebrating his 37th year in the electric power field and is a project director for a new power plant development in California for GenOn Energy. Lawhn and his wife, Dorla, reside in Henderson, Nevada (courtesy Ronald J. Will).

in 1967 was similar to the previous trip, except this time my luggage arrived without delay. We saw the new 1968 Corvette and took an enjoyable train trip into Canada to see a Shakespearean play. We were also introduced to High Density Polyurethane Foam as a model body material. I sure wish I'd had that foam in 1961.

When my model appeared on the giant screen at the awards banquet in Detroit indicating I had won a National Styling Scholarship of $1,000, I felt a tranquil satisfaction about all my Guild efforts. Despite the award, I had one year of college under a full scholarship remaining and did not use it. After the trust holding the scholarship learned I had graduated, I received a check for the full amount. The Regional Award trips remain the most memorable part of my Guild experience. I will always remember the quality and dignity with which the trips were conducted.

Despite my improved craftsmanship and eventual success, there was one aspect that persistently challenged me. Creating smooth, narrow scored lines for body cuts required more time and patience than I could afford. I never found a tool that enabled me to score neat lines in a short amount of time. Eventually, I just cut lines that were too wide, filled them with wood filler then ran a sharp blade through the filler before it hardened. I eventually achieved clean scored lines, but it required multiple passes over every line.

Gratitude

As a boy, I loved to visualize conceptual car designs. The Fisher Body Craftsman's Guild enabled me to express my interest in cars in a way most boys could not. It required me to forego many normal teenage activities, but I always felt I was pursuing something very special. It required a lot of work and at times was very frustrating, but I found unparalleled excitement in the challenge of building a Guild entry. The Guild was the catalyst that transformed my simple boyhood fascination with cars into an extraordinary and unique experience that has ultimately become part of automotive design history. I will always be grateful for the privilege to be a Guildsman.

I am also grateful for the friendship along the way of other Guildsmen, like Robert Straight, John Steiner and Michael Reese, with whom I shared ideas and improved my game. I appreciated seeing all the finished models of the top regional model builders at two Guild conventions. The Guild publications provided valuable information, but it wasn't the same as conferring one-on-one with other participants in your community.

My family, friends and teachers in Texas had little interest in the Guild and did not appreciate its beneficial values. There was always someone who would ridicule my models by asking "which end is the front?" implying the design was so eccentric one couldn't tell. I envy the Guildsmen who had the availability of clubs, school shops and other support to enhance their Guild experience.

But, in the four decades since my Guild participation, I have reflected on what the program has meant to me. It certainly provided hours of great enjoyment and yielded memories that will last a lifetime.

The experience honed skills that have proved valuable throughout my professional career. After earning a graduate degree in nuclear engineering from Texas A&M University, I joined a major electric utility in Texas where I was involved in the development of new nuclear, coal and gas power plant projects. Later in my career, I managed support functions for business expansion. Skills developed through the Guild such as planning, work execution, attention to detail and a drive for continuous improvement have been invaluable to my career. I am currently a project director for a new power plant development in California for GenOn Energy and reside with my wife, Dorla, in Henderson, Nevada.

Conclusions

Years later, I repainted the body of the 1966 regional winner silver, keeping the black indicated window areas, in order to better appreciate the enhanced color contrast possibilities mentioned by the Guild judges.

Over the years, my scholarship model has remained in good condition, although my repainted 1966 regional winner did not fare as well. It suffered several chips in the paint, a cracked taillight and the loss of the front bumper, grille and turn signals. In December 2006, I embarked on an effort to restore the model to its original condition.

I stripped the model down to the bare wood and prepared the surface to be repainted its original dark turquoise color. Remembering the judge's comment, I made one deviation from the original — I painted window areas white. The missing or damaged parts were recreated in the same manner in which I originally crafted them. The restoration was completed in February 2009. It reminded me how much enjoyment I had building Guild models — as well as how much work was involved!

Notes

1. *Model Car and Science*, October 1967, p. 43.

27

Blood, Sweat and Fumes: Twin Rivalry Equaled Twin Scholarships

Tom W. Graboski and Terry P. Graboski (1966–1968)

In the 1950s you were either a Ford family or a GM Family. We were a Ford family, so getting any Fisher Body Craftsman's Guild information by visiting a GM showroom was out of the question. On our eleventh birthdays (1959), family friends gave us (the twins, Terry and Tom) an information package on the Guild competition, but we were too busy making models of airplanes, army tanks and trucks and then blowing them up with cherry bombs to notice. The information was discarded somewhere. After saving our allowances, we bought our first 1:25 scale AMT plastic model car kits. We got hooked on model building, the models and trophies multiplied to fill our den, and subsequently, this nurtured our Guild success which served as a segue to our fine arts studies and careers in design.

In 1962, Tom at age 14 won the first of what was to be a room full of trophies for model cars: second place at Miami's legendary Orange Blossom Hobby Shop's model car contest. Soon we both had over 50 trophies in the den of the house where we did homework and built customized model cars. Building all those 1:25 scale plastic model cars proved to be good practice for building Guild models.

For years, we had gone to the annual auto shows when they came to Miami. By bugging our parents, we also got to attend many of custom car and hot rod shows, seeing custom cars like Ed Roth's *Beatnik Bandit* and *Outlaw* and others like *El Matador* and *Silhouette*. Then building model cars got serious for us! We started customizing the models using plastic wood putty and candy apple paint jobs out of the little spray cans. We were getting pretty good at it.

In 1966, high school graduation loomed and college prospects looked grim due to the lack of resources. We were age 18, and the Guild's promise of a scholarship began to seem like a good idea. We located the General Motors Guild information we had, sent in for a kit of rules and regulations, and began to investigate model materials and sketch ideas. We were seniors in a commercial art program at Miami Central High School, but there were no workshop facilities available. No one at the school was familiar with the Guild, either. Orange Blossom Hobby Shop offered some help with tools and equipment, but what we didn't know was huge.

1966 Entries (Age 18)

We carved our first Guild models from balsa wood because we could not find basswood as described in the informational packet, and used a confiscated 3-inch switchblade taken from a young Tenderfoot Scout as our carving tool. It had a very thin blade of poor steel, but if sharpened continuously, it cut the balsa better than anything we had at the time. Detail trim was carved into the body with old dental tools. The only other tool we used was a power Dremel tool.

The carving was done by eye using only a few basic templates to achieve symmetry. Using automotive lacquers, the paint jobs were yellow for Tom's model and 1965 Mustang Poppy Red for Terry's. The black indicated windows areas were painted with an aerosol spray can. Painting the colors was okay, but getting the black to dry without fogging in the humid Miami air took work and ingenuity.

Tom's 1966 model design was a three-wheel suburban run-about, and Terry's was a coupe with the four wheels in a diamond-shaped wheel base configuration. The models were finished on time and sent to Detroit in early June. In August, we received a letter, a certificate, and a check for $100 for a Senior Division Second State Award for Tom and a check for $50 for a Senior Division Third State Award for Terry. This was better than trophies! Now we had to get serious for the 1967 competition.

When they interviewed us about our 1966 wins, our mother was mentioned in *The Voice,* a local Catholic newspaper from the St. Vincent de Paul Parish, stating her Guild philosophy: "Although she admits that the balsa wood from which they now make their

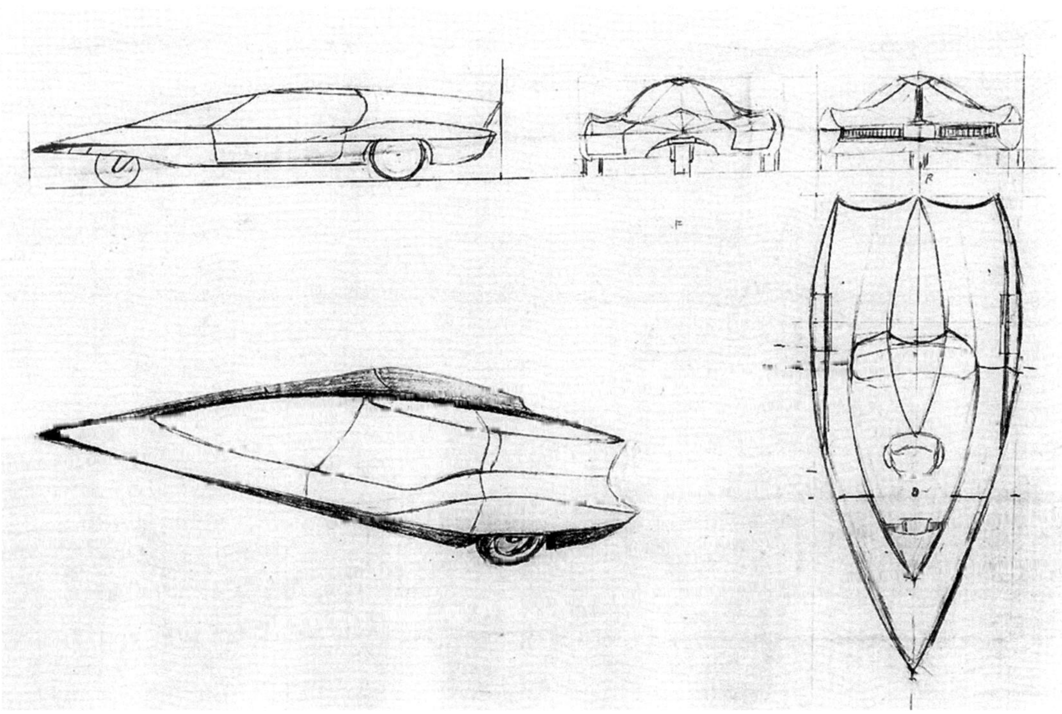

Design sketch for an Open Competition model featuring a triangular wheelbase drawn by Terry P. Graboski for the 1966 Guild competition (courtesy Terry P. Graboski).

Terry P. Graboski (ages 17–18) of Miami, won the 1966 Senior Division Third State Award for Florida ($50) with this entry. Carved from balsa wood, the 2-door coupe had a diamond-shaped wheel configuration and qualified for the Open Competition. It was painted a 1965 Mustang "Poppy Red" automotive lacquer, with a black indicated windshield area. The detailed trim was carved into the body using dental tools and a power rotary Dremel tool. Symmetry of the model was maintained by the "rock-of-the-eye" method and a few basic templates (courtesy Terry P. Graboski).

models sticks terribly to the rugs throughout the house, she believes parents should encourage and work with their children in whatever it is they are interested in doing."

1967 Entries (Age 19)

By now we were both fine art majors at Miami Dade Community College. We still had no shop or power tools. Getting a block of balsa of sufficient size was also a problem. One night, an insurance man came by our house to talk to our parents. We showed him

Tom W. Graboski (ages 17–18) won the 1966 Senior Division Second State Award for Florida ($100) with this design. Carved from balsa wood, the three-wheel, suburban run-about was painted yellow with black indicated windshield areas. It was designed for the Open Competition with two wheels in front and one wheel in the rear. All detailed trim was carved into the body using dental tools and a rotary Dremel tool. Symmetry was maintained by the "rock-of-the-eye" method and a few basic templates (courtesy Tom W. Graboski).

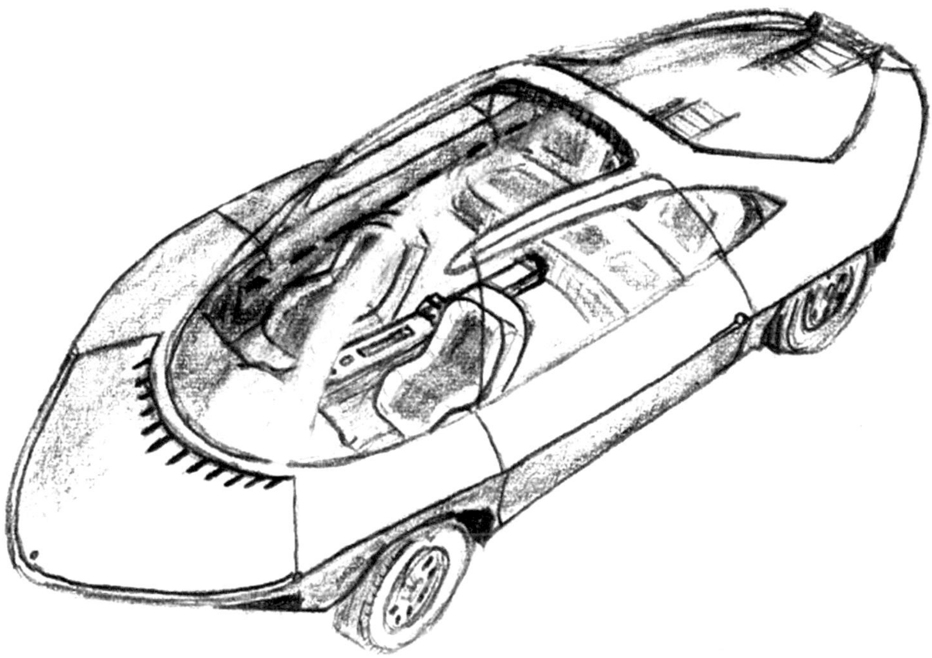

Design sketch of a model with a large glass canopy area, drawn by Terry P. Graboski for the 1967 Guild competition (courtesy Terry P. Graboski).

our 1966 models and lamented the balsa wood situation. Two days later when we came home from school there was a 12-inch round, 5½-foot long log of balsa wood sitting on our front porch. It seems the insurance man had once started a large model plane project but later abandoned it. We were in luck.

Getting the log planed to a rectangular block took lots of calls to local cabinet shops. Finally, we found a man who opened his immaculately clean shop on a Saturday and ran the log through the planer to give us a usable shape, although it left what looked like a 5-foot high pile of shavings on the floor. He also gave us a set of used, very small wood carving tools with which to carve our models. We were on our way!

Terry's 1967 model design was a metallic blue teardrop-shaped coupe with a large, louvered rear deck, solid-top construction and carved from balsa wood. The soft balsa wood surface was sealed with a hard fiberglass resin finish. We had learned about inlaying sterling silver wire into grooves for window trim and using X-Acto saws to cut in door, hood and trunk opening grooves. We even had some of the brass bumper trim pieces chrome-plated. Our craftsmanship was way better than in 1966.

Terry's 1967 metallic blue, two-door sports coupe model was designed for the open competition. The rear boasted a narrower track width, where the body was tapered, compared to the front track. It received 99 out of 100 points for design originality. Chrome window trim was simulated using silver wire glued into grooves and sanded flat. Terry's model, which took six months to complete, featured concealed head lights, "Baby Moon" wheel covers, and a louvered rear deck for cooling the rear engine power plant. The rear racing stripe was used to conceal damage the model incurred from a last minute fall from the work bench.

Tom's 1967 model was a light metallic green sports coupe with solid-top construction and carved from balsa wood. It was a very aerodynamic, a wedge-shape from the side view

Tom W. Graboski (ages 18–19) won the 1967 Senior Division Second State Award for Florida ($100) with this design. This 2-seater sports coupe was carved from balsa wood, sealed with hard polyester resin and painted a light metallic green automotive lacquer with black indicated window areas. Inlaid sterling silver wire was filed flat for window trim. The model was used by GM in an exhibit which toured the Southeast United States to promote the Guild for 6 to 8 months. This model had red stripe tires and outside rear view mirrors (courtesy Tom W. Graboski).

in which the windshield flowed up from the hood to the mid-line of the roof creating a large, expansive area of greenhouse glass. The indicated window areas were all painted black. There were triangular themed front air intakes and complementary triangular-shaped rear taillights.

The building went well until May, when the models were ready for paint. It rained every day that month. The painting routine went something like this: we would arrive home from school by 3 o'clock in the afternoon while our parents were still at work. We would turn on the window air conditioner in the living room, then close every window and door in the house except the one to the den. After an hour or so the air conditioning cooled the house and lowered the humidity. We would then spray paint both cars quickly in the den and run the cars into the living room to dry closer to the air conditioner. The den was then vented by placing a fan in the windows to suck the fumes out. I'm sure we lost a few brain cells painting indoors that way. But the black paint just would not stop fogging. Our Scoutmaster suggested we could use his air-conditioned office to paint the final color. We went there twice on a Saturday, and it finally worked — no fogging.

An unpublished photograph from *The Voice* (1967) of the Graboski twins Tom W. Graboski (left) and Terry P. Graboski (ages 18–19) with his 1966 model entry in-hand. *The Voice* ran an article about Terry P. Graboski (age 19) of the St. Vincent de Paul Parish, who had just won a National Styling Scholarship in the 1967 Guild competition. The winning entry was still in Detroit. This photograph of Tom and Terry was taken at their work bench at home, where they built all their Fisher Body model entries, customized their AMT plastic car model kits and re-engineered their slot cars to go faster (courtesy A.R. Soto, Florida/Catholic of Miami, Archdiocese of Miami).

The models were again sent in on time, and the telegram and letters again arrived: Tom again won the Senior Division Second State Award for Florida of $100, while Terry took Senior Division First State Award $150 and the Southeast Regional Award for Region 8, which encompassed Florida, Missouri, Mississippi, Alabama and Georgia. Terry also received an invitation to attend the 1967 Guild Convention in Detroit. We were thrilled!

Terry's 1967 Guild Convention Trip

The summer of 1967 was a strange year to go to Detroit. Civil unrest and rioting were breaking out all over town. Telegrams arrived from GM assuring our parents that Detroit was a safe place to visit, even if half of Detroit was burning. National Guard troops and tanks were in the streets and buses were being shot at on the freeways. Somehow our parents agreed to let Terry attend.

His convention experience was fantastic. Awaiting him were a new suit of clothes, an official Guild duffle bag and a formal-looking program outlining the convention's events, which included going to Stratford, Ontario, on a special train to see a Shakespeare play, touring the Tech Center, riding in the GM concept cars — the Corvette Mako Shark and Corvair Monza GT's — what more could a young man want?

The 1967 Guild convention was held at the Northland Inn away from the riots, but there was still a curfew over the entire city. Anyone caught out after curfew was in big trouble. But that didn't faze us enterprising Guildsmen one bit. Room service! We called up a pizza joint at 9:30 PM and ordered pizzas to be delivered to the hotel. They arrived after the 10 o'clock evening curfew and it tasted even better knowing that the deliveryman had broken the law.

The best part was seeing the actual models of all the competitors up close and talking to other winners about their building experiences and techniques. The craftsmanship on the winning cars was so good! The best ones seemed to be made by the seniors at the Art Center College of Design. Go figure! How could mere mortals from Miami compete with Art Center's model shop and instructors? It just showed us what had to be done for next year and to try even harder to achieve the top award.

Tom sat with our parents in the house that night in August 1967 listening to the national radio broadcast to hear the overall results from Detroit. Terry was awarded one of the $1,000 National Styling Scholarships.

Tom's 1967 model did not return from competition right away. We received a letter saying it was going on tour with the Fisher Body Craftsman's Guild representative who worked the Southeastern United States territory giving presentations to area high schools. It spent the better part of a year in a box in the back of a station wagon. When we got it back, the model had clearly suffered from the experience, as all the paint was cracked and crazed. But a lot of kids saw that car. One of the judges told Terry at the convention that Tom's model had been considered for a styling award as well.

1968 Entries (Age 20)

Our last Guild competition would be 1968, since we would be too old after that. We didn't know it at the time, but it would also be the last Guild contest for everyone.

Terry P. Graboski (ages 19–20) won the 1968 Senior Division First State Award for Florida ($150) and Regional Award (Florida, Mississippi, Alabama, Georgia) with this teardrop-shaped, 2 door sports coupe, painted dark metallic green automotive lacquer with black indicated window areas. The model was carved and shaped from closed-cell, rigid polyurethane foam, which was coated with a hard polyester resin shell for surface strength. This supported the scored door opening grooves and inlaid sterling silver wire filed flat for window trim. This was an Open Competition model with a narrower rear wheel track width. A white and maroon decal encircles the nose just behind the grille (courtesy Terry P. Graboski).

A year earlier, the Guild introduced a new modeling material to the Guildsmen called closed cell, rigid polyurethane foam. Terry decided he had to have it for the 1967–1968 competition, but GM took a long time to send it to the Guildsmen who sent in $25 for a foam block. It did not arrive until January 1968 — half of the competition year had been wasted.

Tom couldn't wait. There was just enough balsa from that old balsa wood log to make one more car. By Christmas 1967, he had completed a three-wheel suburban car, featuring a triangular wheel base with a single wheel in the front. It also had inlaid silver wire window trim and a pretty good paint finish using a pale turquoise color. Tom's first of two entries for 1968 was completed.

Then the foam arrived. It was a block about 24 inches long by 6 inches high and 8 inches wide. After Terry cut out his basic form, there was a nearly 9-inch piece left over. It was not enough for a full-size car, but Tom was bored just watching his brother carve his model. He started doodling on a mini-size car to fit the small foam piece. It was mid-January 1968 when the Guild process began again for Tom on model entry No. 2.

Terry's 1968 Guild model was a dark metallic green, two-door, teardrop shaped sports coupe carved from closed-cell polyurethane foam. It was designed for the open competition category with a wide front track and a narrow rear track.

Tom took the leftover block and carved a very small, dark metallic green automotive lacquer, urban electric commuter car, similar in proportion to the recent Mercedes-Benz Smart Cars, or micro cars, being imported into the United States some 40 years later. It was an open competition model with a trapezoid-shaped wheel base. Once the body shape was finalized, the closed-cell polyurethane foam was coated with hard polyester resin shell to give the surface strength. This provided a firm foundation for mounting the inlaid silver trim, scoring door, ventilation louvers and hood opening grooves, and preparing the body for a flawless paint job.

The smaller size of the Tom's design created a problem as the GM supplied rubber wheels/tires were the wrong size. That was solved by putting one of the wheels into a power drill and cutting off about ⅜ inch of the radius for what then resembled a wide flat slick. Tom made eight halves of fiberglass, which produced the four complete mini-tires called "Rally Wheels," each with the word "Firestone" printed in place of the whitewalls. Racing decals were applied to the finished paint job, as were outside rear view mirrors.

Painting in South Florida was still a problem as were costs. We both used the same dark metallic green for Terry's 1968 model and Tom's 1968 (second model entry). Fogging and some primer problems plagued the process. The black paint kept fogging, which again put us in the Scoutmaster's office once more. There were a few blemishes, some covered by convenient colored decal strips, and off they went — three models — two of them Tom's and one Terry's.

The Guild letters came in late July. Tom's urban mini-car won the Senior Division Second State Award for Florida ($100) and Terry, again, won Senior Division First State Award for Florida ($150) and the Regional Award, plus a trip to the 1968 Guild Convention. Tom was thrilled for him, but ... that wasn't all!

The second trip to the convention was less memorable for Terry. He had seen the models from the Art Center College of Design students and had pretty much made up his mind that that was where he wanted to attend college. Then, the shocker! The Guild was over! GM had killed it. There were no more sponsorships and no more competitions. We were too old to compete, but the thought of it not being there for the next generation of dreamers was very sad. They said that the kids no longer wanted to build models and that the competition's quality had dwindled.

Tom and his parents again waited anxiously for the radio show from Detroit in August 1968 and while Terry did not place in the money for the top senior awards, when they announced the scholarship winners, Tom's name was announced. The urban mini-car had won a National Styling Scholarship award worth $1,000. We were now even.

One of the judges at the 1968 convention told Terry that his model had placed fifth

Terry P. Graboski with his 1967 National Styling Scholarship model. He built Guild models in 1966, 1967 and 1968. He graduated from Art Center College of Design in 1971 with a bachelor of fine arts degree in advertising and lives in Santa Monica, California, with his wife, Constance Beck, and son Paul. He formed a business partnership with his wife called Beck and Graboski Design specializing in architectural signing, way finding and environmental graphics (courtesy Terry P. Graboski).

Tom W. Graboski with his 1967 Senior Division Second State Award for Florida, model in-hand at MFA Boston. He graduated from the Art Center College of Design in 1971 with a bachelor of fine arts degree in interior design and lives with his wife, Gina, in Coral Gables, Florida, with their two daughters, Morgan and Alexis. He is a graphic designer specializing in architectural graphics and signage and is the President of Tom Graboski Associates, Inc. (courtesy Tom W. Graboski).

in the senior scholarship competition. No award but quite an honor nonetheless. Terry was not eligible for another National Styling Scholarship, as GM only allowed one per competitor.

Still, Terry started Art Center College of Design in 1968 with his Fisher Body Convention roommate Larry K. Eby, who had participated in the Guild from 1962 to 1968, winning a 1967 National Styling Scholarship ($1,000) and a 1968 Second National Senior Scholarship worth $4,000. Tom started a year later after going to the University of Michigan and then transferring to Art Center. We graduated from Art Center together in 1971.

Neither of us pursued careers in automotive design. Both of us became graphic designers specializing in architectural graphics and signage. We have our own businesses on separate coasts with Tom in Miami, Florida, and Terry in Santa Monica, California. Tom lives in Coral Gables with his wife, Gina, who is also a designer, and their two teenage daughters, Morgan and Alexis.

Terry lives in Santa Monica with his wife of 38 years, Constance (who is from Detroit), and their 21-year-old son, Paul. Terry established the Beck and Graboski Design in Santa Monica in 1978 with his wife-partner specializing in architectural signing, way finding and environmental graphics. His clients have included Walt Disney, Universal Studios, Twentieth Century–Fox, Los Angeles County Museum of Art, the City of Los Angeles, and the City of Santa Monica. Terry has taught classes at the Academy of Art in San Francisco and the Otis College of Art and Design in Los Angeles. He still keeps his three Guild models in his office for clients to enjoy. In the recent past his Guild models were part of a concept car exhibit at the Petersen Automotive Museum in Los Angeles.

Forty years after building his award-winning mini-car, Tom measured the dimensions of his 1968 urban vehicle when preparing to attend the 2008 Fisher Body Guildsman's Reunion in Boston. He found it mirrored the exact proportions of the new Mercedes-Benz Smart Cars!

Both of us are still cars nuts. Although collectors of antique toy cars, Tom is the proud owner of a 1949 MG TC, while Terry owns a 1969 Datsun 510 auto x-racer. The gems of our collections, however, are the seven Guild models we made together in that little den in our house in Miami and all the memories twin brothers can share from three years, five state awards, two regional awards, and two national scholarships. We can still smell the lacquer.

28

"Even If I Had Wanted to Avoid It, Design Would Have Come Looking for Me"

Spencer L. Mackay (1966–1968)

My father was an industrial designer, and I was born with designer genes. Despite that fact, my independent nature led me to resist the pull of design precisely because that was what my father did for a living. The Fisher Body Craftsman's Guild was my tipping point to a very satisfying 36-year product design career, the majority of which was at the corporate executive level. For many years, I worked side-by-side with my dad as a partner in the firm he started.

For years, I took my father's design skills — and that genetic influence — for granted. There was no getting away from it, as our home was filled with many opportunities for creativity. My three brothers, two sisters and I grew up drawing spaceships and monsters, building tree houses, making our own fun and exploring our world. I have memories of my dad taking my brothers and I to his office when he had to finish a rush project on a Saturday. He would give us a cup of "Prismacolor" colored pencils and a pad of paper, and we would keep ourselves occupied for hours drawing everything from submarines to aliens.

One time when I was about 5 years old, I remember running out of ideas to draw and asking my dad what I should work on. I think, as a bit of a joke, he suggested I design car hubcaps. I tackled the assignment with gusto. He saved the sketches and showed them around his design office with some amount of pride.

When I was 4 years old, we moved from Detroit, Michigan, to Anaheim, California, in 1954 and then to Tarzana, California, near Los Angeles in 1956. It turned out our next-door neighbors had a boy one day older than I, and we became fast friends. His interest in building plastic model car kits proved contagious. To this day I wonder if I might not be genetically damaged due to all the plastic body putty, model glue and flavors of Testor's spray paint I inadvertently inhaled such as "Candy Tahitian Orange." It was the era of Ed "Big Daddy" Roth, the Rat Fink and wild custom hot rods.

One morning in 1965, while attending Portola Junior High at age 15, I heard that there would be an all-boys assembly. I truly thought they might be telling us about the "birds and the bees," but it turned out to be so much better! Two guys from General Motors gave a slide presentation on designing and building a 1:12 scale model car and showed us images of futuristic transportation concepts. Around 800 boys left that assembly flying high. I suppose it also reflected the attitudes of the era that no girls were invited. That evening I told my dad about the assembly, and we sent in the entry form. When a box from Detroit arrived with a cool blue booklet, *Design and Build a Model Car*, sketches, scale underlays, and a set of 4 molded solid vinyl tires, I was beyond excited. I fondled those tires for about a week, smelling them and circulating them in and out of the box several times a day.

1966 Entry (Age 15–16)

It seemed natural to start doing sketches of a sports car. I don't know what the percentages were between sedans and sports cars entered, but I can't imagine too many teenage boys getting excited about 4-door family sedans. I had a lot to learn about surface development, but I plowed ahead with enthusiasm. I had access to my dad's model shop and all the tools, not to mention his sage advice. He looked at the project as a chance to spend some quality time with me. It was always my design and my car, but he would challenge my thinking and help me understand the design process.

Having access to automotive modeling clay, I built a clay armature, or buck, consisting of shellac-painted wood and then smeared on the hot beeswax-based clay. Next, I started refining the design in three dimensions with my hands and clay tools. At some point the design had changed so much that I began running out of clay and exposing the buck. After some quick surgery, I was able to get a design I was happy with, namely a 2-door sports coupe for the Guild's open category.

After reviewing the suggestions in the Guild's "how to" booklet about carving poplar for the car, I decided to go a different route and make plaster molds off the clay. This was a technique that I had seen used in my dad's shop, and it looked like it would save time. I still remember making the molds in a type of plaster called "Hydrocal" and—after drying, sealing and waxing the molds—pouring the car itself in "Hydrostone," a harder denser plaster. I believe these plasters were used in the dental industry for models of people's teeth. Some of the Guild models from the year before were done with complete interiors, but since no extra points seemed to be awarded for the huge amount of effort, I decided to skip that and go with a painted greenhouse (called a "solid top" in Guild parlance).[1]

The body of the model made from "Hydrostone" was primed and painted with automotive lacquers. I sealed the plaster with shellac and then painted it with automotive lacquer-based primer. The body was painted a color called Vermillion, which I'd never heard of until then. I selected it from some color chips, and the orangey-red looked great on my car. Window trim was added around the dark indicated window areas. I used chrome Chart Pack tape. Turned aluminum wheel covers and shaped aluminum bumpers were added after polishing them to a high brilliance, red acrylic taillights were polished and fitted and even a license plate with my initials (SLM) was added for a finishing touch. I rubbed out the paint and waxed it with carnauba paste wax. I built a bulletproof plywood shipping crate, nestled the plaster model in foam rubber, and sent it off to Detroit. After that point, I put the Guild in the back of my mind and got on with my regular life.

That summer, I had a job as a counselor in a Boy Scout camp in the mountains. I think it was some time in August that I was called into the director's office and shown a very official letter with GM's Fisher Body logo on it. I couldn't believe that I had won the Junior Division First State Award for California ($150) and the Regional Award. Region 20, which consisted of all of California, was one of the toughest competitive areas, given

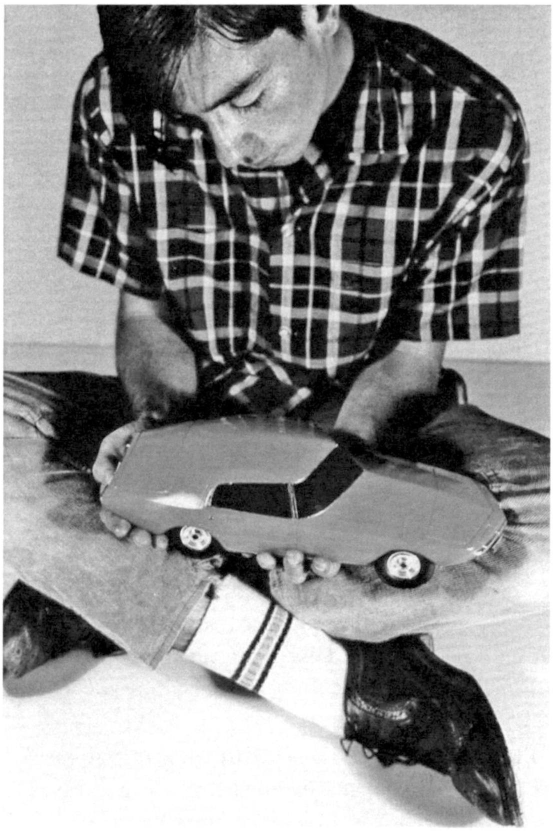

Spencer L. Mackay (age 16) holds his first Craftsman Guild model car, an Open Competition, 2-door sports coupe which won the 1966 Junior Division First State Award for California ($150) and Regional Award. Region 20 encompassed all of California. The car was designed and developed using a beeswax and sulphur-based clay called Chavant Clay. It is still used today by professional sculptors in automotive design studios. Hydrocal plaster molds were taken from the clay model and the car model itself was cast from Hydrostone, a hard dental plaster. The plaster model was dried and sealed with shellac, coated with a lacquerbased primer and painted Vermillion, a red-orange automotive lacquer color. The wheels were turned from aluminum and polished. The bumpers were shaped aluminum. Some chrome tape was used (courtesy Spencer L. Mackay).

the southern part of the state's custom car culture. I have a hard time now remembering how my car-building schedule corresponded to my school semesters. I just remember working to the Guild deadline and packing and shipping the car.

That year, the Guild's trophy had been redesigned. I noticed when I received one that the center emblem looked like it was a re-purposed Cadillac badge with the U-shaped wreath trademark. I remember thinking that was a smart use of an available part for a low quantity application.

1966 Guild Convention Trip

I left camp a week early and took my first ever airline flight on a DC-8 jetliner to my old hometown. Once I arrived in Detroit, it was quite a sight to see all Guildsmen with their matching light blue blazers and navy slacks. We looked pretty snappy. Three of us all bought navy blue Greek fisherman hats in a nearby store. If the Guild would have given us those, we probably wouldn't have worn them, but since we had the idea we thought they were pretty cool.

The Guild had the resources to treat the regional winners like rock stars and pamper us with educational side-trips. They loaded us all on a train — my first train ride complete with a dining car and porters, too — and we rode in style to Kalamazoo to the stamping plant where they were making fenders, hoods and other parts. We also visited Great Lakes Steel, where we saw red-hot ingots turned into thousand foot-long sheets of rolled steel.

Of course, the high point of the trip for us was the visit to the GM Styling Center. Not only did we get to tour the design studios and talk with the designers, we got to line up and ride in some of the concept cars. I rode in the Stingray I and II, the Mako Shark, the Monza GT and a few others. We saw the Firebird I, II and III series turbine concept cars as well but weren't offered rides in those cars. Regrettably, my 1966 design did not win a GM scholarship, but I brought home a lot of great memories.

1967 Entry (Ages 16–17)

Upon returning from Detroit and fresh off the success of a regional win on my first try, I started planning my second model car almost immediately. It was to be a 2-door, sports coupe for the Guild's open category, and I decided to go much wilder with the style. Designed in clay, this model featured flaring side panels behind the doors, wraparound taillights and graceful sloping contours integrating the greenhouse without notches. A mid-engine design was supported by side air scoops for cooling and combustion air. The front wheel track width was narrower than the rear. Metal parts were fabricated from aluminum and polished, while the paint job was a pale yellow acrylic lacquer—a 1967 production Cadillac color. There were dark-indicated window areas, and chrome "Chart Pack" tape for the window trim.

I stuck with the same design and fabrication techniques that I had used on the first model car, but, unfortunately, this time the car arrived at GM for judging with damage. Despite what I thought was another bulletproof shipping box, a scoop area in front of the rear wheel had broken and detached along with the metal scoop grille.

I received a Senior Division Honorable Mention Award for California and $25 cash. It speaks to the generosity of the Guild judges that, even though the car was damaged, I also received a National Styling Scholarship worth $1,000 based on the merits of the design.

Spencer L. Mackay (left) of Tarzana, California, and Karlin Wong of Torrance, California, wait at the Los Angeles International Airport for their flight to Detroit for the 1966 Guild convention. Note their left lapel Guild name tags with the Napoleonic Coach graphic. Karlin Wong was the 1966 Senior Regional winner and Spencer Mackay was the 1966 Junior Regional winner. Both represented Region 20 (the entire state of California). While at the convention, Mr. Wong won a 1966 National Styling Scholarship worth $1,000 (courtesy Spencer L. Mackay).

Spencer L. Mackay (age 17) holds his 1967 Guild model entry right before it was packed and shipped to Detroit. The 2-door sports coupe, Open Competition model won a 1966 Senior Division Honorable Mention award for California ($25) and a National Styling Scholarship ($1,000) award. It featured flaring side panels behind the doors, wraparound taillights and graceful sloping contours integrating the greenhouse and the body. The mid-engine design was supported by side scoops for cooling and combustion air, and the front tire track width was narrower than the rear tire track. It was designed and fabricated similar to the 1966 model—cast from a master clay model. The pale yellow acrylic lacquer paint job was a production Cadillac color from that year and had dark indicated window areas. Metal parts were fabricated from aluminum and polished. The model arrived in Detroit broken despite careful packing and became Spencer's last plaster car for that reason (courtesy Spencer L. Mackay).

It was then that my career objectives started to come into focus at the age of 17. Bolstered by this prestigious award, I told the editor of *Model Car and Science* magazine that I planned a career in industrial design.[2] I thought that if I could get recognition at this high level for something I really passionately enjoyed, perhaps I could succeed in a design career.

Reseda High School in Tarzana, California, also received a trophy that year, and I was taken by surprise at the awards assembly when I went to the podium to receive the recognition. I assumed the trophy for the school was a public relations move by GM, as the principal knew nothing of my project and was as surprised as I that I was receiving the award at school.

Since I was not a regional winner that year, I did not make the trip to Detroit. However, I did get the chance to meet Guildsman Stewart Reed (1964 Regional winner, 1968—4th National Senior Scholarship, $2,000 award) and Bruce Claypool (1967 National Styling Scholarship, $1,000 award) in Burbank, California, to review their Guild models. They were students at the Art Center College of Design and their model cars were so awesome. I was humbled to have received any recognition, especially since Stuart had been the one to look me up and suggest we meet.

1968 Entry (Ages 17–18)

When I started my third model car, I did not realize it would be the last year of the competition. It probably would have been my last year anyway, but it was sad to see a great institution retired due to budgetary considerations.

Learning from my past experience and the damage to my second model, I decided to follow the Guild's new recommended model building practice of using light weight, rigid polyurethane foam and sealing the surface with Liquitex modeling paste. This was supposed to be a faster way to build a Guild model. In the end, I think it might have had more

Top: Spencer L. Mackay's (age 18) Guild entry which won the 1968 Senior Division First State Award for California ($150) and Regional Award. The model was a 2-door sports coupe designed for the Open Competition. A clay master model was used and transferred via sections and templates to closed-cell, rigid polyurethane foam — a new faster and quicker model building method being promoted by the Guild. The model was surfaced with a hard thin layer of Liquitex modeling paste and then coated with Bondo polyester body filler. The body was painted a metallic green color and the window areas indicated with a very dark green color. Bright work was highly polished aluminum. The large V-shaped windshield provided forward and side vision and was split down the middle with a structural support member. A bold piece of aluminum brightwork dominated the centerline of the model from the cowl area to the rear end (courtesy Spencer L. Mackay).
Bottom: Rear view of Spencer L. Mackay's (age 18) 1968 winning Guild entry. There were no rear windows in this design, but the rules allowed for this in the Open Competition, if the contestant explained on the special features card that a video camera rear view system was hypothetically present. The personalized license plate combines Mackay's initials and a 3 signifying his third Guild competition. The trim was machined and hand-shaped aluminum. The taillights were machined from acrylic sheet with foil behind them to make them more reflective. Also note the "M" for Mackay on the central bird "badge" just above the vanity license plate (courtesy Spencer L. Mackay).

polyester "Bondo"™ on it than modeling paste, but it made a light car with sufficient durability. It arrived in Detroit undamaged. I still did a clay model, but used sections and templates to transfer the design to the foam and cut it using files and sandpaper.

This was another 2-door sports coupe, designed for the Guild open category, painted metallic green with a large dark area indicating the windshield area. Most people think this area was black, but it was actually a very dark green to better harmonize with my car's color. There were no rear or side windows. Guild rules allowed a contestant to get away with no rear windows in that category if the Special Features card on the bottom of the model explained that a video camera would substitute for the driver's rear view vision. The V-shaped windshield, which accommodated forward and side vision, was split down the middle with a structural support member. A bold sculpted piece of aluminum brightwork dominated the centerline of the model from the cowl (bottom of the windshield) to the rear end, where the video camera was located.

Although minimally employed on the front and rear of the model, the brightwork was machined, hand-shaped aluminum. The large triangular taillights were machined from red acrylic sheet and aluminum foil placed behind them to make them reflective. The taillight lenses had been recessed into the body, and I shaped a very delicate aluminum rear end ornament and a recessed license plate with the number SLM-3. There were complementary green accents around the wheel covers and the tire side-walls.

The third model car won a Senior Division First State Award for California and Regional Award, and in the process, I earned another trip back to Detroit. There were no blazers that year, however, and it was evident that the budget had been cut back, as there were fewer side trips as well. My biggest disappointment in learning about the competition's demise was the feeling that I would no longer have the chance to "do it again, only better." Theoretically, I still had several more years of eligibility (ages 18–19, 19–20, and 20–21), but I actually had learned upon arriving at the convention that it would be the last.

A few years later, I received a call from Los Angeles Pierce College, a junior college that I had attended, letting me know that my 1968 trophy in their trophy case was about to be thrown out and asking if I would like to have it. It was similar to the one Reseda High School received in 1967. I don't know how many Guildsmen have the bigger "school version" of the trophy, but I have one alongside the three regular trophies.

After graduating from Art Center College of Design in 1974 with a major in product design, I continued on my industrial design career. I worked as a designer at Creative Design Affiliates (1975–1976), senior designer at Emerson Johnson Mackay (EJM), Inc. (1976–1981), vice president of EJM, Inc. (1981 -1997), and I am currently the owner and president of M Industrial Design (formerly EJM, Inc.). As you might suspect, the "M" stands for Mackay. In the original name of the company, the Mackay was my father. I suppose the "M" in M Industrial Design is me. The company is a global strategic design and innovation company that I've been a part of for 34 years.[3] I've worked with our team of designers on projects in the areas of consumer products, home electronics, medical equipment, and fitness equipment. Two of our specialties are structural packaging for cosmetics and food applications as well as innovation and ideation workshops with clients worldwide.

I know for a fact that a few students at Portola Junior High sent in their Guild enrollment cards and received their kits after that fateful fall 1965 assembly — no girls, birds or bees — but as far as I know, I was the only one to pursue and complete a model car. I think I put in almost 300 hours on that 1966 model car with too many design revisions to count.

I have wonderful memories of time spent consulting with my father and asking his

Spencer L. Mackay, in a recent photograph holding his 1968 Region 20 Senior Division award winner. His three cars are still on display in his home office. Ultimately, he translated his love of design and model building into a career in industrial design and is currently president and owner of M Industrial Design in Burbank, California (courtesy Spencer L. Mackay).

advice as well as great memories of those three competitions. As proof, I still have the three original model cars on display in my home office. I never did repair model car number two, and I just show the good side. But even more than the memories, I have the perspective to see the huge influence participating in the Fisher Body Craftsman's Guild competitions had in my life. It gave me the confidence to quit resisting the attraction of design as a career and make the decision to pursue it as a lifetime endeavor. It gave me a respect for attention to detail and craftsmanship — a hallmark of my profession. I feel privileged to be an alumnus of such a unique and storied group, and still feel like a lucky kid.

Notes

1. In the 1950s, the total 450 score sheet points were re-distributed to accommodate the work load of an open convertible type model, but this faded into disuse.
2. *Model Car and Science*, October 1967.
3. The Web site address for Spencer L. Mackay's company, M Industrial Design, is www.idbyM.com.

29

The Last Winner

Grant L. Onnie (1967–1968)

First, I want to thank the editor of this book for his time and effort to preserve the memory of a little known but honorable slice of the General Motors legacy. In the context of the automotive design world, you might imagine that being honored for my efforts as a young amateur was one of the high points of my entire life — something like winning an Academy Award. Who would have known that a little over four decades later, GM would be recovering from a bankruptcy? Today, as I reflect on those glory days, it is both ironic and sad for me to see so many folks suffering from the poor business planning and possibly the greed of a few. But that is another story. For now, here is my "Craftsman's Guild" story.

As a teen, I designed and built two model cars for the GM guild competition. My first example tied for Michigan's Senior Division 1st Place Award in 1967. Luckily, the next year, my attempt became the Senior Division's National First Place winner of the General Motors model car competition. But how had I arrived there? At the time, only the competition's judges knew that it would be the last year that GM would sponsor the event and of note, it would affect the outcome of the judging, an interesting part of this story as you will see.

Winning a running race means passing the finish line first. Winning the model car competition was more like a beauty pageant. Let's face it, some years, blond and sleek is in and some years it isn't.

Let me take you back to 1958. As an eight-year-old curious and happy-go-lucky Italian American boy, I would rush home after school, drop off my school books and run to my father's basement barbershop, a block away, on Oakland Avenue, where I shined shoes for a living. I recall wanting to look like Ricky Nelson, but was made to wear a Princeton style haircut. I didn't mind it much if I wasn't looking in the mirror.

My father's shop was just across the street from the Chrysler Motors headquarters in Highland Park, Michigan, a thriving inner city suburb then, surrounded on all sides by the Motor City. My dad's clients were mostly automotive executives, engineers, designers and lab technicians. I shined as many of their shoes as I could, listening intently as those very tall men in crisp white shirts and ties discussed their business and design issues. It must have been right there that I was infused with ideas and dreams about cars and design, all ears, buffing away, pockets filled with quarters, and fingernails as black as could be.

At thirty-five cents a shine plus a generous tip, I soon had what I thought was a lot of money; money that I would spend on HO scale model trains every Saturday at the hobby store in the neighborhood. For the next few years, if I wasn't at the basement barbershop, or dodging the bullies in the neighborhood, I was busy in a corner of my basement at home, where I would go to escape, and live in a world where balsa wood truss bridges and plaster of Paris mountains filled my railroad world, a landscape of which I was in complete control.

I soon discovered the master-apprentice relationship when on one warm July Sunday evening, I wandered off (not a fearful bone in my body) while my family listened to the wonderful free sounds of the Detroit symphony orchestra. Eight-year-olds could do that then. I stumbled upon a building, not far from the fairgrounds concert shell, and peering in a window, I couldn't believe my eyes; I had found the "old timers," a group of men, the Detroit O-gauge model railroad club.

Lucky me! Spending time there, every Sunday, watching those old pros build with their amazing skill over scratch built miniatures raised the bar, allowing me to leap as a model maker, to understand the importance of precision, even picking up some trade secrets. I took them home and applied them to my own little train layout. I was falling in love. It was the kind of love that a young one gains from accomplishment, learning new skills, and then getting the praise that sustains a young person and helps them to grow. That, after all, was part of what the Guild competition was all about.

Fast forward to my twelfth year. My obsessions turned from trains to pretty girls, really hot cars, the hit Motown songs of the early sixties and of course the Beatles and my new electric guitar. No longer any need for files or needle-nose pliers or sandpaper. The model railroad became a dusty remnant of boyhood. My first two years of high school were all about "The Inmates," my rock and roll band, Debbie Sue, my first real sweetheart; my dad's '55 hand-me-down DeSoto (which I eventually totaled); and the cruel politics of high school. Everything had changed except for the fact that I still had little control over my hairstyle. Oddly enough, just when I no longer wanted to look like Ricky Nelson, but more like a Paul McCartney, my haircuts started to look more like Ricky. As the front man in a rock band, better Ricky than Bob Gibson in a Princeton.

So it goes. By 16, my love life was in ruins and the rock band was breaking up over a mishap with a case of Schlitz beer. I tossed the sharkskin suit and convinced my mom to buy me a peacoat. Sound familiar? That fall (1966), the beginning of my junior year, there was an assembly in the high school auditorium for "young men only," where I listened intently as a couple of cool suits from General Motors explained (with props and a slide show) that we could enter into a competition to build a 12th scale model car and win a college scholarship. I didn't have any singular love interest, and had nothing but time on my hands. I realized that I had some model building skills and the $5,000 scholarship reward was alluring. Maybe I could attend a university after graduation instead of going to the local junior college. I brought my application form home and mentioned it to my dad at dinner that night. He swallowed a forkful of rigatoni and said, "Send it in, son! Who knows, you could win it all," not knowing that the second time around I really would.

A couple of weeks later my official entry came in the mail with all of the dimensional guidelines, four official miniature rubber tires, rules, deadlines, everything needed to begin. It was October 1966 and the model car was due the following June 15. Eight months seemed like plenty of time back then, but as you must know, time is relative.

The circumstance of my first attempt was much more fortuitous than I was aware of at the time. Did I mention that Mozart didn't become a great composer playing stickball in the playground everyday, nor did Willie Mays learn his baseball skills by locking himself in a piano room? Destiny was at work. My environment offered advantages that a kid from Kansas City did not have. That is half the battle in life. Even though my dad was only a barber, he had access to the best and brightest minds that a major car company had to offer. He was able to arrange consultations with real professional stylists, who looked over my drawings and sketches, and I learned all I needed to know at the time about proportioning, occupant packaging, aerodynamics, spray painting, the sizzle of a good wheel design and a long list of other topics.

I had access to a lab that could chrome brass fittings for me, and I especially remember a little bespectacled German man, proprietor of a tiny tool making shop somewhere on a side street in Detroit. After understanding my pinch, he made for me a special one-off, miniature ban steel, circular saw blade (a tool that did not exist anywhere in the world until he made it to my specifications). This allowed me to make what I had sketched out on a napkin, my flashy oversized aluminum wheels.

On top of that, a complete wood and metal shop was at my disposal at the high school where Mr. McClellan, who was built like a machine himself, a hearty reliable type, taught me how to use a Bridgeport and a lathe. Until then, metal seemed awfully hard to me; too difficult to shape. Of course the entry package offered methods and material suggestions but it all really started with clay, real professional modeling clay over a wooden buck, and some clay tools from the art shop on Woodward. For me, that was it. I became a hermit, obsessed. Night after late night, week after week, no homecoming dance, no prom, no work, weeks becoming months, I shaped the body of my little car until I was happy with it, cutting in different window shapes and crouching down to view the sculpture as if I were scaled down to a person little enough to hop in and drive my future car away.

My first car actually came in second at the state level, losing out to a son of a General Motors employee.[1] The competition rules stated that if a GM employee's kid won, then the second place winner would also be awarded as 1st place winner, something like a tie. I think the idea was "fair play." Maybe the contest administrators thought that the GM family's kid might have had an unfair advantage.

That was another stroke of luck; or was it fate? Isn't it funny how things come to you when you're reaching out? It allowed me to attend the four-day national convention at the General Motors Technical Center. One hundred and four boys and young men met there every July: 50 first place junior winners (one from each state) up to 15 years old, and 50 senior winners, 16 years old to around 22, or a college senior's age.[2] Four guys were from overseas, part of the internationals in the competition.[3] We were all supplied with matching green blazers and ties, given a tour of the styling studios and allowed to ride in concept cars. It was a real treat for us to have an inside look at how cars were conceived and developed for real. On the fourth evening, there was an extravagant banquet, where the CEO of GM handed out the awards. Even though I did not place nationally the first year, what was most important to me then, as I remember now, was the opportunity to examine the winning senior level models.

The judging criterion was 50 percent styling and 50 percent craftsmanship. The judges were styling executives and pro model makers. So I figured the look of the car was subjective, but the craftsmanship was measurable. Believe me, I studied the winners very closely the paint, the trick of seamlessly gluing two wheels together to achieve a wider track, like a slick, the scoop opening treatments; I soaked up everything I needed to know by asking the winners questions about technique, and that curiosity really paid off the next year.

What can people gain by reading memoirs from faceless men about childhood experiences? Are these stories about winning? Are these stories about the journey? Are they about the love of the arts and crafts? Are they lessons aimed at the virtues in life? Whatever they are, I think for me, in the end this one's about honesty, devotion to something outside of oneself, and finally about the miracle of finding something to do that you love doing and recognizing and fulfilling a possibility that is laid before you.

Here comes the honest part of the story. Of course I entered the next year and won the whole darn thing, beating out college seniors from the Pasadena Art Center, the Harvard of industrial design schools. But it was more luck. The judges knew that it was the last year of the competition and they decided to award the car that was most far reaching and future looking, whereas traditionally, they awarded the cars with the best craftsmanship, more conservative looking. That's right. My model car was not the most precise and best built

In the foreground is the model made by Grant L. Onnie (age 18) of Highland Park, Michigan, which won the 1968 First National Senior Scholarship ($5,000) award. In the background is the model made by David C. Constance of Flint, Michigan, which won the 1968 First National Junior Scholarship ($5,000) award. Carved from high density polyurethane foam and coated with a Bondo polyester skin, Onnie's Model No. 1512, a 2-door sports car designed for a mid-chassis power plant, featured upper and lower air scoops behind the doors and was painted a custom blended, deep pearlescent Caribbean Blue. It was affectionately named Innova II by its teen designer-builder and was inspired by the Italian DeTomaso Mangusta automobile. Constance's model was made from high density polyurethane foam, coated with fiberglass resin and painted a deep pearlescent blue with black indicated window areas. The two angled and layered spoked wheels were fabricated from straight pins with two-stage clock knobs at the center. The polished aluminum bumpers were filed, formed, and shaped from bar stock to fit the body. The window frames were inlaid pieces of fine polished aluminum. Coincidentally, the 1968 Buick Riviera incorporated a similar scooped rear end design (courtesy Ronald J. Will).

car. I knew it immediately when I looked over the 1968 finalists and was very surprised that I won first place because I actually liked other cars over mine. I guess I was the pretty blond that year.

Oh yes, if I were the judge, I would not have chosen my entry for first place.[4] That is the slightly painful truth but the truth is what you should know. This is not to say that my car was not dazzling. My bladed brass wheels were killer thanks to that saw blade. My paint was deep Caribbean blue thanks to a guy at Chrysler who supplied me with a special pearlescent paint from his lab. My window moldings were actually chrome dental brace wire thanks to a tip from a winner the year before (whose father was a dentist, not a barber). Moreover, I was standing on the shoulders of design giants' creations, like the DeTomaso Mangusta, an Italian sports car that remains one of my favorites of all time. I want to thank you, Debbie Sue, wherever you are, for kicking me to the curb that year. It allowed me to focus my energy, love and desire on car design and dare to dream of what could be.

Notes

1. Grant L. Onnie won the Senior Division Regional Award for Michigan in 1967 and an all-expenses paid trip to the Guild convention just a few miles across town from his home in Highland Park. His Model No. 1394 was a plaster casting, painted red with an orange tint and a pearl finish. It is best described as a stylish, domesticated version of a "Big Daddy" Don Garlits' dragster, with a huge axle to cowl ratio and a small, compact rear passenger compartment area. The wide track-width front tires were independent of the body and featured streamlined, teardrop shaped pontoons or tire nacelles. The long, narrow body ended with an enclosed cockpit, a narrow rear axle and a pair of high vertical side air scoops sticking out from the body in front of each rear tire. There appeared to be massive air scoop in the front as well.

2. In discussing the 104 youths, Onnie makes reference to the Guild's Miniature Model Napoleonic Coach Competition held in the 1930s in which a first place junior and a first place senior coach builder were selected from each state to participate in the Guild convention in Detroit. When Onnie participated in 1967 and 1968, the Guild had 20 regions (groups of states) from which a top junior and senior model car builder were selected for the Regional Award (an all-expenses paid trip to Warren, Michigan). This created 40 regional winners who attended, rather than 104. This was the program format used during the late 1940s, as well as the 1950s and 1960s.

3. In 1968 three European Guild winners attended the convention. They included Robert Kohler (age 20) from Geneva, Switzerland; Otto M. Schaumberger (age 20) from West Germany; and Malcolm Webster (age 18) from Melbourne, Victoria, Australia. Source: *GM World*, September-October 1968 issue, and a program booklet, "Fisher Body Craftsman's Guild, Scholarship Award Banquet, Monday, July 22, 1968, 6:30 P.M., Fisher Body Central Offices Auditorium, Warren Michigan."

4. Grant Onnie's 1968 Senior Division First Place National Scholarship Award winning design (Model No. 1512) was a 2-door, deep pearlescent Caribbean Blue sports car inspired by the famous Italian house of automotive style and design named DeTomaso. Envisioned to be powered by a mid-chassis mounted engine, it featured upper and lower scoops behind the doors. It was constructed from high density polyurethane foam with a Bondo polyester skin. It was named Innova II, from the word "innovation," and on the bottom of the model for the Guild judges to see were the words "innovation that is paramount to the auto world today." Many years ago in Detroit, after a "show and tell" for office friends, the 1968 top national award winning model, the "Last of the Mohicans," mysteriously disappeared. Anyone with information is asked to contact the author.

30

Party by Fisher: 2008 Guild Reunion Exhibition in Boston

John L. Jacobus (1961–1966)

One of the admirers of the first Fisher Body Craftsman's Guild book, an advocate and collector of Guild models and memorabilia, was Frederic A. Sharf, vice president of the board of directors at the Museum of Fine Arts (MFA) in Boston. I had mentioned to him in an exchange of notes, half-kidding and half-serious, to let me know when he wanted to have a large Guild model car exhibition and retrospective at the museum. He quickly embraced the idea, and, by November 2007, had confidently adopted the concept, checked the museum's calendar of events, and suggested three potential weekends in August 2008. Naturally, I was thrilled by the prospects of this opportunity.

Later, by phone, he said he considered the Guild models to be pieces of fine art suitable for display in a museum given their high level of design and craftsmanship. I thought about that comment and wondered whether the old art masters Monet, Renoir and Rembrandt would be happy rubbing shoulders, ever so briefly, with such pieces of automotive art, handcrafted by car-crazed American teenagers.

I contacted Guildsmen from across the country using the mailing and marketing lists from the first Guild book.[1] Soon, I received assurances that a large number were interested in dusting off, repairing and even restoring their highest-award Guild model designs and coming to Boston for a weekend. Over the next eight months, I worked with interested Guildsmen and MFA staff on the physical aspects of the exhibit installation and lay-out, security precautions, and an exhibit catalogue. I also directed a Guildsmen-prepared public information and education program. The project was developed iteratively, month-by-month, via email and communications with roughly 50 to 60 patient and cooperative participating Guildsmen.

Mr. Sharf volunteered to design, produce and publish a spectacular high-quality color catalogue documenting all the models on display. The PowerPoint presentations for the public information and education program were organized and coordinated online by Samuel T. Kjellman, who lived nearby in New Hampshire. The program was to be presented four times by the Guildsmen during the two-day exhibition, giving the static display of models context and meaning.

Background

The MFA, like most museums, is a non-profit, so it has to attract paying customers through the front doors with new programs and activities. As such, the staff tries to present collections of works that appeal to a wide variety of enthusiasts. The MFA's mission statement written in 1870, indicates it is chartered to promote "design" as well as "fine arts"—and this occasionally gets exploited. The Guild would be another one of those exceptional cases in the automotive genre, but perfectly timed to coincide with a lull in summer programming just after the hugely successful "El Greco to Velazquez" exhibition ended on July 27, 2008, and GM's 100th corporate anniversary in September 2008.

2008 Guild-MFA Boston Reunion

The name of the Guild-MFA show was *1950s and 1960s Dream Car Model-rama*, a two-day exhibition and retrospective of award-winning models from the Fisher Body Craftsman's Guild representing the years 1947–1968. It was held August 9–10, 2008, to coincide with the 100th corporate anniversary of General Motors Corporation, now the New G.M. or General Motors Company.,

MFA Model Cluster No. 1. at the main entrance with the Napoleonic Coach centerpiece surrounded by early 1950s competition model cars. The 1947 Napoleonic Coach won a Regional Award and was made by Maurice Beliveau. The five model cars in the foreground, left to right, were made by Ronald C. Hill (1950 National Scholarship), Charles Foreman (1952 Regional), Warren Bakken (1953 National Scholarship), Arnold Joslin (1953 National Scholarship) and Charles Foreman (1954 State). Left side of table from the top: rear views of models made by Richard R. Sylvester (1956 Styling Scholarship), Richard R. Sylvester (1957 National Scholarship), William A. Keyser, Jr. (1952 National Scholarship), and Paul Tatseos (1957 Styling Scholarship). Right side of the table from the top: models made by George R. Chartier (1954 State Award), Oskar Heininger (1954 Regional Award), and Milton J. Antonick (1955 National Scholarship). Top of the table (partially obscured by 1947 Napoleonic Coach) are the models made by George E. Anderson (1955 State Award) on the left and Adrian A. Bruno (1955 National Scholarship) on the right (courtesy Ronald J. Will).

Partial view of MFA Model Cluster No. 2 with model entries from mid–1950s to 1960s. The prominent left foreground model, with the dual white canopies, was exhibited by Ace Edward Fogarty (1960 National Scholarship). The exhibitors, listed clockwise around the edge of the cluster from this point, included Thomas L. De Fazio (1960 National Styling Scholarship), Bill Scott (1959 National Scholarship), John Sparnicht (1958 National Styling Scholarship), Bill Scott (1958 State), John Sparnicht (1959 Regional), Paul Tatseos (1958 National Scholarship) and Ronald C. Pellman (1958 Regional). In the bottom, to the right of center, is the dark, double bubble canopies model exhibited by Robert Aikins (1960 National Styling Scholarship), the white body with double bubble canopies above that was Ronald C. Pellman (1960 National Scholarship), behind Pellman is William F. Marks (1957 National Scholarship), and to the left of Marks is William R. Molzon (1959 National Scholarship) (courtesy Ronald J. Will).

The night before the MFA shows, exhibitors and their guests gathered at the MidTown Hotel to share stories and memorabilia and to get acquainted. Approximately 110 Guildsmen, family and friends attended the 2-day event. Forty-seven of the attendees were disciples of GM design executive Harley J. Earl some 40 to 50 years earlier. Overall, Guildsmen exhibitors ranged in age from 58 to 80 years. About 34 percent of the Guildsmen-exhibitors were from the New England area; the majority, or about 64 percent, hailed from across the country, including Arizona, California, Florida, Michigan and Washington.

The model "Dream Car" installation was located on MFA's second floor Gund Gallery Gift Shop area — approximately 21 feet wide by 95 feet long. The floor layout consisted of four tables, 6 feet by 6 feet, with a cluster of models displayed on each table, that were roped off to keep the models out of reach of the public but close enough for a visual inspection. There were about 16 models per cluster and an identification placard for each model stating the Guildsman's name, age, original city and state, as well as year and type of award.

Full view of MFA Model Cluster No 3. with models 1960 to 1963. (1) Gerald A. Simone (1962 Regional), (2) Paul Bonfilio (1961 State), (3) Anthony V. Simone (1961 National Scholarship), (4) Samuel T. Kjellman (1961 State), (5) Ronald J. Will (1961 National Scholarship), (6) Ronald J. Will (1960 Regional), (7) Eugene Simone (1960 National Scholarship), (8) Robert G. Atkins (1963 National Scholarship), (9) Ronald R. Steinhilber (1962 Styling Scholarship), (10) Michael R. "Bobby" D'Mura (1962 Styling Scholarship), (11) Samuel T. Kjellman (1962 State Award), (12) Thomas L. Covert (1962 National Scholarship), (13) Harry E. Schoepf (1962 State), (14) Ronald R. Steinhilber (1963 Regional), (15) Harry E. Schoepf (1960 State), (16) Paul F. Bonfilio (1962 Regional) and (17) Peter G. Wiinikka (1960 National Scholarship) (courtesy Ronald J. Will).

A fabulously preserved 1947 Napoleonic Coach, a Regional award-winner, greeted the patrons and visitors as they entered the hall. A remnant of the original technical scratch-building FBCG program run by Walter C. Leuschner from 1930 to 1947, this miniature model coach was exhibited by 80-year-old Guildsman Maurice Beliveau and his brother Marcel of Bedford, New Hampshire.

A series of six story-boards (24 inches by 30 inches), mounted at eye height, lined one of the walls depicting the Guild's history using large color images, bold graphics and pieces of memorabilia. Additionally, Guildsmen were assigned "low-key" vigilance duties at the MFA exhibit hall, to circulate and to keep an eye on things. Some Guildsmen even stood watch both days from 10 o'clock in the morning until 5 o'clock at night near the models they had made, as an extra precaution.

We actually had "double" security at the MFA and the MidTown Hotel exhibitions as Tony Mazzola (too young to have ever been a Guildsman) and his wife Julie Steinhilber of Malden (now Medford), Massachusetts, organized their scale model enthusiast friends (Ray Berthiaume, Charles Burdick, Bob Montgomery, Stuart Marcus, Terry Rumery, Diane Rumery and John Terosian) from the Boston Metropolitan area into "plain clothes" security details. Since they were plastic model car builders and enthusiasts, they blended into the crowds at the MFA Boston and MidTown Hotel gatherings and could "talk shop" as well as any Guildsman. They looked like Guildsmen, and talked like Guildsmen, but were actually on security duty.

Auto Designer Ronald C. Hill with his 1950 First National Junior Scholarship award winning model in-hand at the MFA Boston event. Originally from Highland Park, California, Ron Hill graduated from the Art Center School in 1954 with a bachelor of fine arts in transportation design and worked as chief designer at GM for Buick, Pontiac, and Chevrolet. He was an assistant director of design at Opel, Vauxhall and GM do Brazil. More recently, he was the chair, Transportation Design Department, Art Center College of Design (courtesy Ronald J. Will).

One of the main ideas behind the exhibition was to have a cross-section of Guildsmen, the creators of the model "Dream Cars" on display, mix with the patrons and their guests, answer questions and share their Guild experiences. This was to be a kind of "living history" experience for the patrons. A number of Detroit auto designers–Guildsmen were on hand to meet and greet the public as well. Those who were inspired by the Guild as teenagers to pursue auto design careers (current or previous design position indicated) included:

Robert G. Aikins (Ford designer, Ford Light Trucks, Ford Motor Company)
George F. Anderson (GM designer, Chevrolet/Oldsmobile Studios; Wolverine Race Car)
Milton J. Antonick (Studebaker, AMC and Chrysler Designer)
Thomas L. Covert (lead designer, Hummer SUV-SUT program, GMC)
Ronald C. Hill (chief designer, Buick, Pontiac, Chevrolet at GMC; assistant director of design at Opel, Vauxhall and GM do Brazil; chair, ACCD Transportation Design Department)
Frederick J. "Bud" Magaldi (design manager, Mustang Studio, Ford Motor Company)
William R. Molzon (Advanced GM Design, Advanced Chrysler Design and Calty Design)
Bill Scott (executive design director, GMC)
Paul Tatseos (chief interior designer, Oldsmobile, Pontiac, Buick and Chevrolet, GM Design)
Ron Will (GM Design, senior designer, Corvette/Camaro Studio, and manager of product planning and design, Subaru of America, developed Subaru Outback and Forester concepts.)

Afterward, one Guildsman's wife, Virginia Tatseos, wrote on her real estate blog on Dec. 27, 2008:

> My husband, Paul Tatseos, won three times on different levels of the competition, giving him a good start on his education at the Art Center College of Design in California. He went to work at General Motors design staff in Warren, Michigan, as did many of the other competitors. Other winners went on to careers at other auto companies, industrial design, engineering, medicine and many other careers. Approximately 60 winners, many who [had] never met before, brought their cars to the museum for a public showing. The turnout was far more than I expected. Hundreds of people over the weekend viewed the cars and talked to their designers, builders and creators. One young man [Sam Simmons] was so enthralled with the designs he was allowed to hold my husband's cars. The father (Jimi Simmons) said: You and Paul really made all the difference for my son's experience there. By the end of the day he was totally comfortable going up and talking to some of the guys about the models they built and what they did later in life.

The 65 model cars exhibited were representative of the FBCG competitions from 1947 to 1968, with 50 percent being the top national scholarship winners (indicated with a N) or national styling scholarship winners (indicated with an S), 27 percent being regional award winners (indicated with an R) and 23 percent being state award winners (indicated with a K). It was a once-in-a-lifetime opportunity for Bostonians, as the models had never really seen the light of day except perhaps among a few GM corporate employees or other Guildsmen. These were the "best of the best" or the "creme de la creme," selected by Harley J. Earl, Bill Mitchell and Chuck Jordan, GM executives who ran the Guild model car design competition between 1937 and 1968.

The model identification placards at the exhibit included the contestant's age at the time of the competition — information that brought gasps from the audience when considering the sophistication and intricacy of the scratch-built model before them. In addition, many model designs, although made 40 to 50 years ago, appeared very contemporary — a sure sign of good design. Taken out of mothballs and dusted off for the first since their

Auto designer Paul Tatseos with his 1957 National Styling Scholarship award winning model at the MFA Boston event. Originally from Boston, Tatseos graduated from the Art Center School in 1961 and went to work for General Motors Design Staff. For a majority of his GM career he was chief designer at the Oldsmobile, Pontiac, Buick and Chevrolet Interior Studios. From 1996 to the present he represents Paul Tatseos and Associates, Inc., a design consulting firm (courtesy Virginia and Paul Tatseos).

Detroit debut, many of the models had been restored or re-conditioned to their original appearance. Several models had been re-created from scratch using closed cell polyurethane foam (e.g., Milton J. Antonick's 1955N and Bud Magaldi's 1959K). A few changed colors, compared to the original entry.

Tony Mazzola and Julie Steinhilber exhibited the restored 1960–1964 Guild models made by Julie's cousin, Ronny R. Steinhilber, of Wayne, New Jersey (deceased 2000). Ronny had been very successful in the Guild competition: 1960 Junior First State Award, 1961 Senior First State, 1962 National Styling Scholarship, 1963 Regional Award and 1964 Regional Award. Guildsmen will always remember the re-styled, aerodynamic family van design Ron created in 1964, some 20 years in advance of the Chrysler Dodge/Plymouth minivan. The proportions were very similar to today's minivans. An adult plastic model car builder and enthusiast, Tony had previously restored the five models he inherited and was able to show them off in the MFA Boston exhibition in honor of Ronny R.

Meghan Simone, daughter of Guildsman Tony Simone who took the 1961 First National Junior Scholarship, wrote on her blog posted Dec. 27, 2008: "So much I could write, but I did get goose bumps and was so proud of my father and uncle [Gene Simone] when I walked in and saw their cars on display."

Technology Saves the Day

The latest computer technology had come to the aid of the Fisher Body Craftsman's Guild many times and supported its preservation, such as in organizing the 2004 reunion at the GM Tech Center, writing and marketing the 2005 FBCG book, and maintaining mailing and marketing lists.

It came to our aid again in preparing for the Boston exhibit. We agreed in advance to orchestrate and produce a public information and education program to help educate the MFA patrons and their guests about GM's Fisher Body Craftsman's Guild and the *Model-rama*

Sam Simmons (age 10) of Newton, Massachusetts, an MFA Boston guest, holds Paul Tatseos' green 1958 Third National Senior Scholarship award winning model. Sam builds $\frac{1}{24}$ scale model cars, enjoys show cars and anticipates working on a more advanced $\frac{1}{12}$ scale model kit of a V-8 engine (with moving parts) with his father in the future. His father, Jimi Simmons, stated "the Guild show was a wonderful experience for him to see men like Paul Tatseos, who turned their passion for cars into careers" (courtesy Virginia and Paul Tatseos).

exhibition. Samuel T. Kjellman volunteered to organize and coordinate the PowerPoint presentations over a 4-month period and did an outstanding job. The Guildsmen-narrated presentations, with follow-up question and answer sessions, were offered four times in the MFA's first floor Mabel Louise Riley Conference Room. The topics included (1) Craftsman's Guild History; (2) Harley J. Earl and the Emergence of Automotive Design; (3) Gordon M. Buehrig, Designer of Classic Automobiles of the 1930s; (4) Eugene (Bob) T. Gregorie, Chief of Ford Motor Styling 1935–1946; (5) Raymond Loewy, Pioneer Industrial Designer 1926–1986; (6) Virgil M. Exner, Jr., and Virgil M. Exner, Sr.; (7) Guildsmen's Personal Stories, (8) Principles of Aesthetics and (9) Design Tools Da Vinci Would Use Today (e.g., Alias, rapid prototyping, stereo-lithographic techniques).

Ace Edward Fogarty with his 1960 Second National Junior Scholarship award winning model in-hand at the Boston event. A Pratt Institute graduate, he and his wife Bonnie (also a Pratt graduate) are toy inventors and formed a partnership called Ace Edward Fogarty Associates. Over 38 years they have been granted numerous patents and consulted for various toy companies such as Hasbro and Mattel. Their best known toys are Gloworm and Glow Friends. The Fogartys have been inducted into the Hasbro Inventors Hall of Fame (courtesy Ronald J. Will).

This was the Guildsmen's "interpretation," or historic perspective, on the Harley J. Earl art form of emotional design rather than rational design that was famous in the 1950s and evident in the static models on display.

Our excellent public information and education speakers included Samuel T. Kjellman, Robert W. Lawhn, Bill Scott, Ron Hill, Richard Earl, Charles Foreman, Thomas L. De Fazio, Richard Sylvester and Milton J Antonick. Breakfast speakers included Charles Foreman, Tony Mazzola, and Richard Earl (Harley J. Earl's grandson). Richard Earl was the keynote speaker at the Colonnade Hotel banquet buffet breakfast Sunday morning. Mr. Foreman took to the podium and discussed how they located Guildsman Stanley F. Mott (1949–1953), and Mazzola made a special tribute to Guildsman Ronny R. Steinhilber (1960–1964).

Based on the presentations in the first floor Mabel Louise Riley Conference Room and the spectacular display of models upstairs, the Craftsman's Guild was very professional and left a positive impression on the public.

High Tech Model Building

At the MFA, one of the models auto designer Milton J. Antonick exhibited was a re-creation of his 1955 First National Senior Scholarship winner — a convertible sports car with a two-tone red and white color scheme. The original prize-winning model had been sold to GM for a few hundred bucks when Milton was 19 years old and is believed to have been kept in a Fisher Body Division collection for years. It is theorized that this unique collection of models was later sold at auction to automobilia collectors when the Fisher Body Division was organized out of existence in 1984, although that has yet to be proven.

The body-shell of the new M. J. Antonick 1955 replica model, built especially for the MFA exhibition, was milled from polyurethane material in April 2008 using the data generated by Alias computer software. The input data for this process was taken from old surviving photographs, sketches and drawings of the 1955 model entry still in the possession of the Antonick family. An instructor and expert on this software, Antonick did all the Alias work himself. Automobiles are being styled and designed using this computer software today.

Rapid prototyping and stereo-lithographic techniques were used to re-create some of

Top left: **Michael R. "Bobby" D'Mura, retired dentist, with 1962 National Styling Scholarship ($1,000) award winning model entry in-hand at the MFA Boston. Originally from Flagstaff, Arizona, Dr. D'Mura built eight Guild models from 1957 to 1964 and won six Regional Awards. His father built a 1933 Napoleonic Coach for the Guild competition and his brother John and cousin Joseph Wayne built award winning Guild model entries as well.** ***Top right:*** **Michael A. Pietruska, industrial designer, with his 1965 Second National Junior Scholarship award winning model he displayed at the MFA Boston. He grew up in the atmosphere of the family's Stamford, Connecticut, upholstery business and nine brothers and sisters. Pietruska attended the Art Center College of Design and studied environmental design. Some of the design projects he worked on included office furniture, exhibits and toys and games with Milton Bradley. Today he is an independent designer (courtesy Ronald J. Will).**

the interior model parts, and of course, many aluminum parts were shaped with a file and good old-fashioned "elbow grease." The 2 7/16 inch outside diameter resin-cast replica wheels, reproductions from the Guild's 1954–1957 period, were obtained from the author and another FBCG project.

Milton Antonick disassembled his new 1955 model in front of a group of attendees at the MidTown Hotel to demonstrate the meaning of Guild-inspired craftsmanship. Their jaws dropped to the pavement as he snapped it all back together. In the end, Milton's replica 1955 Guild model flawlessly and seamlessly took its place on the MFA display tables along with all the others. No one knew that only four months earlier, a computer-controlled machine had milled, cut and shaped the body surfaces of the model. Antonick's 1955 turned out to be a stunning model and an awesome story.

The original Milton J. Antonick model, a two-door sports convertible design, was hand-carved from basswood and painted a Venetian Red and Shoreline Beige color scheme. Mr. Antonick won the 1955 Senior Division First National Scholarship Award worth $4,000 at age 19. This model had parts shaped from aluminum (front grille, gear shift and center mounted taillight escutcheon), parts shaped from brass (HO gauge track for the windshield trim, front bumper Dagmars and headlight bezels), and parts carved from copper (front

Top left: **Kenneth A. Kelly, marketing manager, with 1965 National Styling Scholarship award winning model entry he displayed at the MFA Boston. He was also the 1966 Third National Senior Scholarship award winner. Kelly works in advertising, marketing and public relations for VBP, Inc., and Van Steel, which manufacture and market Corvette high-performance equipment.** ***Top right:*** **Harry E. Schoepf, industrial designer, with his 1960 Senior Division First State award for New Hampshire ($150) model in-hand which was shown at the MFA Boston exhibition. Schoepf made six Guild models from 1957 to 1962, winning a National Styling Scholarship in 1961. He is an antique car restoration expert, particularly the Model T Ford, and has been working on a technically accurate scale 1913 Model T delivery wagon machined from scrap pieces of brass (courtesy Ronald J. Will).**

and rear bumpers using high-voltage buss bars). After being chrome-plated, many of the brass and copper parts and pieces became the model's "brightwork."

High Quality Photoshop Images

The high quality MFA catalogue produced to memorialize the Model Car Exhibition was another case where computer software, namely Photoshop, came to the Guild's rescue.[2] The museum wanted high resolution images with two images per page for an 8½ inch by 11-inch page format. Increasing the size of the model car images provided by Guildsmen required re-rendering surfaces at the pixel level, adding color, adjusting lightness or darkness, and rendering missing parts and pieces as well as making overall enhancements of image color and quality for consistency among the images. This required hundreds of hours of work by Guildsmen Ronald J. Will and William R. "Bill" Molzon. A single image could take eight hours to restore. The MFA color catalogue went to press July 1, 2008, so both men were very busy. Fortunately, Ron and Bill are both auto designers and experienced Photoshop users, so they were able to competently complete this task. Again, the Guild was saved and preserved. The catalogue was a fitting tribute to General Motors' 100th corporate anniversary, which took place Sept. 16, 2008. About 200 copies of the baby-blue covered catalogue were sold that weekend at the MFA bookstore.

Top left: Gary P. Card with 1967 Senior Division Second State for Rhode Island ($100) award winning model in-hand at the Boston Guild retrospective. Card made seven Guild models from 1962 to 1968. He received a bachelor of science degree in industrial technology from Roger Williams University in Bristol, Rhode Island, and pursued marketing and product management. He is an industrial precision instrument company manager. *Top right*: Auto designer Milton J. Antonick with the reproduction of his 1955 First National Senior Scholarship ($4,000) award winning model at the MFA Boston event. As an employee at Studebaker, Antonick worked with Raymond Loewy on the Avanti design. He was hired by Plymouth as a body studio manager and worked on the Plymouth Valiant and Plymouth Barracuda designs. He worked for Chrysler Corporation until 1980. Antonick became vice-president in charge of Design at William Schmidt and enjoyed a Ford-Nissan minivan project designing with the Japanese. He has taught at the College for Creative Studies in Detroit (courtesy Ronald J. Will).

High Tech Preservation of 2008 Reunion

The 2008 Guild Reunion Exhibition at the MFA Boston has been preserved in images and words on the Guild's official Web site, www.fisherguild.com. Items that have been posted include a press release, the MFA Catalogue with color images of models, photos of the MidTown Hotel informal exhibits and MFA preparations, photos of the formal MFA Exhibit, a promotional flyer designed by Guildsman Allen T. Weideman, images of Guild-MFA exhibit area patrons and visitors, and newspaper articles written about the Guild-MFA exhibition as well as blogs by Brett Snyder, Virginia Tatseos, and Meghan Simone describing the Guild-MFA retrospective.

Conclusions

The Guild-MFA collaboration was a highly successful, well-attended event, with two days of good, solid traffic flow. All the public relations and media contacts paid off with weekend ticket revenues consistent with previous years and months. Catalog sales were

Auto designer William R. "Bill" Molzon with his 1959 Second National Senior Scholarship ($4,000) award winning model. Molzon graduated from General Motors Institute (GMI) in 1960 and the Art Center School in 1963 with degrees in engineering and industrial design, respectively. He has enjoyed a career in automotive design at GM Design Staff, Chrysler Design Staff, Rhor Industries, Calty Design Research (aka Toyota) and Molzon Design — his industrial design consulting business (courtesy Ronald J. Will).

brisk, and three out of the four public information and education sessions were packed or overflowing. Guildsmen interacted with patrons all day on both days, creating a good ambient buzz of anticipation and excitement. Bostonians expressed appreciation for having the Guild exhibition in their community. The MFA professional staff was pleased, as were the Guildsmen who had traveled from all over the United States. There were no broken models, no model accidents or saboteurs. All ended well on a good note of cooperation. All the models displayed were exquisite like glistening jewels, having been cleaned, buffed, waxed, repaired and reconditioned with lots of "TLC" for their historic outing. Old Guild friends and competitors became reacquainted, and new friends were made. The design gods and goddesses were happy. And, as we left for home, the Masters and Mona were smiling.

Acknowledgement of MFA Professional Staff

The Guild acknowledges the Museum of Fine Arts' professional staff for taking good care of the Guildsmen, their families and friends and for making the Guild-MFA exhibition a success. They put their reputations on the line with the fine arts community, and included, among others: Frederic A. Sharf, vice-president, board of directors, MFA; Katie Getchell, deputy curator; Danielle Berger, manager of touring exhibitions; Linda Chernoff, director of special events; Barbara Martin, interpretive/educational relations; Adaleta Maslo-Krkovic, new media producer, A/V PowerPoint expert; and Rick Davis A/V — PowerPoint expert.

Acknowledgement of Guildsmen, Family and their Guests

This also acknowledges all the Guildsmen, their family members and guests who contributed to the MFA exhibition with their original, restored or reconditioned Guild model entries and who participated in the historic, "once-in-a-lifetime," 2-day event. The award key code is as follows: N = National Scholarship award winner; S = National Styling Scholarship award winner; R = Regional Award winner; and K = State Awarded winner.

- •Robert Aikins and his wife, Ro, 2 models (1960 S, 1963 N), Farmington Hills, MI.
- •George and Elaine Anderson, 1 model (1955 K), St. Clair Shores, MI.
- •Milton J. Antonick, 2 models (1955 N, 1964 N), Royal Oak, MI. Exhibited his brother Mike Antonick's 1964 N model, from Mt. Vernon, Ohio.

•Warren Bakken and Lynn Philips, 1 model, (1953 N), Seattle, WA.
•Maurice Beliveau and brother Marcel, 1 model (1947 R Napoleonic Coach), Bedford, NH.
•Gordon Bertrand, 1 model (1955 R), Brookfield, MA.
•Paul and Alba Bonfilio, 2 models (1961 K, 1962 R), Little Neck, NY.
•Adrian "Tony" Bruno and his wife, 1 model (1955 N), Rolling Meadows, IL.
•Gary and Ann Card, 1 model (1967 K), Coventry, RI.
•Tom and Margaret Covert (and their son Adam), 1 model (1962 N) Richmond, MI.
•Joan K. Creamer, former GM exterior designer, no model, East Greenwich, RI.
•Tom and Sally De Fazio, 1 model, (1960 S), Cambridge/Watertown, MA.
•Michael R. "Bobby" D'Mura, DDS, and his wife, 1 model (1962 S), Peoria, AZ.
•Ace Edward Fogarty and Bonnie, 1 model (1960 N), Sarasota, FL.
•Charles and Claudia Foreman, 2 models (1952 R and 1954 K), Arlington, TX.
•Jeffrey Goldstein, guild collector, volunteer, Warwick, RI, brought Peter G. Wiinikka's 1960 N model shipped from Hillsboro, OR.
•Tom W. Graboski and Gina Goodin, 2 models (1967 K, 1968 S), Miami, FL.
•Gary and Heather Graham, no model available (1954 N), Bellingham, WA.
•Oskar Heininger, 1 model (1954 R), Framingham, MA.
•George and Lee Herzog, 1 model (1958 R), Beavercreek, OH.
•Tom Hibschman, 1 model (1965 K), Kittery Point, ME.
•Ron and Phyllis Hill, 1 model (1950 N), Newport Beach, CA.
•Gene and Tomi Izuno with family, 1 model (1954 R model), Orinda, CA.
•John Jacobus, 1 model (1966 K) Silver Spring, MD.
•Joan and Arnold Joslin, 1 model (1953 N), Loveland, CO.
•Kenneth and Cheryl Kelly, 2 models (1965 S and 1966 N), Treasure Island, FL.
•Bill and Joan Keyser, 1 model (1952 N), Victor, NY.
•Samuel and Gayle Kjellman, 3 models (1961 K, 1962 K, 1964 R), Weare, NH.
•Robert and Dorla Lawhn, 1 model (1967 S), Henderson, NV.
•Bud and Mary Ellen Magaldi, 3 models (1960 K, 1965 S), Canton, MI, and brought George Chartier's (1954 K) from Northville, MI.
•Bill and Christine Marks, 1 model (1957 N), Wilmington DE.
•Tony Mazzola and Julie Steinhilber, 3 models (1962 S, 1963 R, 1964 R), Malden, MA, and Bostonian friends, in memory of Ronny R. Steinhilber, Guildsman and Pratt Institute graduate.
•Tristram Walker Metcalfe III, 2 models (1963R, 1964R), Northampton, MA.
•James W. Miga, 1 model (1968 K), Boston, MA.
•Bill Molzon, 1 model (1959 N), Clarkston, MI.
•Richard A. Norton and his wife, Helen, no model available (1948 N), Forest Park, IL.
•Ron Pellman and Susan Clark, 2 models (1958 R, 1960 N), Sarasota, FL.
•Mike Pietruska, 1 model (1965 N), Southwick, MA.
•Harry E. Schoepf, 2 models (1960 K and 1962 K), Goffstown, NH.
•Bill and Mary Scott, 3 models (1957 R, 1958 K, 1959 N), Cape Cod, MA.

John L. Jacobus has studied and collected Craftsman's Guild memorabilia since 1974 and got his big break in 1984 working on a Guild exhibit for the Smithsonian's National Museum of American History (NMAH). Using phone books as a search tool, he contacted many former top Guild contestants by mail hoping to motivate them to donate their award winning Guild models to the permanent collection. This culminated in a public exhibit (a large glass case) of 10 model cars and one Napoleonic Coach which were on display in the NMAH's Hall of Transportation from October 1991 until Spring 2001.

- •Gene and Debra Simone, 2 models (1960 N, 1961 N for Tony Simone), North Smithfield, RI.
- •Gerry and Carol Simone, 1 model (1962 R), Middletown, RI.
- •John Sparnicht, 2 models (1958 S, 1959 R) Reno, NV, and guest Robert Sparnicht from CA.
- •Richard Sylvester and Freddie, 2 models (1956 S, 1957 N), Los Angeles, CA.
- •Paul and Virginia Tatseos, 2 models (1957 S, 1958 N), Birmingham, MI.
- •Ovid O. Ward, 1 model (1966 N), Edgartown, MA (Martha's Vineyard).
- •Philip Wescott, no model, FBCG PR and advertising manager, field rep, Newark, DE.
- •Ron and Pat Will, 2 models (1960 R, 1961 N), Fountain Hills, AZ.

Press and media coverage of the event were extensive, resulting in high attendance over the weekend.[4] It was very impressive to see how the local media rallied around this museum event and provided press coverage with newspaper articles, subway magazine and newspaper announcements, as well as Web site blurbs and blogs. A total of 47 Guildsmen participated from across the country, exhibiting some 65 Guild models which ranged from state level winners to top national scholarship winners. Some Guildsmen worked through the planning process, but could not attend.[5] Overall, the model retrospective and reunion project was highly successful. Inquiries and invitations from other museums interested in a possible Craftsman's Guild model exhibition and retrospective have been received. This kind of attention makes the Guildsmen smile and their hearts fill with pride.

Notes

1. Internet technology allowed for a rediscovery of the Guild era and a resurfacing of the competitors and their model cars, through search engines, Web sites, e-mail, online auctions and other venues. This led to the book *The Fisher Body Craftsman's Guild: An Illustrated History,* by John L. Jacobus (McFarland, 2005).

2. The catalog *Fisher Body Craftsman's Guild 1930–1968: Model Car Exhibition, Museum of Fine Arts, Boston, August 9–10, 2008*, by John L. Jacobus and Ronald J. Will with Frederic A. Sharf, contains 93 high resolution images.

3. The Web site was created February 23, 2004, and is managed by Egret Net Web Design of Louisville, Kentucky. The site is performing far more efficiently these days after a March 2007 redesign, with 20,779 unique visitors (38,011 counting repeat visitors) as of January 14, 2011. Almost six years of marketing data has been collected so far. The Fisher-Guild site continues to "go global" with primary visitors from Canada, Australia, United Kingdom, Japan, Switzerland, Germany, Brazil, Ireland, France, Ghana, Italy, New Zealand, Netherlands, Finland, the Russian Federation, Argentina, Czech Republic, India, Poland, Togo, Denmark, Sweden, Norway, Turkey, Hungary, Israel, Belgium, and elsewhere.

4. Coverage of the event was extensive and included the following: (1) *Artnet* magazine; (2) Association of Professional Model-Makers (APMM); (3) "Dream On" article by the *Metro Boston News*; (4) Fisher Body OnLine advertisement; (5) advertisement or announcement, Hemmings *Classic Car* magazine; (6) "Impulses Section: Smart Cars/Driving Ambition" by Maggie Behringer, *Improper Bostonian* magazine, July 30–August 12, 2008; (7) article or announcement in *Manchester Union Leader*; (8) *MAMA Sez*, May 2008, online newsletter, Maryland Automotive Modelers Association, Greenbelt, Maryland; (9) MassCar Model Car Club online advertisement; (10) *Preview MFA Boston*, July–August 2008, Gallery Activities (MFA donors/membership magazine); (11) www.mfa.org, Upcoming Exhibitions, MFA Guild Catalogue for online bookstore; (12) www.fisherguild.com, MFA-Guild press release and images; (13) www.men.style.com/gq, online version of *Gentlemen's Quarterly* (GQ); (14) www.carofthecentury.com, Richard Earl, editor, (15) artdailynews@yahoo.com enquiry by Ignacio Villarreal; (16) "1950s & 1960s 'Dream Car' Model-rama," Society of Automotive Historians, *SAH Journal*, Issue 236, July–August 2008, p. 15; (17) *Scale Auto Magazine* (Calendar of Events), Kalmbach Publications; (18) "Auto Design Boy Wonder Lives Dream," interview with Bud Magaldi, GM Tech Center News and Oakland Tech Center News, July 7, 2008, by Christine Snyder, Staff Reporter, *Detroit Auto Scene*, Warren, MI; (19) "When Design was King: Scale-Model Prototypes Come from Detroit's Golden Age," *The Providence Journal*

(Projo.com), Peter T.C. Elsworth, Staff Journalist, filed August 13, 2008, Providence, RI; (20) "Gallery to Show Dream Cars," by Anne-Gerard Flynn, Reporter, *The Republican Newspaper*, Springfield, Massachusetts, Sunday, June 27, 2008, interviews with Massachusetts Guildsmen Michael Pietruska and Tristram W. Metcalfe III, see www.masslive.com/living/republican/index.ssf?/base/living/-2 ½ the online version of *The Republican* newspaper; (21) Bill O'Brien's AutoNotes, *The Washington Times*, New York Avenue, Washington, D.C.; (22) Vintage Motor Sports; http://www.vintagemotorsports.com/race-calendar.asp (listed in racing event calendar); (23) "Car Designer Fulfills his Boyhood Dream," interview with Bud Magaldi by Dan Murphy, *Patriot-Bridge*, Thursday, July 31, 2008, Charlestown, Massachusetts.

5. These included George R. Chartier (1953 and 1954 State Awards), Allen T. Weideman (1957 and 1958 Regional Awards), Harrell C. Lucky (1956, 1957 and 1958 Regional Awards), Peter G. Wiinikka (1960 National Scholarship), Joseph B. Arnold (1961 National Styling Scholarship), and Mike Antonick (1964 National Scholarship).

In Memoriam: Charles M. "Chuck" Jordan

Ronald J. Will and John M. Mellberg

On December 9, 2010, the Fisher Body Craftsman's Guild lost one of its greatest supporters. Charles M. Jordan passed away at the age of 83. "Chuck," as his many friends called him, was considered to be the "ultimate" Guildsman. He started, as we all did, with a dream of designing and building his own model car for the Guild competition. Chuck achieved his dream at the age of 19 by taking the top Guild award, first place in the Senior Division for 1947. Two short years later, after graduation from MIT, he started his career at GM Styling as a junior designer. From there, his talent and drive propelled him all the way to the top of the automotive design world. At the age of 26, he was chief designer of the Special Projects Studio. Then at 30, Chuck earned the illustrious position as chief designer of the Cadillac Studio.

In spite of his busy schedule, he remained true to the Guild and took time out to serve as a design judge for many years. By 1962 he was named executive in charge of exterior design for GM. Even in this position, you can see Chuck Jordan in period pictures always enthusiastically talking with the new Regional Guild winners each year. He knew the importance of letting these young aspiring designers express themselves through the Guild. Over 25 Guildsmen would eventually become a part of Chuck's design team at GM. Chuck achieved the top design position as vice president of design at General Motors in 1986.

Even after his retirement in 1992, he still felt the need to help young designers follow their dreams. He again freely gave of his own time,

Charles M. "Chuck" Jordan (October 21, 1927–December 9, 2010), photographed at the 2004 Guild reunion in Warren, Michigan. He was the 1947 First National Senior Scholarship ($4,000) award winner in General Motors' Fisher Body Craftsman's Guild model car competition (courtesy Ronald E. Van Gelderen).

teaching students at two California high schools the art and craft of designing and modeling future car designs. I can still hear him teaching me to look for the perfect surface and design, as he would say, “the form should be not too flat and not too fat.” He was always searching for the perfect balance in design. He achieved it not only in car design, but through his personal contributions to the Guild and with his own family life.

The Guild has lost one of its own, truly a giant of design in the auto industry. He will be long remembered and sorely missed.

Contributed by Guildsmen Ronald J. Will (1961 National Senior Scholarship winner) and John M. Mellberg (1966 National Senior Scholarship winner), former GM Styling (Design) employees who worked for Chuck, knew him well and were his devoted friends.

Appendix: Basic Guild Policies and Procedures

1. The Guild competition year was congruent with the public school year (September to June).
2. The contestant's division, or pool of competitors, was determined by the birth year as specified by Fisher Body in their announcements and advertising.
3. All the model design and construction work had to be done by the contestant.
4. The model submissions had to be constructed from a hard, durable substance, not clay.
5. The models were scored anonymously and identified using a 4-digit number system.
6. The models were scored and evaluated in two categories (1) Design and (2) Craftsmanship by three design judges from GM Styling and three industrial arts teachers from the Detroit School System. Quantitative, objective scores were used to winnow and sort the models down to a manageable number, and then, qualitative subject criteria were applied by design management to select the top scholarship winners. An advisory board of educators consulted on the latter.
7. Similar to the miniature model Napoleonic Coach competition, the same car model could be entered year after year (but, hopefully, with successive design and craftsmanship upgrades or improvements).
8. A contestant could win a Styling Scholarship ($1,000) but once. Although they had won a university trust fund for styling, they remained eligible to compete in future competitions.
9. Top National Scholarship winners (the top eight university trust fund recipients) were ineligible for future competitions.

Glossary

ACCD— Art Center College of Design (name beginning in 1966-67 for the ACS).

Acetate— A type of clear, flexible plastic used by the Guildsmen to make plane-curved windshields for their models without heating or stretching. This material was readily available in the hobby shops. A litmus test for this material was that it could be bonded together with acetone or fingernail polish remover.

ACS— Art Center School, Los Angeles, California (the original school name), or "Art Center" for short.

Alias— Brand name for the CAD/CAM computer software used in the automobile design business today for 3-D concept models, rapid prototyping, visualization, sketches, illustrations and animation. Auto designer and Guildsman Milton J. Antonick used this software package to design and build a replica of his 1955 national award winning model for the MFA Boston exhibition. He also taught the subject at the College for Creative Studies in Detroit, Michigan.

Axle-to-cowl ratio— Distance between the front axle and the windshield cowl or bottom edge of the windshield (e.g., long axle to cowl ratio means a long hood design line and a short axle to cowl ratio means a short hood design line). Some of the classic cars from the 1930s had long axle-to-cowl ratios to create a long, lean, low look, whereas Chrysler's H-Body cab-forward designs have short axle-to-cowl ratios.

Backlight— Rear window of an automobile body.

Bauhaus School of Design— A school of design philosophy or ideology founded by Walter Gropius which synthesized technology, craftsmanship and design aesthetics. As indicated by the essays (C. H. Stewart, S. T. Kjellman and R. J. Will), this approach to aesthetics was taught at Pratt Institute, Syracuse University and the Institute of Design at the Illinois Institute of Technology. The Bauhaus taught that "less is more," meaning a cleaner, simpler design is a stronger, better design. The other essential rational design principle was that "form follows function." Harley J. Earl, vice president GM Styling and an automotive marketing genius, not formally schooled in art or design ideologies, followed the design philosophy "form follows emotion."

Beltline— This is an automotive design term defined by the intersection of the "greenhouse" with the top of the "fenderline." Greenhouse, canopy or "upper" are other automotive design terms used to designate the area above the beltline of the automobile body which contains the windshield, side windows and rear window glass, roof and roof structure. The "lower" refers to the body of the automobile.

Bondo— A white or reddish pink colored polyester-based filler used at auto body repair and paint shops to fill dents. Bondo or similar plastic body filler requires a catalyst to harden and cure the material before sanding.

Bubble top or clear canopy— Guildsmen would vacuum-form or stretch a pre heated sheet of Plexiglas over a wooden pattern to make a clear plastic bubble top or clear canopy shape for their models. Harley J. Earl's Firebird experimental car series with fighter jet bubble canopies over the driver and passenger inspired the Guildsmen to use this futuristic design element. Some can also be seen in the Guild models from the late 1940s.

Cab-forward —An automotive body style where the canopy or greenhouse encroaches into the traditional front engine compartment space. In the 1940s, the Guild had kids experimenting with rear engine designs and this produced the "cab-forward" look or what is termed a short axle-to-cowl ratio.

CAD/CAM— computer aided design/computer aided manufacturing.

Carnauba wax —A hard commercial automotive paste wax like Simoniz or Turtle Wax, used by Guildsmen to complete the lustrous paint finish on their models. This was applied as the "finishing touch" after the cured lacquer paint was worked, by-hand, using rubbing compound. Called Brazil wax or palm wax, carnauba is derived from the leaves of the palm (*Copernicia Prunifera*), a plant native to northwest Brazil.

"Cast the model in clay..."— Jargon or expression used by some essayists for making plaster of Paris molds from a clay master model.

CCS— College for Creative Studies (formerly Center for Creative Studies and before that the Society of Arts and Crafts) located in Detroit, Michigan.

Chavant clay— A brand of professional clay used by automotive sculptors in the scale model and full-size modeling process made from beeswax and sulphur. This is a semi-hard clay that has to be carved as opposed to the soft gray hobby shop Plasticine clays. This clay could be recycled by heating, softening and aggregating small chips or scraps. Chavant and Red Bank are two competitive brands. Each auto company uses its own preferred formulation of Chavant clay for its studio models.

Closed-cell polyurethane foam or **high density polyurethane foam**— Easy to carve and sand, the polyurethane rigid foam was covered with a thin layer of Liquitex modeling paste, polyester resin, dental plaster or Bondo. This provided a rock hard surface to carve grooves for the door, hood, truck, and concealed headlight openings as well as to paint. The technique was developed by GM Styling and promoted by the Guild technical staff as a construction method that promised a quicker way to build a Guild models in the 1967-1968 competition year. The underlying assumption was that the Guild would continue beyond that point. Many top scholarship award winning models were made using this method. The strategy behind the quicker, faster building methods, and expanded Junior and Senior age intervals, was to keep the Guild growing and flourishing.

Compound surfaces— The sheet metal of an automobile body is curved in two directions in what are called compound surfaces as opposed to flat or slab surfaces. The designer uses the reflection of light to design and control the shape of these surfaces during the surface development process.

Dremel tool— This is a high speed power rotary cutting tool with a variety of thin blades and attachments used for fine, delicate cutting of wood or metal as well as sanding, drilling and polishing.

DuPont or Duco— Refers to the brand of lacquer automotive paints and products produced by E. I. DuPont de Nemours of Wilmington, Delaware.

Fiberglass —Individual synthetic glass-like fibers, or a cloth of woven synthetic fibers, impregnated with liquid polyester resin or epoxy resin and used to form molded parts by hand lay-up or vacuum bagging. Generally a low production volume manufacturing technique.

"Flop"— Light and dark separation of colors to show off a form most advantageously, as in the surfaces of model cars.

Ford— Ford Motor Company, Dearborn, Michigan.

Gel coat— The first step in the fiberglass lay-up process, a thin coating of catalyzed polyester resin painted on the female mold surface(s). This gel coat provides the outer smooth surface of the final parts before the cloth layers are applied inside. The mold would be coated, first, with a silicone parting compound or release agent. Polyester impregnated fiberglass fibers, or plain fiberglass cloth, would be applied when this surface became tacky.

"The General"— An affectionate name for the former General Motors Corporation used by automotive writers and journalists.

General Motors Company LLC or the New G.M.— Alternative new names for the downsized company that emerged from the 2008 bankruptcy of General Motors Corporation.

GM Technical Center— Located in Warren, Michigan, at the corner of 12 Mile Road and Van Dyke Avenue, this was the epicenter of General Motors' advanced automotive design and technology. The Fisher Body Central Engineering Offices adjoined the property, as did GM's Ternstedt Division. During the mid–50s and 60s, the

Craftsman's Guild competition was conducted at the Fisher Body auditorium, GM Styling (Design) Staff facilities and under the Styling Dome at the GM Tech Center.

GMI— General Motors Institute or General Motors Institute of Management (now called Kettering University). Many top future industry managers and executives (automotive and otherwise) are produced by Kettering University (named after GM's Charles F. Kettering) from their excellent 5-year co-op educational program, which features alternating semesters of work and study.

Greenhouse— An automotive term for the area above the beltline of the automobile body which contains the windshield, side windows and rear window glass, roof and roof structure.

Guild News— Official newsletter of the Fisher Body Craftsman's Guild (1946–1953) distributed free of charge to Guild enrollees.

Guildsman—Official newsletter of the Fisher Body Craftsman's Guild (1953–1968) distributed free of charge to Guild enrollees, which reached a distribution of 600,000 anually in the 1960s.

HO gauge brass track— Model railroad track sold at hobby shops and easily available to the Guildsmen. The common hobby railroad track gauges were N, HO, S (American Flyer), O (Lionel) and G gauge track (outdoor garden trains that employ self-powered engines rather than electric rails to convey power.) Guildsmen liked the HO gauge track parts because they were brass, the right scale and easily adaptable to making brightwork and trim pieces. Pieces of HO brass track could be assembled with lead solder or silver solder. The brass could then be polished and chrome-plated.

Honorable Mention— A state level award for Guildsmen with a $25 cash value. The State awards were first, second, third with four or five Honorable Mentions awards.

Hydrocal— A powder to liquid plaster, like plaster of Paris, used by the dental profession in making prosthetic teeth molds.

Hydrostone— A powder to liquid plaster, like plaster of Paris, used by the dental profession in making prosthetic teeth molds.

IIT— Illinois Institute of Technology, Institute of Design, 350 North LaSalle Drive, Chicago, Illinois.

Junior Division— A Guild competition pool, ages 12 to 15 years old originally, then later expanded to 11 to 15 years old. Using this approach, those contestants with comparable modeling skills and abilities would compete against one another.

Liquitex modeling paste— A liquid to dry, acrylic polymer emulsion or surface preparation painted on framed canvas surfaces to create a hard, textured surface receptive to oil or acrylic paints from an artist's pallet. The 1966 Guild "how to" booklet suggested that Guildsmen paint the entire high density, polyurethane foam models with several coats of modeling paste called "Liquitex" thinned with water (about one-part water to six parts modeling paste). This coating would permit sufficient groove clearances for scribed lines and moldings without cutting into the rigid foam and it provided a substantial paint surface for the model.

MFA Boston— Museum of Fine Arts, Boston.

MIT— Massachusetts Institute of Technology, Cambridge, Massachusetts.

National Scholarship Award— also called GM university trust fund, are sometimes designated or coded in this book with the year and the capital letter N. Up to and including 1955 the national scholarship award amounts were First—$4,000, Second—$3,000, Third—$2,000 and Fourth—$1,000. Starting in 1956 each of these award levels was increased by a $1,000 increment. Ten National Styling Scholarships were added in 1956 as well, worth $1,000 each.

National Styling Scholarship— Recipients of this award ($1,000) for Excellence in Design are sometimes designated in this book with the year and the capital letter S. Beginning in 1956, ten National Styling Scholarships were awarded annually. Based on the "Design" portion of the score sheet (at least 90 percent of available points had to earned), models were selected regardless of the overall score sheet grade, or State Award level, to receive this university trust fund. An Honorable Mention or Third State Award level model was eligible.

Notchback— Jargon for a sedan body-style where the rear window approaches the rear deck or trunk lid at an obtuse angle of 110–120 degrees, creating what is termed a "notch" when viewed from the side, as opposed to a fastback, hatchback or bustleback rear end body design.

Open category or open competition—This competition classification, inaugurated in 1963, allowed for the study of unusual passenger compartment seating configurations, unique methods of passenger entry and exit, accommodation of passenger comfort and convenience, unique tread patterns or wheelbase configurations, unique suspensions and wheel/tire designs. This was another smart marketing ploy by Guild management to expand enrollment numbers and broaden youth interest (similar to expanding the junior and senior age intervals or increasing the number and dollar amounts of scholarship awards). The idea was to remove design restrictions and open up new channels of styling creativity and design experimentation into the automobile of the future. Embraced by the Guildsmen, this new category paid off with a plethora of new innovative models. Guildsmen literally redefined the automobile as we knew it.

Pattern—A piece of wood carved and shaped to scale, or cast in plaster, and used to form model parts and accessories such as a clear plastic windshield, clear canopy, taillights or bucket seats.

Plastic Wood—An air-dried, cellulose fiber filler for chinks, divots and other wood carving surface blemishes, with a hard, smooth surface when sanded. It was available at most hardware stores. Very few models arrived in Detroit without their fair share of plastic wood.

Plexiglas (also plexiglass, "plex")—Plexiglas is a brand of acrylic plastic and plexiglass is the common, generic name for acrylic plastic. This was a very stiff material, even in thin sheet form, and had to be heated to be made more flexible and malleable. The hard shiny exterior surface of the material and variety of colors was the attraction for Guildsmen. The Guildsmen learned how to heat and stretch the heated plastic sheet form of the material over wood or plaster patterns to make windshields and bucket seats. "Plex" rods were also used for headlights, and clear red "plex" flat stock for taillights.

Pratt Institute—Pratt Institute of Art and Design, New York City, New York.

Pumice powder—A slurry of pumice powder and water was used on the final cured lacquer surfaces of the model to give it the polished sheet metal look. Pumice powder was derived from the light volcanic rock that is formed during eruptions and resembles a sponge because it consists of a network of gas bubbles. Pumice powder was a more coarse abrasive than rottenstone.

Regional Award—Models receiving a regional award are sometimes designated or coded in this book by the year and a capital letter R. GMC and Fisher Body divided the country into 20 regions. Each year one Junior Division and one Senior Division winner was selected from each Region for a Regional Award—consisting of an all-expenses paid, 4-day trip to the Guild convention in Detroit. Models were organized and judged by these geographic regions and by age divisions (Junior and Senior), to create "pools" of equivalent competition.

RIT—Rochester Institute of Technology located in Rochester, New York.

Rock-of-the-eye method—Technique used by Guildsmen to achieve model symmetry by eye using a steel ruler or a pair of dividers, or both.

ROTC—The Reserve Officers' Training Corps was available to many college students. A military reserve stipend helped pay their college expenses, in return for active and reserve duty commitments. Upon graduation, they became commissioned officers. The Navy, Army and Air Force participated as a way to recruit young, qualified officers.

Rotten stone or rottenstone—A fine water-abrasive used to polish automotive lacquer paint and was applied by the Guildsmen to obtain a lustrous, mirror finish. The painted surfaces were polished with a water-rottenstone slurry on a soft cloth.

RTV rubber—Room temperature vulcanizing (RTV) rubber which the Guildsmen could buy in liquid form at the hardware store. Guildsmen in the late '40s and early '50s were encouraged to paint layers of this material on their master clay models to make a flexible rubber mask or elastic rubber mold. To maintain its shape, however, an egg crate pattern of plaster of Paris struts needed to be applied over the rubber mask's surface. Commonly called RTV today, this is generally marketed as a silicone-derivative product.

Rubbing compound—A commercial fine or extra fine abrasive emulsion used by Guildsmen to "rub-out" a cured lacquer paint surface with a soft cloth in the palm of their hand. This abrasive was readily available from auto supply stores for a few bucks and was the "last step" before the application of a hard wax. This combination provided the deep mirror finish surfaces seen on the models. This fine abrasive emulsion removed fine scratches, oxidation and watermarks and essentially "moved" or "worked" the paint surface to remove airborne debris, blemishes, or orange peel (over spray). It could be applied to solid, metallic or clear coat finishes. Guildsmen learned to be careful with this abrasive as they risked rubbing through to primer color if not judicious and cautious. The colloquial action verbs derived from this activity were "rubbed-out," or "compounding," or "compounded."

Scale fidelity—A score sub-category under Craftsmanship, referring to the conformance of the model entry to Drawing A, B, and C maximum-minimum specifications, depending on the body style.

Senior Division—A Guild competition pool, ages 16–19 originally, then later expanded to 16–20. Using this approach, those contestants with comparable skills and abilities would compete against one another.

Solid-top construction—A model construction method where the occupant greenhouse is carved or formed from a solid material as opposed to a convertible model or a model with an open interior, windshield or clear plastic canopy.

Spit-polish—A military-style shoe polish technique that was used to bring out the luster in a hand-rubbed lacquer paint job.

State Award—Models receiving a State Award are sometimes designated or coded in this book with the year and the capital letter K. The state awards were either First State, Second State, Third State or Honorable Mention with a cash amount of $150, $100, $50 and $25, respectively. There were four to five Honorable Mention Awards, designated for each State for 7 or 8 cash awards per State depending on the year. In the 1960s, with teenagers from 50 States and the District of Columbia competing, there were about 816 State cash awards designated per year. In the late 1940s and early 1950s, there were 686 State cash awards, and by 1957 they were expanded to 784 State cash awards. A contestant could earn multiple awards: a State Award (cash), a Regional Award (trip to Detroit) and a National Scholarship Award (GM university trust fund) for one design entry.

SUT—sport utility truck

SUV—sport utility vehicle

Talcum powder—Ordinary household baby powder, or dad's talcum powder, was mixed with the polyester gel coat by Guildsmen to create a softer, more malleable finished model car surface so door, hood, rear deck, and retractable headlight opening cut-lines or grooves (1⁄32 inch by 1⁄32 inch) could be made. This made it possible to carve the surface with handmade cutting tools. A cured, virgin polyester resin surface would have been too hard and difficult to cut.

Templates—A cardboard or flat pattern material that could be cut to match the finish side of a clay or wood model to transfer the form to the opposite side.

"Upper" and "lower"—Automobile design jargon referring to the automobile surfaces above the beltline (canopy or greenhouse) and below the beltline (the automobile body).

VMCC—Volvo Monitoring and Concept Center, Camarillo, California.

Windshield cowl or cowl—Part of the automobile body at the base of the windshield which contains a series of air vents or louvers for passenger compartment air intake and a platform or storage area for the windshield wipers.

X-Acto tool kit—A wood carving tool set consisting of a plastic handle and an interchangeable variety of sharp, flat or curved knife blades.

Bibliography

Books

Bayley, Stephen. *Harley Earl.* New York: Taplinger, 1991.

_____. *Harley Earl and the Dream Machine.* New York: Knopf, 1983.

Bonfilio, Paul F. *Fallingwater: The Model.* New York: Rizzoli, 2000.

Brinkley, Douglas. *Wheels for the World: Henry Ford; His Company and a Century of Progress.* New York: Viking, 2003.

Byrne, John A. *The Whiz Kids: Ten Founding Fathers of American Business—And the Legacy They Left Us.* New York: Doubleday, 1993.

Chana, William F. *Over the Wing: The Bill Chana Story.* San Diego, CA: WFC, 2003.

Drake, Albert. *Fifties Flashback: A Nostalgic Trip.* Tucson, AZ: Fisher Books, 1998.

Earl, Harley J. *Correspondence Course: Detroit Institute of Automotive Styling (DIAS).* Detroit, MI: Harley Earl Corporation, ca. 1945.

Farrell, Jim, and Cheryl Farrell. *Ford Design Department: Concept and Show Cars 1932–1961.* Roseburg, OR: Jim and Cheryl Farrell, 1999.

Gandini, Marzia, and Bruno Alfieri. *Shiro Nakamura Nissan Design.* Milan, Italy: Automobilia s.r.l., 2003.

Georgano, Nick, and Nicky Wright. *Art of the American Automobile: The Greatest Stylists and Their Work.* New York: Smithmark, 1995.

Goddard, Jolyon. *Concise History of Science and Invention: An Illustrated Time Line.* Washington, DC: National Geographic Society, 2010.

Graboski, Tom W. *Fisher Body Craftsman's Guild Reunion—Retrospective—Exhibition: Museum of Fine Arts, Boston, August 9–10, 2008.* Coral Gables, FL: Tom Graboski Associates, Inc., Design, 2008.

_____. *Reunion, June 26–27, 2004, Detroit, Fisher Body Craftsman's Guild 1930–1968: An Organization for the Development of Craftsmanship and Creativity Among Boys.* Coral Gables, FL: Tom Graboski Associates, Inc., Design, 2004.

Green, George W. *Special Use Vehicles: An Illustrated History of Unconventional Cars and Trucks Worldwide.* Jefferson, NC: McFarland, 2003.

Grist, Peter. *Virgil Exner, Visioneer: The Official Biography of Virgil M. Exner, Designer Extraordinaire.* Dorchester, England: Veloce, 2007.

Halberstam, David. *The Reckoning.* New York: Morrow, 1986.

Hirshberg, Gerald P. *The Creative Priority: Driving Innovative Business in the Real World.* New York: HarperBusiness, 1998.

Jacobus, John L. *The Fisher Body Craftsman's Guild: An Illustrated History.* Jefferson, NC: McFarland, 2005.

_____, and Ronald J. Will, with Frederic A. Sharf. *Fisher Body Craftsman's Guild 1930–1968 Model Car Exhibition, Museum of Fine Arts, Boston, August 9–10, 2008.* Lynn, MA: Velocity Print Solutions, 2008.

Janicki, Edward. *Cars Detroit Never Built: Fifty Years of American Experimental Cars.* New York: Sterling, 1990.

_____. *Cars Detroit Never Built: Fifty Years of American Experimental Cars.* New York: Sterling, 1995 (revised edition).

Jensen, Todd A. *Automotive Web Sites.* Jefferson, NC: McFarland, 1999.

Lamm, Michael, and David Holls. *A Century of Automotive Style: 100 Years of American Car Design.* Stockton, CA: Lamm Morada, 1996.

Langworth, Richard M., Jan P. Norbye, and the Auto Editors of Consumer Guide. *The Complete History of Chrysler Corporation 1924–1985.* New York: Beekman House, 1985.

Lewis, David L., and the Auto Editors of Consumer Guide. *100 Years of Ford: A Centennial Celebration of the Ford Motor Company.* Lincolnwood, IL: Publications International, 2005.

Loewy, Raymond. *Industrial Design.* Woodstock, NY: Overlook Press, 1979.

McVey, Michael. *The Men of Ford: A Tribute to Ford Motor Company.* Flint, MI: McVey Marketing and Advertising, 2000.

Noakes, Andrew. *The Ultimate History of the Sports Car: From Early Enthusiasts' Racers and Performance Supercars.* Bath, England: Parragon, 2005.

Sharf, Frederic Alan. *The Harry Bentley Bradley Col-*

oring Book: American Automotive Design Innovator. Boston, MA: Frederic A. Sharf, 2008.

_____. *Richard H. Arbib, 1917–1995: Visionary American Designer.* New York: Art Publications, 2006.

_____, and Brigham and Women's Hospital. *American Automobile Art 1950–1970: Original Design and Advertising Drawings.* Newbury, MA: Newburyport Press, 2004.

Sharf, Frederic Alan, and Danielle Berger. *American Automobile Art 1945–1970: Drawings from the Great Age of American Car Design.* Newbury, MA: Newburyport Press, 2007.

Sharf, Frederic Alan, and Richard Arbib. *Future Retro: Drawings from the Great Age of American Automobiles,* Boston, MA: MFA Publications, 2005.

Stein, Jonathan A., et al. *The Art and Colour of General Motors.* Philadelphia, PA: Coachbuilt Press, 2008.

Wayt, Hampton C., and Petersen Automotive Museum. *Driving Through Futures Past: Mid-20th Century Automotive Design.* Aikens, SC: Kythe, 2005.

Weber, Louis, and the Auto Editors of Consumer Guide. *Corvette: Sports Car Superstar,* Lincolnwood, IL: Publications International, 2005.

_____. *100 Years of the American Auto.* Lincolnwood, IL: Publications International, 2003.

Wingrove, Gerald A. *The Complete Car Modeller.* New York: Crown, 1979.

Internet Articles

Billingsley, Eric. "When Kids Designed Cars," *Innovation: America's Journal of Technology Commercialization.* June–July 2009. http://www.innovation-america.org/archive.php?articleID=543.

"Fisher Body Craftsman's Guild Reunion at Eyes on Design 2004," *Car Design News,* http://archive.cardesignnews.com/news/2004/040614fisher-body/index.html.

Jordan, Charles M., and John M. Mellberg. "Fisher Body Craftsman's Guild: Cars of Tomorrow by Boys of Today," *2004 Eyes on Design,* June 26–27, 2004. http://wiki.gmnext.com/wiki/index.php/Cars_of_Tomorrow_by_Boys_of_Today.

McLellan, Rob. "Fisher Body Craftsman's Guild," *The Automotive Chronicles,* October 2005, www.automotivechronicles.com, published by McLellan's Automotive History, Houston, TX.

_____. "Fisher Body Craftsman's Guild: An Illustrated History," *The Automotive Chronicles,* October 2003, www.automotivechronicles.com, published by McLellan's Automotive History, Houston, TX.

_____. "Fisher Body Craftsman's Guild Reunion," *The Automotive Chronicles,* November 2004. www.automotivechronicles.com, published by McLellan's Automotive History, Houston, TX.

Mellberg, John M. "The Fisher Body Craftsman's Guild Model Car Competition," http://www.fisherguild.com/articles/010/index.php.

"Party by Fisher," *Time,* August 31, 1931. http://www.time.com/time/magazine/article/0,9171,929777,00.html.

Prinz, John. "Fisher Body Craftsman's Guild Dream Car Competition," *CarSeek,* http://www.carseek.com/articles/fisher-car-design.html.

Simone, Meghan. "Simone Featured at Fisher Body Craftsman's Guild Special Exhibit at the Museum of Fine Arts, Boston," *Meghan Simone Wedding Videography* blog, posted December 27, 2008. www.megsimone.com.

Snyder, Brett. "Fisher Body Craftsman's Guild Reunion: A Report by Brett Snyder," *Andrew F. Johnson Gallery* blog, posted September 2008. http://fisherguild.com/fisher-guild-reunion/2008/01/index.php.

Tatseos, Virginia. "Fisher Body Craftsman's Guild 2008 Reunion," *Virginia Tatseos Real Estate* blog, posted December 27, 2008.

Print Articles

"Design Award," *JET Magazine,* April 4, 1963.

"Fisher Award," *Maryland Terrapin,* University of Maryland, October 1966.

"Four Reach Finals of Science Contest," *Gazette Community News,* Silver Spring, MD, July 31, 2005.

Giovannini, Joseph. "Automobile Design: Icons of a Golden Age; Post–World War II Classic and Concept Car Designs Take a Front Seat in Florida," *Architectural Digest,* May 2008.

Hutchings, Helen V. "Guilding: Helen V. Hutchings Finds Bright Leadership Mettle in the Fisher Body Craftsman's Guild," *Auto Aficionado,* May-June 2007.

"Influencing Automotive Design 1930–1968: Fisher Body Craftsman's Guild," *Eleventh Annual Car Classic — Leaving Our Century in Style,* sponsored by the Milestone Car Society of California, June 20, 1999.

"Intel Science Talent Search," *The Washington Post,* July 31, 2005.

Jacobus, John L. "GM's Fisher Body Craftsman's Guild: Carving Out a Career in the Design World — 1950's Style," *Wood Carving Illustrated,* Issue 38, Spring 2007.

_____. "Once and Future Guildsmen: A Fisher Guild Scrapbook, 1930 to 1968," *Automobile Quarterly,* 25:2, CBS Magazines, Division of CBS, Newport Beach, CA, Kutztown Publishing, Fall 1987.

_____. "Young Model Makers Took Guild Trip,"

The Collector's Guide to Automobilia, John Gunnell, Editor, Krause Publications, Iola, WI, 1994.

_____, and Skip Geear, "Boys Just Wanted to Have Fun: Remembering General Motors' Fisher Body Craftsman's Guild (1930–1968)—Nurturing the Creative Male Soul," *Automotive History Review,* Issue No. 34, Spring 1999.

"John Jacobus Wins First Prize in Model Car Competition," *Poly Press,* Baltimore Polytechnic Institute, Baltimore, MD, October 27, 1961.

Lassa, Todd. "In Memoriam: Charles M. Jordan," *Motor Trend,* March 2011.

McDonald, Don. "Fisher Foresees the Future," *Motor Trend,* November 1956.

McLellen, Dennis. "Former Vice President of Design at General Motors, Charles M. Jordan 1927–2010," Obituaries Section, *Los Angeles Times,* December 19, 2010.

_____. "Vice President of Design for General Motors, Charles M. 'Chuck' Jordan, 83," Obituaries Section, *The Washington Post,* December 20, 2010.

"Money Rides on Model Cars," *Mechanics Today,* November 1947.

"The Open Road for Boys," author and magazine unknown, October 1933.

Petti, Michael. "Car Designer's Test," Michael Petti, *Colonia Corner,* 8:7, July 2009 (www.coloniacorner.net).

Purdy, Steve. "Recaptured Youth: A Reunion of the Fisher Body Craftsman's Guild," *Collectible Automobile,* December 2004.

"Siemens-Westinghouse Competition for Excellence in Math, Science and Technology," *Gazette Community News,* Silver Spring, MD, December 17, 2003.

"Teen-Age Model Car Designers," *Sunday Sun Magazine, The Baltimore Sun,* Baltimore, MD, August 13, 1961.

"2 Local Teenagers Win Car Body Design Awards," *Baltimore–News American,* Baltimore, MD, August 6, 1961.

"2 Negros Cop Prizes in Auto Craft Competition," *Jet Magazine,* August 1, 1963.

"2 Youths Win Honors in Fisher Body Contest," *The Baltimore News American,* Baltimore, MD, July 24, 1964.

Van Gelderen, Ronald E. "Tribute to Auto Designers," *The Self-Starter,* Cadillac-LaSalle Club of America, November-December 2006.

"Wanted: Boys with Brains and the Urge to Use Them," 1957 Fisher Body Craftsman's Guild Model Car Competition Poster, *Motor Trend,* November 1956.

Weigley, Samuel, and Paul Fraughton. "Napoleon's Carriage Reborn," *The Salt Lake Tribune,* Section E, July 22, 2010. www.tribtown.com.

"Wins Auto Styling Award," *Jet Magazine,* November 25, 1954.

Photo CD ROM

Fisher Body Craftsman's Guild—Maxi CD, marketed by Ronald J. Will, Fountain Hills, AZ. (azwillfh@cox.net)

16 mm Film

1946 Guild Convention Film, 16mm public relations film produced by General Motors Corporation, GM Photographic Section, Detroit 2, Michigan, Copyright 1946.

Real Dreams Industrial Arts Film. 16 mm public relations film produced by General Motors Corporation, GM Photographic Section, Detroit 2, Michigan, Copyright 1958.

Audio Cassette Tapes

Fisher Body Craftsman's Guild Awards Banquet, 1953—Eight (8) National Winners Being Interviewed by Charles F. Kettering, property of General Motors Corporation, Detroit, MI (78 RPM record converted to audio cassette tape courtesy of Wilfred Keagy, from the Wilfred C. Keagy Collection).

Fisher Body Model Car Awards, NBC-TV July 20, 1954, by TV-Time, Off-the-Air Recordings, Woodmere, Long Island, NY. Originally recorded at 33⅓ RPM, converted to DVD by All Things Audio, courtesy of Oskar Heininger from the Oskar Heininger Collection.

Reporters 1955, *Interview with Harley J. Earl,* V.P. GM Styling, Warren, MI.

MFA Catalogue Review

"Fisher Body Guild Exhibit Presented in Color Catalogue," Headlights Section, *Cars and Parts,* February 2009, p. 14.

Museum Brochures

America on the Move, National Museum of American History, Kenneth E. Behring Center, 14th Street and Constitution Avenue, NW, Washington, DC 20013.

The Great Age of American Automobiles, Auto Designer Art Exhibition and Retrospective, February 3 to June 23, 2008, Museum of Art, Ft. Lauderdale, FL.

Automotive Design and Styling Contests (2-Dimensional Design)

Annual Truck and SUV Design Contest, by *Motor Trend's Truck Trend,* http://www.trucktrend.com/future/67/descon/index.html.

Design a New Studebaker Contest, sponsored by Virgil M. Exner, Jr. (1946 National Junior Scholarship winner, FBCG), and the Studebaker National Museum, 201 South Chapman, South Bend, Indiana 46601. http://www.carlustblog.com/2010/03/design-a-new-studebaker.html.

Motor Trend's International Design Contest in cooperation with the California International Auto Show (CIAS). http://www.motortrend.com/auto_news/112_news0311054_descon/index.html.

Peugeot International Design Contest, http://www.motorauthority.com/blog/1027869_20-year-old-romanian-wins-peugeot-design-contest.

Some Scale Model Car Design Contests (Plastic Kits and Scratch-Built)

Annual Model Car Show and Contest, sponsored by the Long Island Auto Replica Society (L.I.A.R.S.), Henrietta Acampora Recreation Center, 39 Montauk Highway, Blue Point, NY, www.liarsmodelcarclub.com.

Classic Plastic Model Car Competition, Contact: Del Paone, President, Classic Plastic Model Club, Lowell, MA, www.classicplastic.org.

GSL International Scale Vehicle Championship and Convention, 10291 South 1300 East, PMB #131, Sandy, UT 84094. Contact: Mark S. Gustavson, msgsl@xmission.com. Also see http://www.gslchampionship.org/News/GSLPreview-1-2011.pdf, http://www.GSLChampionship.org and http://www.TheModelCarMuseum.org.

International Plastic Model Societies (IPMS/USA), Model Car Club Directory, http://www.ipmsusa.org. Contact: National Contest Committee, NCC @ipmsusa.org.

MassCar Model Car Exhibition and Contest, Rustcraft Road, Dedham, MA, www.masscar.com.

Mid-Atlantic NNL (Scale Model Car Competition and Convention), sponsored by the Maryland Automotive Modelers Association. Contacts: Norman F. Veber, Replicas and Miniatures Company of Maryland, 317 Roosevelt Avenue, S.W., Glen Burnie, MD 21061, and Timothy Sickle, Editor, *MAMA's Diner* Newsletter, 15905 Ark Court, Bowie, MD 20716. www.mamaboyz.org.

NNL East (Scale Model Car Competition and Convention). Contact: Tim Geiger, President, Tri-State Model Car Club, 2 Princeton Place, Hazlet, NJ 07730. www.tristatecarclub.

NNL Western Nationals, Scale Model Car Competition and Convention. Contact: Andy Kellock, Event Chairman, P.O. Box 110691, Campbell, CA 95011–0691. www.nnlwest.org.

Scale Model Car Magazines

Car Modeler: Your How-to Modeling Resource 2003, Scale Model Special Issue, Kalmbach Publishing Company, 21027 Crossroads Circle, P.O. Box 1612, Waukesha, WI 53187–1612.

Contest Cars 2004, Scale Model Special Issue, Kalmbach Publishing Company, 21027 Crossroads Circle, P.O. Box 1612, Waukesha, WI 53187–1612.

Fine Scale Modeler, Matthew W. Usher, Editor, Kalmbach Publishing Company, 21027 Crossroads Circle, P.O. Box 1612, Waukesha, WI 53187-1612.

Kustom and Hot Rod Models, Don Graham, Publisher/Editor, 23 West Alexander Avenue, #78, Merced, CA 95348 USA.

Model Cars Magazine, Gregg Hutchings, Editor, 3419A Hinahina Street, Honolulu, HI 98616.

Scale Auto, Jim Haught, Editor, Kalmbach Publishing Company, 21027 Crossroads Circle, P.O. Box 1612, Waukesha, WI 53187–1612.

Scale Model Collector International, Peter Emery, Editor, SAM Publications, Media House, 21 Kingsway, Bedford, MK42–9BJ, United Kingdom.

Toy Cars and Models, Merry Dudley, Editor, F+W Publications, 700 East State Street, Iola, WI 54990-0001.

Auto Design and Automotive History Web Sites

Automotive Hall of Fame, Dearborn, Michigan, www.automotivehalloffame.org.

Automotive Oral Histories, Automobile in American Life and Society, Benson Ford Research Center, Dearborn, MI, by David Crippen, Henry Ford Archives Staff, http://www.autolife.umd.umich.edu/Oral_histories.htm.

Car Body Design, Automotive Design and Engineering, http://www.carbodydesign.com.

Chevrolet Design Centre (Worldwide), http://www.chevroletdesign.com.

"The Fisher Body Craftsman's Guild," *Jitterbuzz.com, Lindy Week Review*, http://www.jitterbuzz.com/fisher.html.

Fisher Body Online, edited by James Sponseller, former Public Relations and Advertising Director (circa 1984), Fisher Body Division, General Motors Corporation, Detroit, MI.

Futurliner News (GM Futurliner Restoration Project), Dean G. Tryon, Editor, 2516 Lauralford Lane, Wake Forest, NC 27587, www.Futurliner.com.

Joe Bortz's 1955 LaSalle Motorama Show Car Restoration Project, Video by "Rides with Chuck," LaSalle II, 1955 GM Motorama Concept Car, EV Conversion, Part 1, http://www.youtube.com/watch?v=pnE0-ZVX4Ho, and An EV Conversion, Bortz LaSalle Roadster, Baby's First Steps,

Part 4, http://www.youtube.com/watch?v=PWhjx Ckw-34&NR=1&feature=fvwp.
The Museum of Automotive Art and Design, developed by Walter Miller, Syracuse, NY, www.moaaad.org.
Official Fisher Body Craftsman's Guild book Web site, edited by John L. Jacobus, http://www.fisherguild.com.
Official Harley J. Earl Web site, edited by A. Richard Earl, http://www.carofthecentury.com.
Society of Automotive Historians, www.autohistory.org.
Stewart Reed Design, Transportation and Industrial Design Consultant, http://www.stewartreeddesign.com/bios.php.

About the Contributors

George E. **Anderson** is a career automobile interior designer who was trained at the Art Center School in Los Angeles and worked for GM Design Staff at the GM Technical Center. Outside of GM, he also found success as a freelance race car designer in 1965 and 1966. He retired from GMC in 2003 after 43 years and makes his home in Farmington Hills, Michigan.

Warren M. **Bakken** graduated from the University of North Dakota with a B.S. in mechanical engineering and worked in the aircraft industry (Lockheed Aircraft Corporation, General Dynamics and Boeing Aircraft). At Boeing he worked as a wind tunnel model designer. He lives in Seattle, Washington.

Paul F. **Bonfilio** trained as an architect and urban planner and retired as commissioner from a major New York City agency. He makes his home in Little Neck, New York. The scale models he made as an architectural model maker and hobbyist can be viewed at the Museum of Modern Art (MOMA) in New York City.

Adrian A. **Bruno** is a former vice president of R&D and currently senior executive at Weber Stephen Products Company (manufacturer of the ubiquitous Weber Grill). He has more than 47 years of experience in product design, management, engineering and marketing and resides in Rolling Meadows, Illinois.

Felix R. **Collazo** majored in architecture and industrial design while in high school. A gifted artist, his architectural renderings adorned the walls of the school. He retired January 2002 from the military at the rank of sergeant first class after 28 years of service between the Army and the Navy. He travels between his homes in Columbia, South Carolina, and Bayamon, Puerto Rico.

Thomas L. **Covert** is a career auto designer and graduate of the Art Center College of Design in 1968 with a B.S. in industrial design. He worked for GM Design Staff at the GM Tech Center (Chevrolet Studio, Advanced Design, Corvette Studio, Pontiac Studio) 13 years and was lead designer on the Hummer H2 brand. He retired from GMC in 2005 and lives in Richmond, Michigan.

Thomas L. **De Fazio** is a retired mechanical engineer and occasional consultant. He holds several degrees from Massachusetts Institute of Technology (MIT), has taught at the Florida Institute of Technology and recently was involved in innovative mobile robotics research work at iRobot Corporation. He resides in Cambridge, Massachusetts.

Charles R. **Foreman** is an advanced composite materials expert who graduated from engineering school and went to work for LTV Aerospace (now Vought Aircraft Industries). He contributed to the development of the B-2 Bomber. Partially retired, he lives in Arlington, Texas.

Terry P. **Graboski** is a 1971 graduate of the Art Center College of Design with a B.F.A. in advertising. His Santa Monica, California, based industrial design firm (Beck and Graboski Design

Office) practices environmental design such as the application of architectural signing, way finding and environmental graphics.

Tom W. **Graboski** holds a B.A. in interior design from the Art Center College of Design (1971) and an M.S. in urban design from the University of Miami School of Architecture (1977). He works in Coral Gables, Florida, and has his own industrial design business (Tom Graboski Associates, Inc.) specializing in architectural graphics and signage.

George G. **Herzog**, while a physics major at Duke University and AF-ROTC student, earned a commission as a U.S. Air Force officer upon graduation and became a program manager at Wright-Patterson Air Force Base (1964–1993) in Fairborn, Ohio. He progressed to deputy director for the system program office contract to modernize the Air Force's training fleet. He is retired and resides in Beavercreek, Ohio.

John L. **Jacobus** holds two engineering degrees, and early on in his career, worked as a design engineer and project engineer at Ternstedt and Fisher Body Divisions of GMC. This led to a 30-year career as an auto safety engineer at the U.S. Department of Transportation in Washington, D.C. A federal retiree, he resides in Silver Spring, Maryland.

Jeffrey A. **Jones** is a master wood carver, master craftsman and career petroleum engineer from Bakersfield, California.

Arnold L. **Joslin** was a summer intern at GM Styling as a college student (1959 and 1960). He obtained a B.S. in mechanical engineering degree from North Carolina State University and became an industrial designer and mechanical design engineer at Hewlett-Packard. He lives in Loveland, Colorado.

Wilfred C. **Keagy** was schooled as an electronic technician and got involved in the design and manufacture of electronic test equipment, computers and aircraft controls. He cleared the land and then designed and built his own home in La Verne, California, brick by brick.

William A. **Keyser**, Jr., is professor emeritus, School of American Crafts, Rochester Institute of Technology, Rochester, New York. He taught furniture design for 35 years (1962–1997) and maintained an independent studio creating custom furniture and sculptures in Victor, New York where he is living today.

Samuel T. **Kjellman**, after graduating from Syracuse University as an industrial design major, went to Detroit to work for Ford Motor Company designing cars (1967–1972), became an appliance designer (1972–1974), worked as a Parker Brothers toy and game designer from 1974 to 1989, and then became an independent product designer (1990 to present). He is from Weare, New Hampshire.

Robert W. **Lawhn** studied physics in college, received a master's degree in nuclear engineering from Texas A&M University and later became the environmental director of Reliant Energy operations in Nevada and California. He recently joined a new energy company, GenOn Energy, and is directing the development of a new power plant in California. He resides in Henderson, Nevada.

Geza A. **Loczi**, after more than 45 years of experience as an auto designer at GM, VW and Volvo, recently retired as the director of the Volvo Monitoring and Concept Center (VMCC) in Camarillo, California. Before he could retire, he was rehired by VMCC as a design consultant. He resides in Moorpark, California.

Spencer L. **Mackay** studied product design at the Art Center College of Design and worked as a senior designer (1976–1981) and later became vice president (1981–1997) of the family's industrial design business, Emerson Johnson Mackay, Inc. (EJM, Inc.) in Burbank, California. Today he is president and owner of M Industrial Design (formerly EJM, Inc.) and designs and packages products for clients all over the world. He resides in Agoura Hills, California.

Frederick J. "Bud" **Magaldi** was a career auto designer at Ford Motor Company. Some of the highlights of his career include managing the 1994 Mustang Design Team and contributing to the design of the 1983 Thunderbird, the groundbreaking 1986 Ford Taurus, the concept for the Ford Explorer Sport Trac (the "Adrenalin" show car), the Lincoln Town Car and the Mustang Mach III. Bud splits his retirement time between Michigan, Maine and Florida.

John M. **Mellberg** began his career on the GM Design Staff, GM Technical Center, then moved on to J.I. Case Corporation where he designed agricultural tractors and construction vehicles. After a period designing aftermarket truck accessories, Mellberg worked for Emergency One, Inc., designing a composite bodied, prototype fire rescue truck called the E-One Daytona. He retired in 2011 after designing new school buses at Thomas Built Buses in High Point, North Carolina. He is now an independent automotive design professional and still an avid modeler.

Walter F. **O'Neill**, Jr., insurance industry executive and antique auto restoration expert, owned and traded hundreds of different antique cars over his lifetime while living in Basking Ridge, New Jersey. He is a familiar face at AACA meets (car shows and flea markets) in Hershey, Pennsylvania.

Grant L. **Onnie**, the last of the top national senior scholarship ($5,000) winners in the Guild competition, attended the School of Arts and Crafts (currently College for Creative Studies) in Detroit, Michigan, and enjoyed a long, successful career in the auto industry. A creative freelance writer, he lives and works in Rochester Hills, Michigan.

Herman I. **Rauth**, with a degree from Kearney State College, was a Nebraska educator for 35 years, and is an amateur stoneware potter from Honey Creek, Iowa.

Joseph R. **Rauth** graduated high school and became a Benedictine lay brother (1955 to 1970), working as a cabinet maker and craftsman on office and church furniture, then became a construction contractor. He is an accomplished artist and stone sculptor from New Haven, Missouri.

Philip J. **Rauth** was a high school administrator for 14 years, and has been a priest and pastor for various Catholic parishes in Nebraska over the years. He retired in 2006 to his home in Omaha, Nebraska.

Roger L. **Schneider** studied industrial design at the Illinois Institute of Technology, Institute of Design, and enjoyed multiple careers. For 12 years he worked on the design of railroad and mass transit control equipment at Westinghouse Air Brake Company. He founded a cultural magazine called *Renaissance: Pittsburgh Showcase Magazine*, the success of which evolved into a broadcasting opportunity at WQED-TV as *Pittsburgh Magazine*. He worked for more than 30 years at American Thermoplastic Company (a national loose-leaf binder manufacturer), eventually rising to plant manager. He lives in Murrysville, Pennsylvania.

William D. **Scott** graduated from Art Center School in 1963 and joined GM Styling Staff at the GM Tech Center. In 1991 he was named the director of design for Oldsmobile Studios (both interior and exterior) and then the Saturn Studios. He became director of interior design for all divisions and for design quality for GM Design (formerly GM Styling). He retired from GM Design in July 2001 as an executive level director of design and lives in Brewster, Massachusetts.

Charles H. **Stewart** studied industrial design at Pratt Institute in New York City and joined GM Styling Staff as an auto designer. He served as a Guild design judge in 1960 and 1961. He retired from GM in 1997 after a 38-year career working in the Chevrolet, Oldsmobile and Cadillac Exterior Studios. He resides in Clawson, Michigan.

Roger D. **Teter**, son of the owner of a Pontiac/GMC dealership in Elkins, West Virginia, attended West Virginia University and completed an undergraduate degree in engineering, and later completed an advanced engineering degree. He worked at Lockheed Missiles and Space in Sunnyvale, California. His late Guildsman brother, Homer Keith Teter, became an auto designer at Ford Motor Company and chairman of the Transportation Design Department at the Art Center College of Design.

Ronald J. **Will** attended the Illinois Institute of Technology (IIT) Institute of Design and joined GM Styling Staff working in the Corvette, Camaro and Cadillac Studios (1966–1976). He designed, built and promoted his own 3-wheel automobile called the "Turbo Phantom" (1976–1980), became a designer at Subaru Research and Design (1981–1991) and retired as manager of product planning and design at Subaru of America (1991 to 2006). Ron is now an advocate for Natural Gas Car, is a member of the Phoenix Auto Press and lives in Fountain Hills, Arizona.

Gordon D. **Williams** is a 1958 graduate of the Art Center School and became an industrial designer and sculptor. After a brief period with GM Styling Staff as a young man, he enjoyed a 40-year career at Gordon Williams Design in San Francisco in the field of commercial/contract interior design. He currently lives in an artists' community in Albuquerque, New Mexico.

Raymond P. **Wykes** was an aerospace engineer involved in advanced composite structures and stress analysis for aerospace and experimental vehicles at Rockwell International, Los Angeles Division. He passed away in January 2011 in Cerritos, California.

Index

Numbers in ***bold italics*** indicate pages with photographs.
A * after an entry indicates a Guildsman.